W9-CKM-526

Marine Molecular Biotechnology

Subseries of Progress in Molecular and Subcellular Biology
Series Editor   Werner E. G. Müller

Progress in Molecular and Subcellular Biology

46

Series Editors   W. E. G. Müller (Managing Editor)
Ph. Jeanteur, Y. Kuchino, M. Reis Custodio,
Robert E. Rhoads, D. Ugarkovic

Volumes Published in the Series

Progress in Molecular
and Subcellular Biology

Subseries:
Marine Molecular Biotechnology

Nobuhiro Fusetani • William Kem

Editors

Marine Toxins as Research Tools

 Springer

Editors

Professor Dr. Nobuhiro Fusetani
Faculty of Fisheries Sciences
Hokkaido University
3-1-1 Minato-cho
Hakodate 041-8611
Japan
discodermia@yahoo.co.jp

William Kem, Ph.D.
Department of Pharmacology & Therapeutics
University of Florida
P.O. Box 100267
Gainesville, FL 32610-0267
USA
wrkem@UFL.EDU

ISSN 1611-6119
ISBN 978-3-540-87892-6   e-ISBN 978-3-540-87895-7
DOI 10.1007/978-3-540-87895-7

Library of Congress Catalog Number: 2008935623

Preface

There has been a steady increase in the number of known marine toxins in the past few decades. The remarkable molecular diversity of these and other marine natural products probably exceeds the diversity of organisms that live in the sea. Earlier pioneering books on marine toxins such as those by Halstead (1965, 1967, 1970) and Hashimoto (1979) greatly stimulated scientific interest in toxins from marine animals and plants. In the past 30 years a major discovery has been that many of these toxins originate in marine bacteria, a diverse group that is only beginning to be investigated by a large number of scientific laboratories.

Even more remarkable than the increased number of known marine toxins is the improved understanding of the chemistry and chemical biology of these molecules, which potently target specific structures and processes essential for life. It is now apparent that there is actually a continuum in the activities of these compounds, from molecules that quickly paralyze by targeting membrane ion channels to molecules that act more slowly by affecting second messenger pathways, gene expression, and cell proliferation or death.

The field of marine natural toxins has become so large that it is difficult to cover it within a single volume. The intent of the current volume is to provide the reader with some selected developments within the field that are providing useful molecular probes and lead compounds for drug discovery. After an overview of the various types of toxic molecules now known, the reader can learn of recent developments in the study of the cone shell peptide toxins, probably the most exciting group of marine toxins to be intensively investigated during the past 20 years (Lewis, Chap. 2). The variety of targets that the conotoxins affect is now quite large and continues to grow beyond the initially researched ligand-gated and voltage-gated ion channels, and now includes biogenic amine transporters and G-protein-coupled receptors as well.

Llewellyn then summarizes recent studies with tetrodotoxins and saxitoxins affecting voltage-gated sodium channels (Chap. 3). These toxins continue to be useful probes of these channels, which turn out to be quite diverse – for instance, some sodium channels displaying resistance to TTX have been shown to be potential targets for developing new medications for the treatment of neuropathic pain, which does not respond to classical analgesics such as morphine. TTX can then be used to block other sodium channels that are also expressed concurrently with the TTX-insensitive channels.

The multiplicity of potassium channels makes the need for selective research probes even greater, so it is fortunate that marine and terrestrial animal venoms seem to contain toxins for any channel that needs to be modified. Initial sea anemone K-channel toxins were found to target delayed rectifier and calcium-activated channels, and these have become useful probes for these channels. In Chap. 4, Diochot and Lazdunski review the properties of a new group of K-channel toxins isolated from sea anemones.

An exciting new area involves the discovery of new probes for kainate-type glutamate channels, as described in the chapter by Swanson and Sakai (Chap. 5). This was enabled by chemical transformations of natural products to extend compound diversity beyond what nature offers. For the first time probes for kainate receptors devoid of significant activity at AMPA-type glutamate receptors have been obtained. These molecules will be extremely useful for dissecting out the respective roles of these two different types of glutamate receptors within the nervous system. They may also suggest new avenues for receptor-selective drug design.

A variety of marine toxins directly or indirectly target the presynaptic machinery that mediates neurotransmitter and neuromodulator release, and these toxins like spider (latrotoxins) and bacterial (botulinum toxins) are becoming useful probes for biochemical dissection of the complex events mediating exocytotic secretion processes. These marine toxins and their use as synaptic probes are presented by Meunier and Molgó in Chap. 6.

Highly cytotoxic metabolites found from marine cyanobacteria, sponges, soft corals, and opisthobranchs are valuable not only as anticancer drug leads but also as research tools. Modes of action studies of these cytotoxins led to the discovery of actin polymerization-inhibiting and -facilitating polyketides as well as tubulin polymerization-inhibiting and -facilitating toxins. Since actin and tubulin are the major components of cytoskeletal system and are involved in basic cellular processes, including maintenance of cell shape, cell division, and intracellular transport, they contribute significantly to our understanding of cellular functions as described by Saito in Chap. 7.

Reversible phosphorylation of proteins, which is mediated by protein kinases and protein phosphatases, is a ubiquitous mechanism for the control of signal transduction networks regulating diverse biological processes. Thus, specific inhibitors and activators of these enzymes are important tools to study biological phenomena. In Chap. 8, Fujiki and Suganuma highlight the involvement of this process in chemical carcinogenesis which was discovered by using marine toxins of the okadaic acid class, specific inhibitors of protein phosphatases 1 and 2A.

Thus, research on marine natural products is providing an impressive platform of unique molecules that are being increasingly exploited as useful molecular probes for understanding myriad biological processes, from control of cell division (and thus proliferation) to modulation of ion channels (and thus chemical as well as electrical signaling). Some of these probes are also serving as "lead" compounds for the design and development of new drugs and pesticides. Prospecting for new compounds coupled with improvements in screening methods for identifying their biological activities should further accelerate progress in this interdisciplinary field.

We hope readers of this volume will gain an appreciation for the potential of this field for facilitating discoveries in biology, chemistry, and medicine.

References

Halstead BW (1965) Poisonous and venomous marine animals of the world, Vol. 1. US Government Printing Office, Washington, DC.

Halstead BW (1967) Poisonous and venomous marine animals of the world, Vol. 2. US Government Printing Office, Washington, DC.

Halstead BW (1970) Poisonous and venomous marine animals of the world, Vol. 3. US Government Printing Office, Washington, DC.

Hashimoto Y (1979) Marine toxins and other bioactive marine metabolites. Japan Scientific Societies Press, Tokyo.

Preface to the Series

Recent developments in the applied field of natural products are impressive, and the speed of progress appears to be almost self-accelerating. The results emerging make it obvious that nature provides chemicals, secondary metabolites, of astonishing complexity. It is generally accepted that these natural products offer new potential for human therapy and biopolymer science. The major disciplines which have contributed, and increasingly contribute, to progress in the successful exploitation of this natural richness include molecular biology and cell biology, flanked by chemistry. The organisms of choice, useful for such exploitation, live in the marine environment. They have the longest evolutionary history during which they could develop strategies to fight successfully against invading organisms and to form large multicellular plants and animals in aqueous medium. The first multicellular organisms, the plants, appeared already 1,000 million years ago (MYA), then the fungi emerged, and finally, animals developed (800 MYA).

Focusing on marine animals, the evolutionary oldest phyla, the Porifera, the Cnidaria, and the Bryozoa, as sessile filter feeders, are exposed not only to a huge variety of commensals but also to toxic microorganisms, bacteria and fungi. In order to overcome these threats, they developed a panel of defense systems, for example, their immune system, which is closely related to those existing in higher metazoans, the Protostomia and Deuterostomia. In addition, due to this characteristic, they became outstandingly successful during evolution; they developed a chemical defense system which enabled them to fight in a specific manner against invaders. These chemicals are of low molecular weight and of nonproteinaceous nature. Because of the chemical complexity and the presence of asymmetrical atom centers in these compounds, a high diversity of compounds became theoretically possible. In a natural selective process, during evolution, only those compounds could survive which caused the most potent bioactivity and provided the most powerful protection for the host in which they were synthesized. This means that during evolution nature continuously modified the basic structures and their derivatives for optimal function. In principle, the approach used in combinatorial chemistry is the same, but turned out to be painful and only in few cases successful. In consequence, it is advisable to copy and exploit nature for these strategies to select for bioactive drugs. Besides the mentioned metazoan phyla, other animal phyla, such as the higher evolved animals, the mollusks or tunicates, or certain algal groups, also produce compounds for their chemical defense which are of interest scientifically and for potential application.

There is, however, one drawback. Usually, the amount of starting material used as a source for the extraction of most bioactive compounds found in marine organisms is minute and, hence, not sufficient for their further application in biomedicine. Furthermore, the constraints of the conventions for the protection of nature limit the commercial exploitation of novel compounds, since only a small number of organisms can be collected from the biotope. Consequently, exploitation must be sustainable; that is, it should not endanger the equilibrium of the biota in a given ecosystem. However, the protection of biodiversity in nature, in general, and those organisms living in the marine environment, in particular, holds an inherent opportunity if this activity is based on genetic approaches. From the research on molecular biodiversity, benefits for human society emerge which are of obvious commercial value; the transfer of basic scientific achievements to applicable products is the task and the subject of *Marine Molecular Biotechnology*. This discipline uses modern molecular and cell biological techniques for the sustainable production of bioactive compounds and for the improvement of fermentation technologies in bioreactors.

Hence, marine molecular biotechnology is the discipline which strives to define and solve the problems regarding the sustainable exploitation of nature for human health and welfare, through the cooperation between scientists working in marine biology, molecular biology, microbiology, and chemistry. Such collaboration is now going on successfully in several laboratories.

It is the aim of this new subset of thematically connected volumes within our series "Progress in Molecular and Subcellular Biology" to provide an actual forum for the exchange of ideas and expertise between colleagues working in this exciting field of "Marine Molecular Biotechnology." It also aims to disseminate the results to those researchers who are interested in the recent achievements in this area or are just curious to learn how science can help to exploit nature in a sustainable manner for human prosperity.

Werner E.G. Müller

Contents

Contributors

S. Diochot
Institut de Pharmacologie Moléculaire et Cellulaire, Centre National de la
Recherche Scientifique, Université de Nice-Sophia-Antipolis, 660 Route
des Lucioles 06560 Valbonne, France

H. Fujiki
Faculty of Pharmaceutical Sciences, Tokushima Bunri University, Yamashiro-cho,
Tokushima 770-8514, Japan
hfujiki@ph.bunri-u.ac.jp

N. Fusetani
Graduate School of Fisheries Sciences, Hokkaido University, Minato-cho,
Hakodate 041-8611, Japan
discodermia@yahoo.co.jp

W. Kem
Department of Pharmacology and Therapeutics, University of Florida,
P.O. Box 100267, Gainesville, FL 32610-0267, USA
wrkem@UFL.EDU

M. Lazdunski
Institut de Pharmacologie Moléculaire et Cellulaire, CNRS, Sophia Antipolis,
F-06560 Valbonne, France
lazdunski@ipmc.cnrs.fr

R.J. Lewis
Institute for Molecular Biosciences, The University of Queensland,
Brisbane 4072, Australia
r.lewis@imb.uq.edu.au

L.E. Llewellyn
Australian Institute of Marine Science, Townsville MC, QLD 4810, Australia
L.Llewellyn@aims.gov.au

C. Mattei
CNRS, Institut de Neurobiologie Alfred Fessard, FRC2118,
Laboratoire de Neurobiologie Cellulaire et Moléculaire,
UPR 9040, 1, avenue de la Terrasse, Gif sur Yvette, F-91198, France
and
Délégation Générale pour l'Armement 7-9 rue des Mathurins, Bagneux, F-92220,
France

F.A. Meunier
Queensland Brain Institute and School of Biomedical Sciences, The University
of Queensland, St. Lucia, Queensland 4061, Australia

J. Molgó
Délégation Générale pour l'Armement 7-9 rue des Mathurins, Bagneux, F-92220,
France

Shin-ya Saito
Department of Pharmacology, Faculty of Pharmacy, University of Shizuoka,
Yada 52-1, Suruga-ku, Shizuoka 422-8526, Japan
synsaito@u-shizuoka-ken.ac.jp

R. Sakai
Faculty of Fisheries Sciences, Hokkaido University, Hakodate 041-8611,
Japan

M. Suganuma
Research Institute for Clinical Oncology, Saitama Cancer Center,
Saitama 362-0806, Japan
masami@cancer-c.pref.saitama.jp

G.T. Swanson
Department of Molecular Pharmacology and Biological Chemistry, Northwestern
University, Feinberg School of Medicine, 303 E. Chicago Ave., Chicago, IL,
60611
gtswanson@northwestern.edu

Marine Toxins: An Overview

Nobuhiro Fusetani and William Kem

N. Fusetani (✉)
Graduate School of Fisheries Sciences, Hokkaido University, Minato-cho,
Hakodate 041-8611, Japan
e-mail: discodermia@yahoo.co.jp

W. Kem
Department of Pharmacology and Therapeutics, University of Florida, Gainesville,
FL 32610-0267, USA
e-mail: wrkem@UFL.EDU

N. Fusetani and W. Kem (eds.), *Marine Toxins as Research Tools*,
Progress in Molecular and Subcellular Biology, Marine Molecular Biotechnology 46,
DOI: 10.1007/978-3-540-87895-7, © Springer-Verlag Berlin Heidelberg 2009

Abstract Oceans provide enormous and diverse space for marine life. Invertebrates are conspicuous inhabitants in certain zones such as the intertidal; many are soft-bodied, relatively immobile and lack obvious physical defenses. These animals frequently have evolved chemical defenses against predators and overgrowth by fouling organisms. Marine animals may accumulate and use a variety of toxins from prey organisms and from symbiotic microorganisms for their own purposes. Thus, toxic animals are particularly abundant in the oceans. The toxins vary from small molecules to high molecular weight proteins and display unique chemical and biological features of scientific interest. Many of these substances can serve as useful research tools or molecular models for the design of new drugs and pesticides. This chapter provides an initial survey of these toxins and their salient properties.

1   Introduction

The world's oceans cover more than 70% of the earth's surface and represent greater than 95% of the biosphere. According to the World Conservation Monitoring Center, ~1.75 million species of living organisms have been described by systematists, but these probably represent just a fraction of the total number of existent species. Estimates of the total number occupying the Earth vary considerably, from 3.5 to 100 million species (Groombridge et al. 2000, cited in Müller et al. 2003). Because marine species are relatively inaccessible compared with terrestrial species, it is likely that currently described marine organismic diversity (<500,000 species) is greatly underestimated several-fold. The phyletic diversity of animals (and bacteria) in the marine biosphere is exceptional: all but one of the 35 principal animal phyla are represented in aquatic environments and 8 of these phyla are exclusively aquatic. Most marine invertebrates that are sessile and soft-bodied have evolved chemical means to defend against predators and overgrowth of competing species. In fact, high percentages of them are poisonous and venomous (Halstead 1965, 1967, 1970; Hashimoto 1979). It should be noted that normally non-toxic animals can become poisonous by sequestering toxins from their prey organisms. For example, many bivalves accumulate toxins from dinoflagellates (see Sect. 3) while ciguateric fishes accumulate ciguatoxins and related toxins through food chains (see also Sect. 3 and Chap. 3). Tetrodotoxins are found in many marine organisms ranging from algae to various fishes; perhaps they are accumulated from bacteria and/or through food chains (see also Sect. 10 and Chap. 3). Opisthobranch mollusks sequester defensive substances from prey organisms including seaweeds, sponges and ascidians (Faulkner 1992). Other marine invertebrates, especially sponges and ascidians, often contain highly cytotoxic compounds that probably originate from symbiotic microbes, namely bacteria, cyanobacteria and dinoflagellates (Brewley and Faulkner 1998; Piel 2004).

A variety of marine animals possess peptides and proteins that are used either for defensive means, catching prey, or both. Two prominent examples are cone shells and sea anemones, whose venoms are complex mixtures of highly toxic peptides with different modes of action (see also Chap. 2 for discussion of the cone shell peptides). It should be also noted that antimicrobial peptides of several types are found in a wide range of marine animals (Andreu and Rivas 1998; Bulet et al. 2004).

2 Cyanobacteria

Freshwater cyanobacterial (blue–green algal) blooms of the genera *Anabaena*, *Aphanizomenon*, and *Microcystis* have been known to produce neuro- and hepatotoxic compounds that cause mass mortalities of birds, wild animals and livestock (van Apeldoorn et al. 2007). Though marine cyanobacteria also form blooms, these have not yet presented serious health and economic problems. It is now known that marine species, especially *Lyngbya majuscula*, produce an amazing variety of bioactive metabolites, including highly cytotoxic compounds (Gerwick et al. 2001; Burja et al. 2001). Many cytotoxic compounds found in marine sponges and gastropods are also of cyanobacterial origin (Burja et al. 2001; Yamada and Kigoshi 1997).

2.1 Microcystins

The hepatotoxic cyclic heptapeptides produced by freshwater cyanobacteria are collectively referred to as microcystins, since they were first isolated from *Microcystis aeruginosa* (van Apeldoorn et al. 2007). Currently 64 microcystins are known; they have the general structure cyclo-(D-Ala-X-D-MeAsp-Z-Adda-D-Glu-Mdha), where X and Z are variable L-amino acids. The first investigated and most prominent member is microcystin-LR (**1**), displaying a mouse LD_{50} of 50–60 µg/kg (i.p.). Microcystins are powerful cancer promoters that act by inhibiting protein phosphatases 1 and 2A (see Chap. 8).

A similar cyclic pentapeptide named nodularin (**2**) having toxicity and mode of action comparable to those of microcystin-LR was isolated from the brackish water cyanobacterium *Nodularia spumigena*. Seven variants of nodularins are known at present. An analogue of nodularin named motuporin has been found in the marine sponge *Theonella swinhoei*. It is likely produced by symbiotic cyanobacteria (see Chap. 8).

It should be noted that microcystins affect zooplankton community composition (Hanson et al. 2005).

1

2

2.2 Antilatoxin and Kalkitoxin

Among a large variety of bioactive metabolites produced by *L. majuscula,* particularly interesting are antilatoxin (**3**) and kalkitoxin (**4**), ichthyotoxins that respectively activate and inhibit voltage-gated Na+ channels (Gerwick et al. 2001). It is intriguing that antilatoxin binds the site of Na+ channels different from that of brevetoxins (Al-Sabi et al. 2006).

3

4

2.3   Alkaloids

2.3.1   Anatoxins

Anatoxin-*a* (**5**) and homoanatoxin-*a* (**6**) are fast-acting neurotoxins produced by cyanobacteria of the genera *Anabaena, Oscillatoria, Cylindrosperum*, and *Aphanizomenon* (van Apeldoorn et al. 2007). They transiently stimulate nicotinic acetylcholine (ACh) receptors by binding directly at the ACh binding sites but ultimately block neuromuscular transmission. Their mouse LD_{50} values were 375 and 250 µg/kg, respectively.

Interestingly, anatoxin-*a*(S) (**7**) produced by *Anabaena* spp. inhibits cholinesterase and its mouse LD_{50} is 31 mg/kg (i.p.).

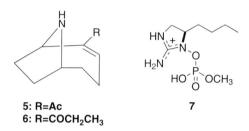

5: R=Ac
6: R=COCH₂CH₃   7

2.3.2   Saxitoxins

Saxitoxins originally discovered from marine bivalves that were infested by dinoflagellates are also produced by cyanobacteria of the genera *Anabaena, Aphanizomenon*, and *Lyngbya* (see Sect. 3.1).

2.3.3   Cylindrospermopsins

Cylindrospermopsin (**8**), an unusual alkaloid possessing a tricyclic guanidium moiety, was first isolated from an Australian strain of *Cylindrospermopsis raciborskii*. Several variants have been reported from various cyanobacteria (van Apeldoorn et al. 2007). Cylindrospermopsin blocks protein synthesis and induces kidney and liver failure (Froscio et al. 2008).

8

2.3.4 Lyngbyatoxins

Lyngbya majuscula is a causative agent of "swimmers' itch," a skin dermatitis well-known in Hawaii, and lyngbyatoxins and aplysiatoxins are responsible for this syndrome. Lyngbyatoxin A (**9**) is actually teleocidin A-1 produced by actinomycetes of the genus *Streptomyces* and is a potent tumor promoter (see Chap. 8). It was also reported to be a causative agent of turtle poisoning (Ito et al. 2002).

2.4 Polyketides

2.4.1 Aplysiatoxins

Aplysiatoxin (**10**) and debromoaplysiatoxin (**11**) were originally isolated from the digestive glands of the Hawaiian sea hare *Stylochaelus longicauda* and were later found to originate from *L. majuscule,* a prey of the sea hare (van Apeldoorn et al. 2007). Some variants of aplysiatoxins are known from marine cyanobacteria (Burja et al. 2001). These compounds are inflammatory agents and tumor promoters (see Chap. 8). They were also reported as causative agents of human intoxication from ingestion of the Hawaiian red alga *Gracilaria coronopifolida* (Nagai et al. 1996).

9

10: R=Br
11: R=H

2.4.2 Other Cyanobacterial Toxins

A large number of cytotoxic polyketides have been reported from marine cyanobacteria of the genera *Lyngbya, Symploca*, and *Scytonema* (Gerwick et al. 2001; Burja et al. 2001). Scytophycin (**12**), an unusual macrolide isolated from *Scytonema pseudohofmanni*, inhibits actin polymerization (Burja et al. 2001; see Chap. 7).

12

3 Dinoflagellate (Pyrrophyta) Toxins

Dinoflagellates, aquatic photosynthetic eukaryotes, are classified as belonging to the kingdom Protoctista (Protista) (Camacho et al. 2007). They produce an array of highly toxic metabolites, many of which are unprecedented in terrestrial secondary metabolites, are involved in human intoxications from ingestion of sea food and cause mortality of marine animals (Yasumoto and Murata 1993; Daranas et al. 2001; Cembella 2003).

3.1 Saxitoxins

Saxitoxins, highly potent neurotoxins, cause paralytic shellfish poisoning from ingestion of bivalves (see Chap. 3). Bivalves such as mussels, scallops and oysters become toxic by sequestering saxitoxins from dinoflagellates, *Alexanidrium catenella, A. tamarense, Gymnodinium catenatum,* and *Pyrodinium bahamense* var. *compress*a. In addition to saxitoxin (**13**) which was first isolated from the Alaska batter clam *Saxidomas giganteus*, more than 30 variants have been reported from bivalves, dinoflagellates, and cyanobacteria of the genera *Anabaena, Aphanizomenon, Nostoc,* and *Oscillatoria* (Llewellyn 2006). Saxitoxin is one of the most toxic variants with an LD_{50} value of $10\,\mu g/kg$ in mice (i.p.). It inhibits voltage-gated Na^+ channels by binding at the external surface of the pore (Site 1) (see Chap. 3).

13

3.2 Polyethers

Since the discovery of brevetoxin B from *Karenia brevis* (formally, *Gymnodinium breve* and *Ptichodiscus brevis*), a wide array of polyethers have been isolated from dinoflagellates as well as mollusks that prey on dinoflagellates and toxin containing fish (Daranas et al. 2001).

3.2.1 Brevetoxins

Brevetoxin B (**14**), an ichthyotoxic constituent of the red-tide forming dinoflagellate *K. brevis,* was the first isolated "ladder-shaped polyether" from nature. This was followed by the more potent polyether, brevetoxin A (**15**). Though not highly toxic to mice (LD_{50} values of ~50 µg/kg), both are potent ichthyotoxins (LC_{50s} 3–15 ng/ml for zebrafish). Fifteen brevetoxins have been isolated from *K. bevis* (Baden et al. 2005). Brevetoxins activate voltage-gated Na^+ channels by binding at Site 5 (see Chaps., 3 and 6).

Brevetoxins and their metabolites are involved in neurotoxic shellfish poisoning (NSP), in which some symptoms are similar to those of ciguatera fish poisonig (CFP) (see Sects. 3.2.2 and 8.4). Like ciguatoxins, brevetoxins accumulate in fish (Naar et al. 2007).

14

15

3.2.2 Ciguatoxins

Ciguatera is a food poisoning that results from eating subtropical and tropical fish at certain times and places. Ciguatoxin was named for a toxic principle first isolated from a Pacific red snapper (*Lutjanus bohar),* which frequently causes this disorder (Hashimoto 1979; Nicholson and Lewis 2006). Later, it was isolated in a purity suffi-cient for chemical analysis from the moray eel *Gymnothrax javanicus* and its structure was shown to be a complex ladder-shaped polyether (**16**) (Yasumoto and Murata 1993;

Yasumoto 2005; Nicholson and Lewis 2006). Actually, ciguatoxin is a metabolite of ciguatoxin 4B (**17**); the latter compound is synthesized in the dinoflagellate *Gambierdiscus toxicus*. Ciguatoxin 4B is metabolized to ciguatoxin in fishes as it passes up the food chain from herbivorous fish that feed upon seaweeds containing the dinoflagellate to carnivorous fish (Yasumoto 2005). So far more than 30 ciguatoxin variants have been isolated or detected from fishes or this dinoflagellate. Ciguatoxin (LD_{50} 0.25 µg/kg in mice) is about 40 times more toxic that tetrodotoxin and activates voltage-gated Na^+ channels after binding to Site 5 (see Chaps. 3 and 6).

16

17

3.2.3 Maitotoxin

Maitotoxin (**18**) was named after the Tahitian name for the surgeonfish *Ctenochaetus striatus* which frequently causes ciguatera in Tahiti (Murata and Yasumoto 2000). Again, the fish accumulates this *Gambierdiscus toxicus* toxin from its prey. Maitotoxn is the most complex polyether to be isolated and the most toxic natural product known (mouse LD_{50} 50 ng/kg). It increases membrane permeability to Ca^{2+} by an as yet unknown mechanism.

18

3.2.4   Gambierol

G. toxicus produces a variety of polyethers with potent biological activities, among which gambierol (**19**) is intriguing because it strongly inhibits voltage-gated K$^+$ channels with an IC$_{50}$ value of 1.8 ng/ml and is toxic to mice with LD$_{50}$ 50 µg/kg (Ghiaroni et al. 2005). Gambierol's mechanism of action is discussed in Chap. 6.

19

3.2.5   Prymnesins

Although produced by the haptophyte *Prymnesium parvum*, prymnesin-2 (**20**) should be mentioned here (Murata and Yasumoto 2000). It is a causative agent of massive fish kills in brackish waters. In fact, its minimal icthyotoxic concentration is 3 nM against the medaka *Tanichthys albonubes*. It is also highly hemolytic. *Prymnesium* extracts' modes of action are discussed in Chap. 6.

20

3.2.6 Other Polyethers

Many other polyether compounds, most of which are of dinoflagellate origin, are involved in human intoxications from ingestion of mollusks. Therefore, they are described in Sect. 8.

3.3 Long-Chain Polyketides

Dinoflagellates of the genus *Amphidinium* produce long-chain polyketides, e.g. amphidinol-3 (**21**), that are highly hemolytic (Yasumoto and Murata 1993). Symbiotic dinoflagellates of the genus *Symbiodinium* produce long-chain polyhydroxylated macrolides, including zooxanthellatoxins, zooxanthellamides, and symbiodinolide (**22**), that activate voltage-gated Ca^{2+} channels (Moriya et al. 1998; Onodera et al. 2005; Kita et al. 2007).

21

22

3.4 Macrolides

Among an array of macrolides produced by dinoflagellates, goniodomin A (**23**) isolated from *Alexandrium hiranoi* (formerly, *Goniodoma pseudogonyaulax*) was reported to be highly ichthyotoxic and to stimulate actomyosin ATPase activity (Daranas et al. 2001). Prorocentrolide (**24**), produced by *Prorocentrum* spp., is a unique macrolide possessing a cyclic imine moiety that is a fast-acting toxin with LD$_{50}$ 400 µg/kg (i.p.) in mice (Torigoe et al. 1988).

23   **24**

4 Macroalgal Toxins

A wide variety of seaweeds have been consumed in Asian countries and, to a lesser extent in some parts of Europe and North America, but only small numbers of human intoxications were reported (Hashimoto 1979). Coralline algae of the genus *Jania* were reported to contain tetrodotoxins (Yasumoto et al. 1986a). Brown algae often contain meroditerpenoids (Blunt and Munro 2007), of which 2β,3α-epitaondiol (**25**) and others isolated from *Stypopodium flabelliforme* were reported to be voltage-gated Na$^+$ channel modifiers (Al-Sabi et al. 2006).

25

4.1 Kainic and Domoic Acids

Several seaweeds have been used as anhelminthics in China and Japan (Hashimoto 1979); α-kainic (**26**) and domoic acids (**27**) were isolated from the red algae *Digenea simplex* and *Chondria armata*, respectively. These amino acids are potent agonists of glutamate receptors (see Chap. 5). Domoic acid is also a causative agent of amnesic shellfish poisoning as mentioned later (Sect. 8.5).

4.2 Polycavernosides

Human intoxications from ingestion of red algae of the genus *Gracilaria* have been reported in several places (Hashimoto 1979; Nagai et al. 1996; Paquette and Yotsu-Yamashita 2007). As mentioned earlier, aplysiatoxin and debromoaplysiatoxin were causative agents in the Hawaiian intoxications (Nagai et al. 1996). Polycavernoside A (**28**) and other congeners were isolated from the *G. edulis* that caused fatal human intoxication in Guam and Philippines (Paquette and Yotsu-Yamashita 2007). This compound is toxic to mice with LD_{99} values of 0.2–0.4 mg/kg. Polycavernoside A triggers an intial extracellular Ca^{2+} entry produced by a mechanism other than activation of L-type voltage-gated Ca^{2+} channels or muscarinic receptors (Cagide et al. 2007). Polycavernosides are believed to be of cyanobacterial origin, perhaps *Lyngbya* spp.

4.3 Other Macroalgal Toxins

A large number of ichthyotoxic terpenoids have been isolated from macroalgae (Blunt and Munro 2007), but their toxicities toward mammals have not been evaluated.

5 Sponge Toxins

Sponges, often referred to as the most primitive multicellular animals, harbor large numbers of symbiotic microorganisms that are frequently thought to be the real producers of bioactive sponge metabolites (Piel 2004). Although a large number of such metabolites have been isolated from marine sponges, few of them have been evaluated for toxicity toward mammals. Significant numbers of cytotoxic compounds of sponge origin target tubulin or actin as mentioned in Chap. 7, while others are potent tumor promoters that inhibit protein phosphatases 1 and 2A as described in Chap. 8.

5.1 Polyalkylpyridiniums

Sponges of the order Haplosclerida often contain 3-alkylpyridinium salts (Andersen et al. 1996; Sepčić 2000), some of which were reported to be toxic to mammals. The first example of this class is halitoxin (**29**) isolated from *Haliclona* spp. which is actually a mixture of oligomeric/polymeric 3-alkylpyridinium salts (Schmitz et al. 1978; Turk et al. 2007). Halitoxin is hemolytic, ichthyotoxic and toxic to mice (LD_{50} 1.4 mg/kg, i.v.). A similar polymeric mixture (**30**) from the Mediterranean *Reniera* (*Haliclona*) *sarai* and referred to as poly-APs (polymeric alkylpyridinium salts) largely consists of two major polymers with molecular weights of 5 and 19 Da. It showed a wide range of biological activities, including hemolytic, lethal (above 2.7 mg/kg i.p. in rats), and acetylcholinesterase inhibitory (Sepčić et al. 1997a, b). More significantly, it forms transient pores in cell membranes like halitoxin (McClelland et al. 2003).

29

30

5.2   Kainate Receptor Agonists

Several agonists of glutamate receptors including dysiherbaine (**31**) have been isolated *Dysidea harbasea* (Sanders et al. 2006) as will be described in Chap. 6.

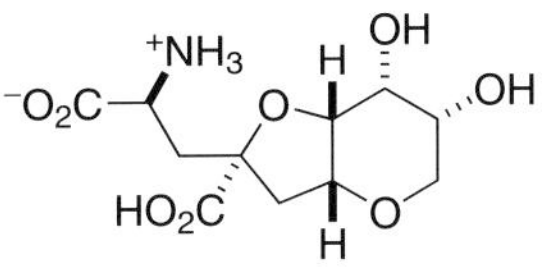

31

5.3   Other Sponge Toxins

Bastadins [e.g. bastadin 5 (**32**)], bromotyrosine-derived metabolites isolated from *Ianthella basata,* are selective modulators of sarcoplasmic reticulum Ca^{2+} channels and behave either as full or partial agonists (Zucchi and Ronca-Testoni 1997). Penaramides (**33**), novel acylated polyamines isolated from *Penares* aff. *incrustans*, inhibit binding of ω-conotoxin GVIA to N-type Ca^{2+} channels at µM levels (Ushio-Sata et al. 1996). Cyclostellettamines A–F [A (**34**)], cyclic 3-alkylpyridines isolated from the marine sponge *Stelletta maxima*, are muscarinic acetylcholine receptor antagonists (Fusetani et al. 1994).

Finally, a variety of isocyanoterpenoids and related terpenoids are known from sponges, especially *Acanthella* spp. They show a wide range of activities including ichthyotoxic and cytotoxic, but their modes of action as well as toxicity toward mammals are unknown (Chang 2000).

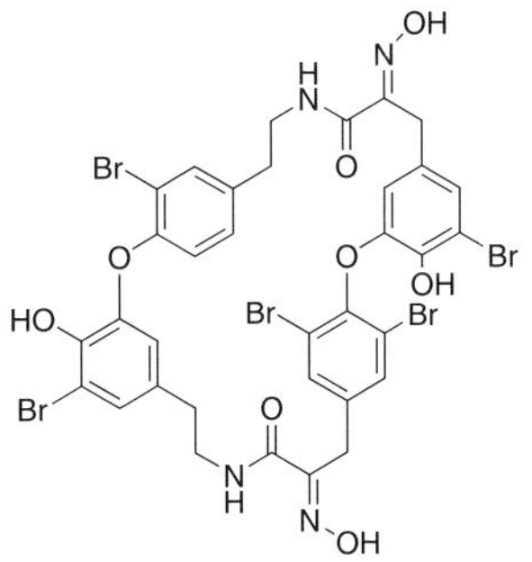

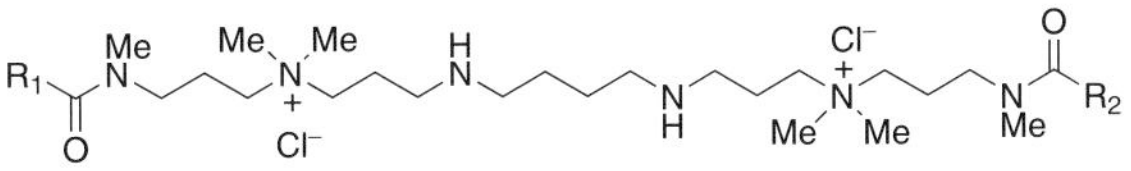

R1, R2=$C_{10}H_{21}$

33

32

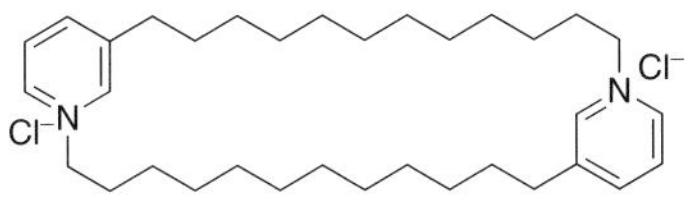

34

6   Cnidarian (Coelenterate) Toxins

Soft corals have been intensively investigated for their secondary metabolites, and a large number of terpenoids, especially diterpenoids have been isolated (Blunt and Munro 2007). Cembranoids, a typical class of soft coral diterpenes, exhibit a wide range of biological activities (Coll 1992). Lophotoxin (**35**) and related cembranoids isolated from gorgonians of the genera *Lophogorgia* and *Pseugopterogogia* are irreversive nicotinic receptor antagonists; lophotoxin is lethal to mice with an LD_{50} value of 8 mg/kg and blocks the binding of cobra a-toxin to nicotinic receptors of BC3H-1 cells with IC_{50} values of $1–2 \mu M$ (Fenical et al. 1981; Culver et al. 1985). Cnidarians possess cnidocysts (previously called nematocysts) that contain peptidic and proteinaceous toxins; these will be described in Sects. 6.2 and 6.3.

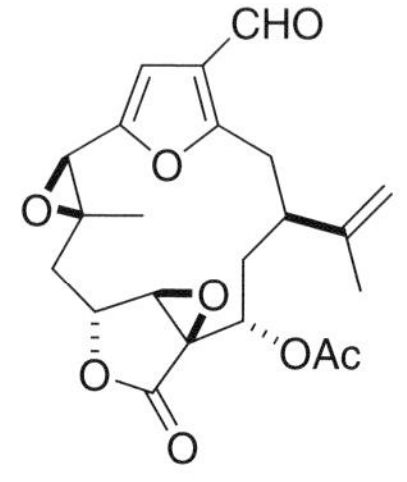

35

6.1 Palytoxins

Palytoxin (**36**), is a highly unusual metabolite occurring in zoanthids of the genus *Palythoa* (Hashimoto 1979; Moore 1985; Uemura 1991). It is highly toxic with a mouse LD_{50} of 0.45 µg/kg (i.p.). Its mode of action has been extensively studied and it has been demonstrated that palytoxin binds to the cell membrane Na, K-ATPase and converts it from an ion pump into an ion channel, which greatly increases membrane Na^+ permeability and results in a large inward current (Artigas and Gadsby 2003; Hilgemann 2003). It should be noted that palytoxin is a novel skin tumor pomoter (Wattenberg 2007; also, see Chap. 8).

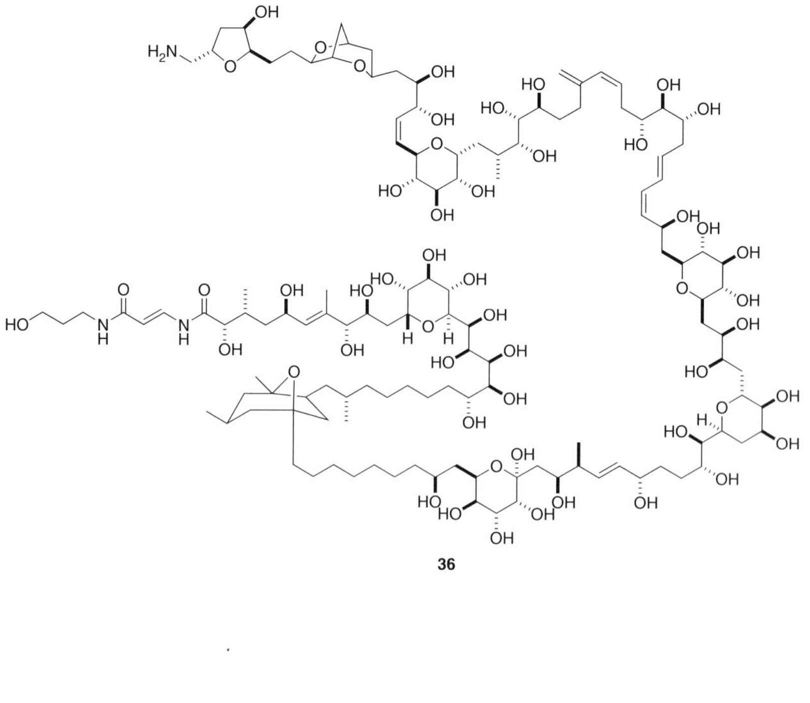

36

Palytoxin and its analogues have been reported to be responsible for such human intoxications as clupeotoxism (Onuma et al. 1999), xanthid crab poisoning (Yasumoto et al. 1986b) and filefish poisoning (Hashimoto 1979). Palytoxin analogues have been isolated also from the red alga *Chondria armata* (Maeda et al. 1985) and the dinoflagellates of the genus *Osteropsis* (Usami et al. 1995; Riobo et al. 2006). Palytoxin is sequestered by a wide range of coral reef animals from zoanthid colonies (Gleibs and Mebs 1999).

6.2 Cnidarian Peptides

The phylum Cnidaria (the three classes are: anthozoans, schyphozoans, and hydrozoans) capture their prey and repel predators via cnidae, also called cnidocysts or nematocysts. Many forms of cnidae have been described and the "cnidome" of each cnidarian species usually consists of several anatomically distinct forms. Each cnidocyst is elaborated within the Golgi apparatus of a cnidocyte, the cell which makes this unique structure. When cnidocytes are activated by a combination of chemical and mechanical stimuli, an arrow-like tube within the cnidocyst rapidly everts to make contact with the organism that stimulated the discharge. Depending on the cnidocyte, the tubule may be used to ensnare the prey (adhesive spirocyst) or, if barbed, to actually penetrate the skin of the victim and inject a paralyzing venom (most of the >30 different types of cnidocytes are thought to have this role). All members of this large, almost entirely marine phylum possess cnidae and most of these are predicted to contain venoms composed of a variety of peptides and proteins (peptides with MW >10,000). While there may be some relatively non-toxic cnidarian species, these would be quite exceptional.

Cnidarian peptides have not yet been investigated in as much detail as the *Conus* peptides, because they are usually of larger molecular size, which makes structure determination and subsequent synthesis for biological evaluation quite challenging. Furthermore, cnidarian toxins are often diffusely distributed over the entire surface of the body rather than being localized within a discrete organ, and this makes toxin collection and purification more difficult. It is likely that the myriad types of toxins produced by cnidarians will ultimately be found to be as diverse as those of the cone shells, a genus producing an amazingly diverse group of peptide toxins.

6.2.1 Na+ Channel Modulating Toxins

These were the first cnidarian peptides purified. ATXs I–III isolated from a European species (*Anemonia sulcata*; Béress et al. 1975) and anthopleurin A from a Pacific species (*Anthopleura xanthrogrammica*, Norton et al. 1976) were initially investigated for their cardiotonic (inotropic) and nerve action potential modifying activities. These "long" peptides consist of a single peptide chain of 45–48 amino acid residues cross-linked by three disulfide bonds. High resolution NMR studies

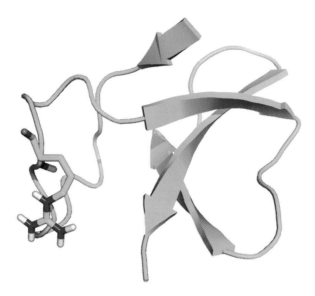

Fig. 1 Ribbon diagram of the solution structure of ShI, a peptide modulating Na channel inactiva-tion from the Carribbean mat sea anemone, *Stichodactyla helianthus* (Kem et al., 1989). The folded structures of the so-called "long" sea anemone neurotoxins are stabilized by several seg-ments of β-pleated sheets and three disulfide bonds; all contain a relatively flexible loop that may also interact with the Na channel (Fogh et al. 1990; Wilcox et al. 1993)

have yielded a number of similar solution structures; the ShI structure (Fig. 1) was of particularly high resolution (Fogh et al. 1990).

Although the existence of 40+ homologous long toxin sequences might seem adequate for assessing the structure-activity relationship for these toxins, most of these sequences are from closely related sea anemone species. Future studies of related toxins from other sea anemone families are expected to provide a better perspective as to the structural variability of these peptides.

These toxins act by slowing the repolarization phase of nerve and muscle action potentials by inhibiting the process of Na^+ channel inactivation (Murayama et al. 1972; Salgado and Kem 1992). When a nerve action potential, which may now last hundreds of milliseconds (like a myocardial action potential) reaches the nerve terminal, massive release of neurotransmitter results, causing hyperexcitability and convulsions. The cardiotonic actions of the long peptides are of potential therapeutic interest; however, the positive inotropic effect is transitory, leads to calcium-loading of the myocardial cell and is often accompanied by arrhythmias. Currently these Na^+ channel toxins are mainly used as molecular probes for studying the gating mechanisms of these channels.

While most sea anemone Na^+ channel modulating peptides are "long" toxins, rep-resentatives of another subfamily, the "short" toxins are found in a few species. These contain 30–32 residues and 4 disulfide bonds. They lack homology with the long toxins. So far, these peptides are only known to affect crustacean and insect nerve action potentials in a manner similar to the long sea anemone toxins (summarized by Honma et al. 2003).

Other anthozoans, particularly hexacorals, represent an almost unstudied group with respect to their Na$^+$ (and other) channel toxins. However, a 12 kDa peptide isolated from an unidentified *Goniopora* species (Hashimoto and Ashida 1973) was shown to delay Na$^+$ channel inactivation (Muramatsu et al. 1985; Gonoi et al. 1986). While its effects were similar in many respects to those of the anemone toxins, its sequence was not homologous to them (Ashida et al. 1987).

6.2.2   Sea Anemone K$^+$ Channel Peptide Toxins

Currently the most actively investigated ion channel modulating peptides are those which affect the gating of various K$^+$ channels, which are more diverse than Na$^+$ channels, in function as well as in number. While K$^+$ channels are important during the repolarization phase of an action potential, they also regulate excitability by controlling resting membrane potential and conductance; they also regulate calcium fluxes in non-electrically excitable cells such as T-lymphocytes as well as electrogenic cells. The initial "short" K$^+$ channel toxins characterized were BgK (Aneiros et al. 1993) and ShK (Castenada et al. 1995), both isolated from Carribbean sea anemones. These basic peptides consist of a 35–37 residue chain whose folded structure is stabilized by 4 disulfide bonds. Each peptide contains two small alpha-helical segments that flank an essential lysyl sidechain that enters and blocks the pore. The mechanism of channel block is nearly identical to what had previously been demonstrated for a scorpion toxin, charybdotoxin (Pennington et al. 1996a; Dauplais et al. 1997). These anemone toxins mainly block "delayed rectifier" K channels that are activated during the repolarization phase of action potential, but they also block an intermediate conductance Ca^{2+}-activated K$^+$ channel found in certain blood cells. Although they are minor constituents of sea anemone extracts, their highly basic character and thermal stability (even at 100°C) facilitated their isolation and subsequent sequence determinations. They were readily synthesized by solid-phase methods (Pennington et al. 1995); this allowed detailed structure-activity analyses, including so-called "alanine" scans, to identify residues critical for ion channel binding (Pennington et al. 1996a, b; Alessandri-Haber et al. 1999). The solution structure of synthetic ShK (Fig. 2) could then be determined by NMR methods (Tudor et al. 1996). Analogs of ShK are currently in development for treating autoimmune diseases because they are able to preferentially inhibit the proliferation of antigen-activated T-lymphocytes. By blocking Kv1.3 channels the membrane potential of the lymphocyte is reduced and this indirectly reduces the Ca^{2+} influx of required to stimulate proliferation (Kalman et al. 1998; Kem et al. 1999; Beeton et al. 2006).

Anemone peptides affecting other types of K$^+$ channels are also known and these will be described in Chap. 4.

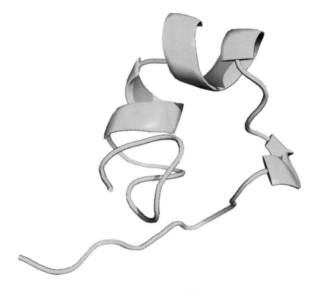

Fig. 2 Ribbon diagram of the solution structure of ShK, a peptide blocking certain K channels from the Carribbean mat sea anemone, *Stichodactyla helianthus*. A short stretch of α-helix is present, but no β-pleated sheet structure is present (Tudor et al. 1996)

6.2.3   A Protein Toxin that Blocks Ca²⁺ Channels

A coral toxin (MW~19,000) blocking Ca channels has been isolated from an unidentified Red Sea *Goniopora* species (Qar et al. 1986). Systematic screening of extracts from other anthozoan species likely will provide other Ca^{2+} channel blockers and modulators.

6.3   Cnidarian Protein Toxins

6.3.1   Actinoporins

Two hemolytic sea anemone toxins, one from European Atlantic and Mediterranean coasts (equinatoxin) and one from the Gulf of Mexico (*Stichodactyla* cytolysin II) were initially isolated (Macek and Lebez 1988; Kem and Dunn 1988). Edman and mass spectrometric sequence analyses and comparison of Florida-collected cytolysin II and Cuba-collected Sticholysin I showed that they are the same molecule ((Blumenthal and Kem 1983; Huerta et al. 2001; Stevens et al. 2002). These toxins are small (~21 kDa) proteins and though they lack disulfide bonds, their folded structures are very stable due to an extensive BB-pleated sheet secondary structure that involves most of their peptide bonds. Because these sea anemone (order Actinaria) toxins structurally resemble bacterial porins by having a predominantly β-sheet

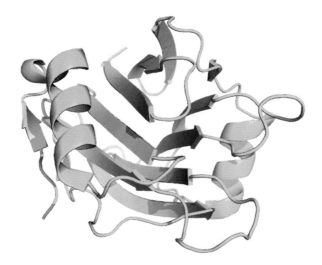

Fig. 3 Ribbon diagram of the solution structure of equinatoxin II, a cytolytic pore-forming protein (actinoporin) occurring in the European sea anemone *Actinia equine*. Although most of the folded structure of this type of toxin is stabilized by β-pleated sheet secondary structure, a helical segment near the N-terminus is essential for membrane insertion and pore formation in association with three other monomers (Hinds et al. 2002)

secondary structure and form stable, relatively large transmembrane pores, they are now commonly referred to as "actinoporins" (Kem 1988). Four monomers of the toxin aggregate to form a functional pore. One of the most interesting properties of the actinoporins, their ability to insert into lipid bilayers membranes and behave like membrane proteins, is greatly facilitated by the presence of small amounts of sphingomyelin. A high resolution X-ray crystallographic structure of equinatoxin monomer reveals an α-helical segment near the N-terminus that other studies (Fig. 3) implicate in membrane insertion and pore formation (Athaniasiadis et al. 2001). X-ray diffraction and EM studies of 2D crytals of sticholysin II adsorbed to a lipid layer have provided a low resolution model of the tetrameric pore (Mancheno et al. 2003).

6.3.2 Scyphozoan and Hydrozoan Toxins

In spite of many attempts, these proteins have been very difficult to isolate as stable, active toxins. However, recent studies with certain scyphozoan and hydrozoan toxins have been successful (Nagai et al. 2000; 2002). These relatively large proteins were sequenced by molecular biological methods. Two additional toxins from the cubomedusan *Chironex fleckeri,* whose sting can be lethal to swimmers, were recently isolated and sequenced (Brinkman and Burnell 2007). These homologous proteins (MW ~45,000) have certain common features.

The first hydrozoan toxin that was isolated and sequenced was a lytic protein called "hydralysin" (Zhang et al. 2003). It was reported that this toxin is not located within cnidocytes. Further cloning revealed several other hydralysins of similar molecular size (27–31 kDa) and sequence (Sher et al. 2005). Recently a new method for isolating toxins from undischarged cnidocysts of fire corals *Millepora* spp., another hydrozoan group, was reported. Three different toxic fractions were resolved. The smallest MW fraction was partially sequenced and its entire sequence inferred by analysis from its cDNA (Iguchi et al. 2007). MCTx-1 is an acidic protein of 222 amino acid residues that lacks phospholipase A_2 activity but is lethal (IC_{50} 79 ng/ml) to cultured mouse leukemia cells and to crayfish (LD_{50} 160 ug/ml). Interestingly, it displays some homology with extracellular matrix proteins called dermatopontins. The successful use of recombinant DNA methods in these investigations will hopefully inspire others to "fish out" sequences of other schyphozoan and hydrozoan toxins in the near future, which will facilitate the production of antibodies for treating victims of their stings and provide insights into the mechanisms of action and evolution of these protein toxins.

7 Worm Toxins

7.1 Annelid Alkaloids

The polychaete *Lumbriconereis brevicirra* secrets a toxic mucus when disturbed; its active constituent is nereistoxin (**37**), a dimethylamine possessing a 1,2-dithiolane ring. Its structure was proposed in 1962 and subsequently confirmed by synthesis (Okaichi and Hashimoto 1962; Hagiwara et al. 1965). It is toxic to mice (LD_{50} 30 mg/kg i.v.) and also displays ichthyotoxic and insecticidal properties. It inhibits nicotinic acetylcholine receptors (nAChRs). Depending on the particular nAChR involved, it (1) competes with acetylcholine at postsynaptic membrane nicotinic acetylcholine receptor sites, and (2) directly blocks the nAChR ion channel. A family of insecticides that includes cartap (the first member, initially developed in Japan) is based on the nereistoxin structure (Hashimoto 1979).

$$Me_2N-\text{\textless} \begin{smallmatrix} S \\ | \\ S \end{smallmatrix}$$

37

7.2 Annelid Peptides and Proteins

Arenicins are novel antimicrobial peptides that were recently isolated from coelomocytes of the lugworm *Arenicola marina*. Two isoforms (MW~2,800) were sequenced. Arenicins act upon Gram-positive and Gram-negative bacteria and fungi (Ovchinnikova et al. 2004). The interesting solution structure of arenicin and its membrane actions have been the subject of recent papers (Lee et al. 2007; Andra et al. 2008).

Glycerotoxin, a very large (~320 kDa) protein from the blood worm (*Glycera convoluta*, from Atlantic coast of France) specifically stimulates the exocytotic release of neurotransmitter from nerve terminals and is thus a useful tool for understanding the presynaptic processes that are involved in neurotransmitter release (see Chap. 6).

7.3   Nemertine Alkaloids

Bacq, a Belgian pharmacologist, discovered the existence of toxins in nemertines, a phylum of carnivorous marine worms that currently includes ~1,000 described species. Many species that belong to the class Hoplonemertinea contain pyridine alkaloids that function as toxins and feeding repellents (Kem 1971; Kem et al. 2001, 2003). Hoplonemertines paralyze their prey, releasing an alkaloidal venom from a proboscis that simultaneously punctures the skin of the prey with a mineralized stylet. These alkaloid toxins allow the worm to rapidly subdue its prey (usually annelids or crustaceans) since they target nicotinic acetylcholine receptors important for body movements.

Anabaseine, 2-(3-pyridyl)-3,4,5,6-tetrahydropyridine (**38**) was the first alkaloid to be isolated and identified (Kem et al. 1971; Kem et al., 1997). The skin and proboscis of a Pacific hoplonemertine, *Paranemertes peregrina*, contains very high concentrations of this compound (Kem 1971). Anabaseine, like nicotine (**39**), stimulates all known nAChRs to some degree. While nicotine preferentially stimulates a brain nicotinic acetylcholine receptor (alpha4beta2) associated with smoking pleasure, anabaseine preferentially stimulates the neuromuscular nAChR and another brain nAChR, the so-called alpha7 receptor (Kem et al. 1997). A 3-benzylidene anabaseine derivative, GTS-21 (**40**), was shown to improve learning and memory in animal experiments and clinical trials (Kem et al. 2008; Kem et al., 2004; 2006). It is the first compound found to selectively stimulate the $\alpha 7$ type nAChR.

Another hoplonemertine found in the northwestern Atlantic as well as Pacific oceans, *Amphiporus angulatus*, contains at least 15 different pyridine alkaloids (Kem et al. 1976). This is one of the largest hoplonemertines known, attaining a biomass of about one gram fresh weight; most hoplonemertines are less than 0.1 g fresh weight. While anabaseine is only a trace constituent of this species, the related 2,3'-bipyridyl (**41**) readily paralyzes crustaceans. The 3-methyl-2,3'-bipyridyl was recently identified (Kem et al., submitted). All eight C-methylated 2,3'-bipyridyls were synthesized and tested for toxicity and ability to inhibit barnacle larval settlement (Kem et al. 2003). Nemertelline (**42**), the most abundant alkaloid, is a tetrapyridyl with similarities to the tripyridyl tobacco alkaloid nicotelline (Thesing and Muller 1956; Cruskie et al., 1995), whose structure is identical to nemertelline without the nemertelline A ring. The biological activities of nemertelline are not yet well understood, though it has been demonstrated to inhibit the settlement of barnacle larvae at relatively high concentrations (Kem and Soti 2001).

41   42

Although the toxins of only a few readily accessible species have been examined in any detail, it is already clear that hoplonemertines contain a multitude of heterocyclic compounds. Future investigations of a wider variety of species should yield many new structures and interesting biological activities.

7.4 *Nemertine Peptide Toxins*

Nemertines that are not hoplonemertines also capture their prey with a long proboscis that wraps around the prey and prevents its escape. Heteronemertines such as the large *Cerebratulus* and *Parbolasia* species are known to produce peptide neurotoxins and/or cytolysins, some of which have been sequenced (Kem, 1976; Blumenthal and Kem, 1976; 1980; Blumenthal et al., 1981). *Cerebratulus lacteus (Cl)*, a relatively common intertidal nemertine along the eastern seaboard of North America produces at least four homologous neurotoxins (MW ~6,000), referred to as B toxins because they elute from a gel column after the larger A-type cytolysins. The B peptides are toxic to crustaceans but not mammals. These are very basic peptides consisting largely of α-helical segments that are folded together and

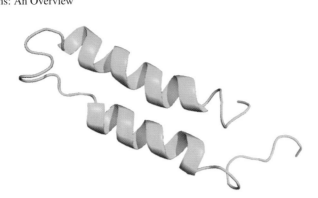

Fig. 4 Ribbon diagram of the solution structure of neurotoxin B-IV, a peptide occurring in the Atlantic heteronemertine *Cerebratulus lacteus*. Two long α-helices connected by a hairpin B-turn and stabilized by four disulfide bonds constitute the scaffold of this elongated toxin, which induces spontaneous action potentials in crustacean nerves (Tudor et al. 1996)

stabilized by four disulfide bonds (Kem 1976; Kem et al. 1990). The folded structure of the most abundant crustacean paralyzing toxin Cl-BIV (Fig. 4) is rather unique for neurotoxins affecting voltage-gated ion channels, consisting primarily of two α-helices separated by a hairpin turn (Barnham et al. 1997).

The *Cerebratulus* B-neurotoxins transiently excite but then block the generation of action potentials in crustacean nerves and thereby cause spastic followed by flaccid paralysis. Their targets are likely sodium channels, as TTX inhibits their contractile action on the isolated crayfish nerve-opener muscle preparation (Kem, unpublished results).

The *Cerebratulus* A toxins are small (~10 kDa) cytolytic proteins localized in the integumentary (skin and proboscis) tissues. Cl-AIII is the most abundant and well characterized isotoxin (Kem and Blumenthal 1978). These peptides are also very basic and contain several disulfide bonds to stabilize their folded structures. A homologous, cytolytic protein was isolated from a very large (>2 m) Antarctic species, *Parborlasia corrugans*. Whereas the Cl-A isotoxins were difficult to chromatographically resolve, separation of the various *Parborlasia* isotoxins was not possible due to their similarity in size, ionization or other characteristics (Berne et al. 2003). Cloning techniques will probably be required to satisfactorily resolve and sequence these toxins. These heteronemertine cytolysins presumably permeabilize membranes by forming pores or by acting as potent detergents. Both the A and B types of heteronemertine toxins may be used offensively as well defensively, even if there is no special skin puncturing mechanism available for facilitating their entry into a victim. *Cerebratulus* feeds on clams as well as annelids; it displays an uncanny ability to slip its proboscis between the two shells of a clam and cause it to open.

Paleonemertines, a third class, also possess cytolytic proteins that have not yet been isolated (Kem 1971, 1994).

7.5   Other Worm Toxins

Two species of flatworms (Miyazawa et al. 1987; Ritson-Williams et al. 2006) and a polychaete (Yasumoto et al. 1986) have been reported to contain tetrodotoxins. Arrowworms (phylum Chetognaths) are abundant members of planktonic predators, among which six species are known to possess tetrodotoxin (Thuesen et al. 1988). Flatworms and arrowworms are likely to use tetrodotoxin to catch prey organisms; arrow worms may sequester tetrodotoxin from *Vibrio alginolyticus* living associated with these worms (Thuesen and Kogure 1989).

8   Mollusks

Bivalve and gastropod mollusks often accumulate a variety of toxic compounds in digestive glands (mid-gut glands) from prey organisms and cause human intoxications known as shellfish poisonings mentioned below (Halstead 1967; Hashimoto 1979; Camacho et al. 2007). In addition, the trumpet shell *Charonia sauliae* and the Japanese ivory shell *Babylonia japaonica*, carnivorous gastropods, and the blue-ringed octopus *Octopus maculosa* accumulate tetrodotoxin (Hashimoto 1979; Mosher and Fuhrman 1984). Cone snails of the genus *Conus* hunt prey organisms, fishes, mollusks and worms, using complex mixtures of toxic peptides as described in Chap. 2.

8.1   Paralytic Shellfish Poison (PSP)

Bivalves accumulate saxitoxin and its derivatives which are sequestered from dinoflagellates and cause paralytic shellfish poisoning as mentioned in Sect. 2.

8.2   Diarrhetic Shellfish Poison (DSP)

Since the first outbreak of diarrhetic shellfish poisoning (DSP) reported from northern Japan in 1976, large numbers of DSP incidents have been reported from Europe, Japan, and other countries. Okadaic acid (**43**) and its derivatives named dinophysistoxins-1 (**44**) and −3 (**45**) as well as pectenotoxins [pectenotoxin-2 (**46**)] were isolated from mussels and dinoflagellates of the genus *Dinophysis* (Daranas et al. 2001; Ciminello and Fattorusso 2004; Camacho et al. 2007) these polyethers except for pectenotoxins inhibit protein phosphatases 1 and 2A strongly. Although the toxicities of okadaic acid and various dinophysistoxins are not strong (LD_{50} values of 160–500 µg/kg, i.p. in mice), their potent tumor promotion activity is of serious concern from the viewpoint of public health as mentioned in Chap. 8. The pectenotoxins cause similar symptoms and are as toxic as okadaic acid, but pectenotoxin-2 is not a tumor promoter; rather, it inhibits actin polymerization (Burgess and Shaw 2001; also, see Chap. 7). It should be noted that more than 20 okadaic acid analogues have been isolated from mollusks, sponges and dinoflagellates of the genera *Dinophysis, Prorocentrum* (Daranas et al. 2001).

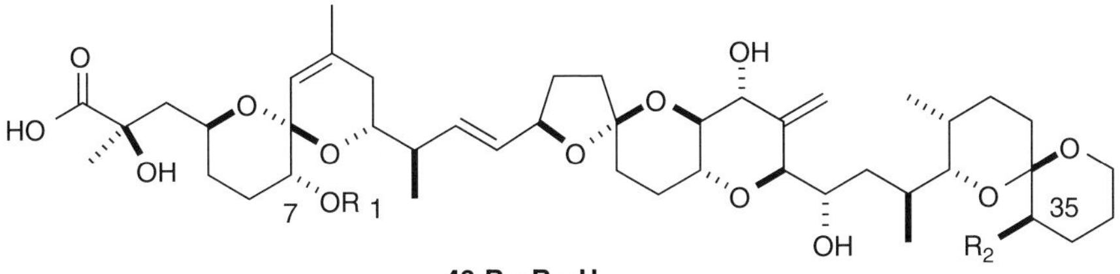

43:R₁=R₂=H
44:R₁=H, R₂=Me
45:R₁=acyl, R₂=Me

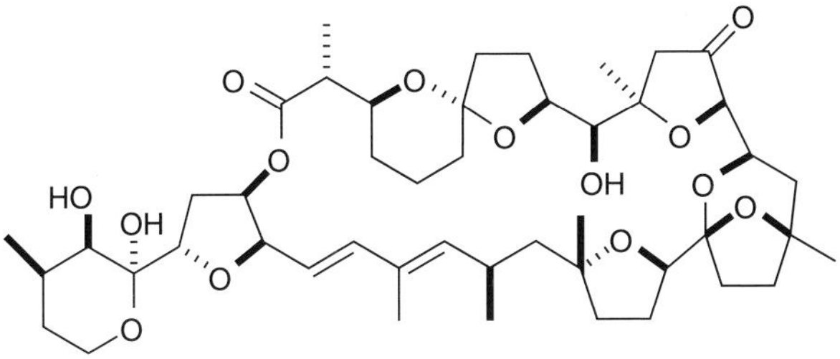

46

Yessotoxin (**47**) is a ladder-shaped polyether originally isolated as a causative agent of DSP from the Japanese scallop *Patinopecten yessoensis* (Bowden 2006). More than 22 analogues of yessotoxins have been isolated either from shellfish or the dinoflagellates *Protoceratium reticulatum* and *Lingulodinium polyedrum*. They kill mice within hours at doses of 100–200 µg/kg and are highly cytotoxic. At cellular levels, accumulation of a 100 kDa fragment of E-cadherin was observed when treated with yessotoxin. It also produces Ca^{2+} influx through some types of Ca^{2+} channels. However, yessotoxins are no longer considered as causative agents of DSP due to their low oral toxicity and lack of diarrhetic acitivity in mice (Ciminello and Fattorusso 2004).

47

8.3 Azaspiracid Shellfish Poison (AZP)

A human intoxication similar to DSP occurred after ingestion of mussels in Europe in 1995. It was named azaspiracid shellfish poisoning (AZP) after azaspiracids, the causative agents isolated from mussels; 11 azaspiracids have been known at present (Virariño et al. 2007). Azaspiracid-1 (**48**), a novel nitrogen-containing polyether, is lethal to mice at i.p. doses of 110–200 µg/kg, highly cytotoxic (IC_{50} values of 10^{-9} to 10^{-8} M), and teratogenic (Ronzitti et al. 2007). Azaspiracids cause severe damage in the epithelium of several organs and induce accumulation of an E-cadherin fragment (as the case of yessotoxins).

48

8.4   Neurotoxic Shellfish Poison (NSP)

During *Karenia brevis* blooms bivalves often become poisonous and cause human intoxications similar to ciguatera, as mentioned earlier. The causative agents are thought to be brevetoxins and their metabolites such as brevetoxin B1 (**49**) (Yasumoto 2005).

49

8.5   Amnesic Shellfish Poison (ASP)

A unique food poisoning characterized by the short-term memory loss occurred by ingesting mussels in Canada in 1987 and was named amnesic shellfish poisoning (ASP) (Mos 2001; Sobel and Painter 2005). The causative agent was identified as domoic acid and it was shown to originate from the diatom *Pseudonitzschia multiseries*. Domoic acid and several congeners are produced by several diatoms of the same genus (Daranas et al. 2001). Mass mortalities of marine animals occur from eating domoic acid containing prey (Mos 2001).

8.6   Pinnatoxins and Related Cyclic Imines

Food poisonings from ingestion of bivalves of the genus *Pinna* are known in China and southern Japan. In fact, four novel cyclic imines named pinnatoxins A–D were isolated from the digestive glands of *P. muricata* collected in Okinawa (Kita and Uemura 2005; Molgo et al. 2007). Pinnatoxin A (**50**) is lethal to mice with an LD_{99} value of 180 μg/kg (i.p.), while pinnatoxins B and C (**51, 52**) are more toxic (LD_{99} 20 μg/kg). Similar toxins named pteriatoxins A–C [A (**53**)] have been isolated from the Okinawan bivalve *Pteria penguin*.

50   **51: 34*S***   **53**
                          52: 34*R*

Spirolides [spirolide A (**54**)] were originally discovered from the digestive glands of mussels and scallops near aquaculture sites along the eastern shore of Nova Scotia, and were later found in the dinoflagellate *Alexandrium ostenfeldii* (Gill et al. 2003; Molgo et al. 2007). Spirolides are fast-acting toxins that may target the muscarinic and nicotinic receptors. Gymnodimine (**55**) is a causative agent of the food poisoning resembling neurotoxic shellfish poisoning occurred from ingestion of New Zealand oysters (Molgo et al. 2007). It is produced by the dinoflagellate *Karenia selliformis*. Gymnodimine is lethal to mice with an LD_{50} value of 96 μg/kg (i.p.) and broadly targets nicotinic acetylcholine receptors (Molgo et al. 2007).

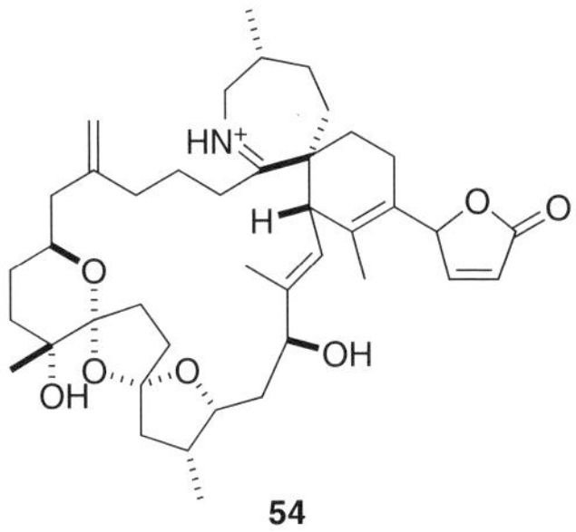

54

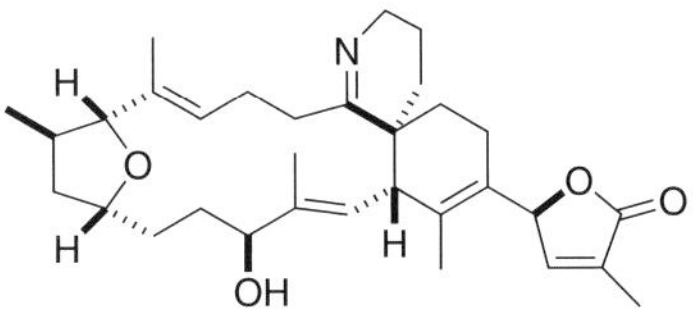

55

8.7   Surugatoxins

The Japanese ivory shell *Babylonia japaonica* is a carnivorous gastropod and is known to accumulate toxic substances such as tetrodotoxin from food (Hashimoto 1979). It caused an unusual food poisoning characterized by failing eyesight. Surugatoxin (**56**) was originally isolated as a causative agent from the digestive glands, but further examination led to isolation of neosurugatoxin (**57**), which is 100 times more active than surugatoxin, inducing mydriasis in mice. The conversion of surugatoxin to neosurugatoxin was experimentally demonstrated (Kosuge et al. 1982). Neosurugatoxin is a relatively potent antagonist at nicotinic acetylcholine receptors (Hayashi et al. 1984). Interestingly, neosurugatoxin was produced by a Coryneform bacterium isolated from sea sediments inhabited by *B. japonica* (Kosuge et al. 1985).

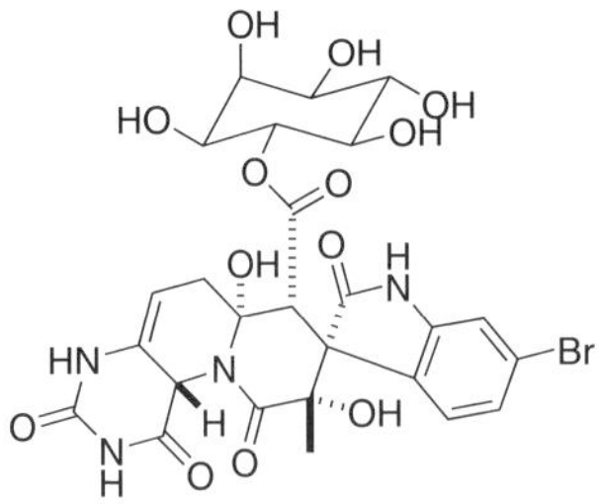

56

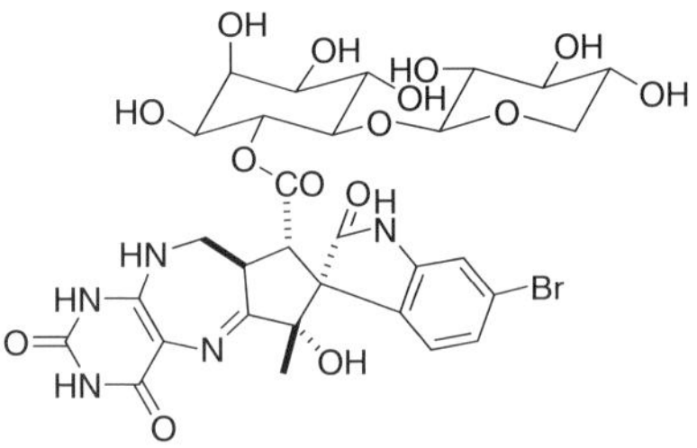

57

8.8 Other Toxins

Several Japanese abalones (genus *Haliotis)* accumulate pyrophaephorbide *a* (**58**) in their viscera in early spring and an unusual food poisoning from eating such viscera was reported in northern Japan (Hashimoto 1979). Pyropheophorbide *a* is a photo-dynamic agent derived from chlorophyll *a*. Thus, abalone poisoning is a rare photosensitization disease.

58

A protein toxin fraction named cephalotoxin that potently paralyses crabs is found in the posterior salivary glands of octopuses and cuttlefishes (Ghiretti 1959; Songdahl and Shapiro 1974). The α- and β-cephalotoxins purified from *Octopus vulgaris* are glycoproteins with respective molecular weights of 91,200 and 33,900 (Cariello and Zanetti 1977). Their mechanism of action is not yet known. A cyto-lytic protein displaying only weak crab paralytic activity occurs in the posterior salivary gland of the Atlantic squid, *Loligo pealii* (Kem and Scott 1980). The blue-ringed octopus *Hapalochlaena maculosa* is unique among octopi in that its posterior salivary glands contain TTX, which is secreted as a venom when it bites an attacker. An additional toxin, hapalotoxin, is apparently more effective in immobilizing prey (Savage and Howden 1977).

9 Arthropods, Bryozoans and Echinoderms

Crabs belonging to the family Xanthidae cause human intoxications in Indo-Pacific regions, while horseshoe crabs sporadically cause food poisonings in South-East Asian countries (Halstead 1967; 1984; Hashimoto 1979). While the xanthid crabs contain palytoxin, saxitoxins, and/or tetrodotoxins (Halstead 1984; Yasumoto 2005), horseshoe crabs often contain either saxitoxins or tetrodotoxins (Halstead 1984; Miyazawa and Noguchi 2001). The origins of these toxins are unknown, except for tetrodotoxins in xanthid crabs that presumably originate from red algae of the genus *Jania* (Yasumoto 2005). It was also reported that the starfish *Asteropecten polyacanthus* contains tetrodotoxin (Mosher and Fuhrman 1984).

"Dogger Bank itch" is an allergic contact dermatitis disease popular among North Sea fishermen caused by sensitization to an allergen from the marine bryozoan *Alcyonidium gelatinosum*. The allergen was identified as (2-hydroxyethyl) dimethylsulfoxonium (**59**) (Carlé and Christophersen 1980). This compound was also found in a sponge (Warabi et al. 2001).

59

9.1   Holothurins and Asterosaponins

The first animal saponin named holothurin was isolated from the sea cucumber *Holothuria vagabunda* in 1955 (Yamanouchi 1955) and the sea cucumber saponins have been generally called "holothurins." The complete structure of a holothurin [holotoxin A (**60**) from *Sticopus japonicus*] was elucidated in 1978 (Hashimoto 1979). Holothurins are triterpenoid saponins composed of a lanosterol-derived aglycone and several sugars (Minale et al. 1993) and show a wide range of biological activities such as hemolytic, ichthyotoxic, antimicrobial (Hashimoto 1979). Asterosaponins [thornasteroside A (**61**) from the crown-of-thorns starfish *Acanthaster planci*] are steroidal saponins derived from starfishes and exhibit a variety of biological activities similar to the holothurins (Hashimoto 1979; Minale et al. 1993). In recent decades a large number of saponins have been isolated from sponges, cnidarians and brittle stars, in addition to sea cucumbers and starfishes (Blunt and Munro 2007).

60

61

9.2   Protein Toxins

From some sea urchins a venomous secretion is elicited when globiferous pedicellariae, flower-shaped structures mounted on stalks found between the spines, are mechanically and chemically stimulated at the same time. Some urchins (*Diadema* species, for example) also have long slender spines that are reputed to be poisonous, but this has not yet been unequivocally demonstrated. Pedicellarian toxins are largely, if not totally, peptidic substances. Investigations on several poisonous European and IndoPacific sea urchins have reported a variety of physiological actions of pedicellarian venom, including hemolysis, inflammation, algesia (pain), neurotoxicity and cardiovascular depression. It is not yet clear that these activities are due to the same peptides. Most activities are retained on concanavalin A chromatographic columns and can be released by the addition of specific sugars to the eluting solution. Two D-galactose-specific sea urchin lectins (SUL-1 and -2) with smooth muscle stimulatory activity were isolated from *T. pileolus* (Nakagawa et al. 1991). SUL-1 (MW~32,000) is mitogenic and induces dendritic cell maturation (Takei and Nakagawa 2006). These lectins could become useful probes for investigating the processes involved in the exocytotic secretion of neurotransmitters and other signaling molecules (Takei et al. 1993).

While most starfish chemically defend themselves by means of their integumentary saponins, the crown of thorns starfish also contain secretes interesting basic peptide toxins (Shiomi et al. 1988a; Ota, 2006). The most lethal, isolated protein (MW~25,000) displayed a mouse i.p. LD_{50} of 0.43 mg/kg. It potently damaged the mouse liver (Shiomi et al. 1988b). More recently the hepatotoxic plancitoxins were thoroughly purified and characterized by cloning. Both homologs were sequenced and shown to be homologous to deoxyribonucleases; their DNAase activities were then demonstrated. Thus, plancitoxin I is the first known example of a toxic DNAase (Shiomi et al. 2004).

10   Urochordates and Fishes

Ascidians (tunicates) contain a variety of cytotoxic metabolites, some of which may be toxic to mammalians (Blunt and Munro 2007). Pictamine (**62**), a quinolizidine alkaloid isolated from the ascidian *Clavelina picta* and lepadin B (**63**), a decahydroquinoline alkaloid from *C. lepadiformis* block neuronal nicotinic acetylcholine receptors at µM levels (Tsuneki et al. 2005), whereas lepadiformine (**64**), a rare decahydro-1*H*-pyrrolo[2,1-*j*]quinoline isolated from *Clavelina* spp. inhibits cardiac muscle K$^+$ channels (Sauviat et al. 2006).

Subtropical and tropical fishes accumulate ciguatoxins and cause ciguatera as mentioned Sect. 3.2.2 (also see Chap. 3). Some other types of fish poisonings have been known, but their causative agents remain to be fully elucidated (Hashimoto 1979).

62   **63**   **64**

10.1   Tetrodotoxins

It is well known that tetrodotoxin (**65**) is widely distributed in puffer fish of the family Tetraodontidae and is believed to be originated from bacteria of the genera *Vibrio, Pseudoalteromonas* and others (Yasumoto 2005). At moment, more than ten variants of tetrodotoxin have been isolated from a wide range of organisms including xanthid crabs, horseshoe crabs, frogs, and newts in addition to puffers and other marine animals that mentioned earlier (Miyazawa and Noguchi 2001). Interestingly, freshwater puffer fish contain saxitoxins, but not terodotoxins (Ngy et al. 2008). Tetrodotoxins bind to Site 1 on Na^+ channels and block sodium influx into muscle and nerve cells (see Chap. 3).

65

10.2   Peptide and Protein Toxins

10.2.1   Fish Integument Toxins

Integumentary secretions of soapfishs, another teleost family, contain hemolytic peptides called grammistins (Hashimoto and Oshima 1972; Hashimoto 1979; Sugiyama et al., 2005). These are linear peptides of 12–28 residues that contain many basic amino acids and possess alpha-helical secondary structures. These peptides display antimicrobial activities against a wide variety of bacteria. Several grammistins were recently cloned (Kaji et al. 2006). Pardaxins (named after the genus *Pardachirus*) are surface-active anti-microbial and hemolytic peptides that are secreted from the skin of the Red Sea sole when it is stressed; these were purified and shown to form pores in liposomes (Lazarovici et al. 1986; Thompson et al. 1986 also, see Chap. 6).

10.2.2 Fish Spine Toxins

Quite a few families of fish have members that possess poisonous spines. Perhaps the most well known are the scorpion fish including the Pacific stonefishs and lionfish. During the past 15 years the proteinaceous toxins isolated from three stonefish, *Synanceja horrida* (stonustoxin), *S. trachynis* (trachnilysin) and *S. verrucosa* (verrucotoxin and neoverrucotoxin) have been purified and extensively characterized (Poh et al. 1991; Ueda et al. 2006; Ghadessy et al. 1996). They are large (~150 kDa) but relatively stable proteins, containing two subunits of similar size. The cloned α- and β-subunits (71 and 79 kDa, respectively) of stonustoxin display ~50% sequence homology. There are 7 and 8 cysteine residues, respectively, in these subunits, and 10 of the 15 residues participate in intrachain disulfide linkages. The remaining five cysteines are uncombined. The stonefish toxins are pore-forming proteins, explaining their ability to lyse erythrocytes (Chen et al. 1997; Khoo 2002). The synaptic effects of trachnilysin will be described further in Chap. 6.

Many other fish contain poisonous spines that inflict painful wounds in fisherman and bathers. Oriental catfish (*Plotosus lineatus*) spine venom contains hemolytic, edema-forming and lethal proteins. The hemolytic protein fraction is of large size (~180 kDa) compared with the lethal proteins (~12 kDa). One lethal protein has been purified and partially characterized (Shiomi et al. 1986). Sting rays also possess poisonous spines and broken spines are often embedded in the surfaces of their predators; their proteinaceous toxins have yet to be purified and characterized.

11 Concluding Comments

Investigations of marine toxins during the past few decades have provided a remarkable diversity of molecules new to science. Besides stimulating chemical interest in unique structures posing many synthetic challenges, these substances often possess such unique targets that they can serve not only as useful research tools, but in some cases can either be useful drug candidates in their naturally occurring forms or as leads for designing analogs that possess even more selective actions. Traditionally, small, non-peptide molecules (MW < ~500) have served as lead compounds for drug design, due to their superior bioavailability and ease of synthesis, but this approach is rapidly being enlarged to include much larger molecules including peptides and nucleic acids. Who would have thought that an extremely lethal, large bacterial protein like botulinum toxin would become a very useful drug for treating various muscle spasms (Cooper 2007). Similarly a sea anemone peptide of 35 residues (ShK, Fig. 1) that blocks an important K$^+$ channel (Kv1.3) in T-lymphocytes has reasonable pharmacokinetic properties and has become a drug candidate for treating host-graft (transplantation) immune rejection and certain autoimmune diseases (Beeton et al. 2006).

Regardless of whether a natural product or its analogue is to be used as a molecular probe or as a drug, target selectivity will generally be of the utmost importance. If the toxin is to be used as a probe in an in vitro system, where additional sites of action are not present, the degree of selectivity is much less stringent. On the other hand, for use as in vivo probe or as a drug administered to the whole organism, the natural product may require considerable structural manipulation (engineering) to improve target selectivity, reduce toxicity and improve pharmacokinetic properties. Natural products including toxins often preferentially interact with one or a few receptor subtypes, but it would be rare for a toxin to affect only a single receptor subtype, as this would limit its ability in nature to neutralize a wide variety of prey or predators.

Historically, marine natural products research has lagged behind the investigation of terrestrial natural products, due to the limited accessibility of most marine organisms. However, the chemical diversity of natural toxins among marine animals is likely to be much higher than for terrestrial animal toxins, as the phyletic diversity of marine animals is greater. Since the chemical diversity of marine organisms is only beginning to be appreciated, it can be anticipated that future studies will reveal a plethora of new compounds of scientific and potential therapeutic interest.

Acknowledgments We thank Professor Ben Dunn, Department of Biochemistry and Molecular Biology, University of Florida for preparing the ribbon structures (Figs. 1–4) from Brookhaven Protein Data Bank structures.

References

Alessandri-Haber, Lecoq A, Gasparini S, Grangier-Macmath G, Jacquet G, Harvey AL, de Medeirosi C, Rowan EG, Gola M, Menez A, Crest M (1999) Mapping the functional anatomy of BgK on Kv1.1, Kv1.2, and Kv2.3. Cues to design analogs with enhanced selectivity. J Biol Chem 274:35653–35661.

Al-Sabi A, McArthur J, Ostoumov V, French RJ (2006) Marine toxins that target votage-gated sodium channels. Mar Drugs 4:157–192.

Andersen RJ, van Soest RWM, Kong F (1996) 3-Alkylpiperidine alkaloids isolated from marine sponges in the order Haplosclerida alkaloids. Chem Biol Perspect 10:301–355.

Andra J, Jakovkin I, Grotzinger J, Hecht O, Krasnosdembskaya AD, Golldmann T, Gutsmann T, Leippe M (2008) Structure and mode of action of the antimicrobial peptide arenicin. Biochem J 410:113–122.

Andreu D, Rivas L (1998) Animal antimicrobial peptides: an overview. Biopolymer 47: 415–433.

Aneiros A, Garcia I, Martinez JR, Harvey AL, Anderson AJ, Marshall DL, Engstrom A, Hellman U, Karlsson E (1993) A potassium channel toxin from the secretion of the sea anemone *Bunodosoma granulifera*. Isolation, amino acid sequence and biological activity. Biochim Biophys Acta 1157:86–92.

Artigas P, Gadsby DC (2003) Na$^+$/K$^+$-pump ligands modulate gating of palytoxin-induced ion channels. Proc Natl Acad Sci USA 100:501–505.

Ashida K, Toda H, Fujiwara M, Sakiyama F (1987) Amino acid sequence of *Goniopora* toxin. Jpn J Pharmacol 43 (Suppl.) Abst. p. 33.

Athaniasiadis A, Anderluh G, Macek P, Turk D (2001) Crystal structure of the soluble form of equinatoxin II, a pore-forming toxin from the sea anemone *Actinia equina*. Structure 9:341–346.

Baden DG, Bourdelais AJ, Jacocks H, Michelliza H, Naar J (2005) Natural and derivative brevetoxins: historical background, multiplicity, and effects. Environ Health Perspect 113: 621–625.

Barnham KJ, Dyke TR, Kem WR, Norton RS (1997) Structure of neurotoxin B-IV from the marine worm *Cerebratulus lacteus*: a helical hairpin cross-linked by disulfide binding. J Mol Biol 288:886–902.

Beeton C, Wulff H, Standifer NE, Azam P, Mullen KM, Penniington MW, Kolski-Andreaco A, Wei E, Grino A, Counts DR, Wang PH, LeeHealey CJ, Andrews BS, Sankaranarayanan A, Homerick D, roeck WW, Tehranzadeh J, Stanhope KL, Zimin P, Havel PJ, Griffey S, Knaus H-G, Nepom GT, Gutman GA, Calabresi PA, Chandy KG (2006) Kv1.3 channels are a therapeutic target for T cell-mediated autoimmune diseases. Proc Nat Acad Sci USA 103: 17414–17419.

Béress L, Béress R, Wunderer G (1975) Purification of three polypeptides with neuro- and cardiotoxic activity from the sea anemone *Anemonia sulcata*. Toxicon 13:359–367.

Berne S, Sepčić K, Križaj I, Kem WR, McClintock JB, Turk T (2003) Isolation and characterization of a cytolytic protein from mucus secretions of the Antarctic heteronemertine *Parborlasia corrugatus*. Toxicon 41:483–491.

Blumenthal KM, Kem WR (1976) Primary structure of *Cerebratulus lacteus* toxin B-IV. J Biol Chem 251:6025–6029.

Blumenthal KM, Kem WR (1980) Primary structure of *Cerebratulus lacteus* toxin A-III. J Biol Chem 255:8266–8272.

Blumenthal KM, Kem WR (1983) Primary structure of *Stoichactis helianthus* cytotoxin III. J Biol Chem 258:5574–5581.

Blumenthal KM, Keim PS, Heinrikson RL, Kem WR (1981) Amino acid sequence of *Cerebratulus* toxin B-II and revised structure of toxin B-IV. J Biol Chem 256:9063–9067.

Blunt JW, Munro MYHG (2007) Dictionary of marine natural products. Chapman & Hall/CRC, Boca Raton.

Bowden BF (2006) Yessotoxins – polycyclic ethers from dinoflagellates: relationships to diarrhetic shellfish toxins. Toxin Rev 25:137–157.

Brewley CA, Faulkner DJ (1998) Lithistid sponges: star performers or hosts to the stars. Angew Chem Int Ed 37:2162–2178.

Brinkman D, Burnell J (2007) Identification, cloning and sequencing of two major venom proteins from the box jellyfish, *Chironex fleckeri*. Toxicon 50:850–860.

Bulet P, Stöcklin R, Menin L (2004) Anti-microbial peptides: from invertebrates to vertebrates. Immunol Rev 198:169–184.

Burgess V, Shaw G (2001) Pectenotoxins – an issue for public health. A review of their comparative toxicology and metabolism. Environ Int 27:275.

Burja AM, Banaigs B, Abou-Mansour E, Burgess JG, Wright, PC (2001) Marine cyanobacteria – a prolific source of natural products. Tetrahedron 57:9347–9377.

Cagide E, Louzao MC, Ares IR, Vieytes MR, Yotsu-Yamashita M, Paquette LA, Yasumoto T, Botana LM (2007) Effects of a synthetic analog of polycavernoside A on human neuroblastoma cells. Cell Physiol Biochem 19:185–194.

Camacho FG, Rodríguez JG, Mirón As, García MCC, Belarbi EH, Chisti Y, Grima EM (2007) Biotechnological significance of toxic marine dinoglagellates. Biotechnol Adv 25:176–194.

Cariello L, Zanetti L (1977) α- and β-cephalotoxin: two paralyzing proteins from posterior salivary glands of *Octopus vulgaris*. Comp Biochem Physiol 57C:169–173.

Carlé JS, Christophersen C (1980) Doger Bank Itch. The allergen is (2-hydroxyethyl)dimethylsulfonium ion. J. Am Chem Soc 102:5107–5108.

Castenada O, Sotolongo V, Amor AM, Stocklin R, Anderson AJ, Harvey AL, Engstrom A, Wernstedt C, Karlsson E (1995) Characterization of a potassium channel toxin from the Carribbean sea anemone S*tichodactyla helianthus*. Toxicon 33:603–613.

Cembella AD (2003) Chemical ecology of eukaryotic microalgae in marine ecosystems. Phycologia 42:420–447.

Chang CWJ (2000) Naturally occurring isocyano/isothiocyanato and related compounds. Fortschr Chem Org Naturst 80:1–186.

Chen D, Kini RM, Yuen R, Khoo HE (1997) Haemolytic activity of stonustoxin from stonefish (*Synanceja horrida*) venom: pore formation and the role of cationic amino acid residues. Biochem J 325:685–691.

Ciminello P, Fattorusso E (2004) Shellfish toxins – chemical studies on northern Adriatic mussels. Eur J Org Chem 2533–2551.

Culver P, Burch M, Potenza C, Wasserman L, Fenical W, Taytor P (1985) Structure-activity relationships for the irreversible blockade of nicotinic receptor agonist sites by lophotoxin and congeneric diterpene lactones. Mol Pharmacol 28:436–444.

Coll JC (1992) The chemistry and chemical ecology of octocorals (Coelenterata, Anthozoa, Octocorallia). Chem Rev 92:613–631.

Cooper G (2007) Therapeutic uses of botulinum yoxin. Humana Press, Totowa, NJ.

Daranas AH, Norte M, Fernández JJ (2001) Toxic marine microalgae. Toxicon 39:1101–1132.

Dauplais M, Lecoq A, Song J, Cotton J, Jamin N, Gilquin B, Roumestand C, vita C, de Medeiros CLC, Rowan EG, Harvey AL, Menez A (1997) On the convergent evolution of animal toxins. Conservation of a diad of functional residues in potassium channel-blocking toxins with unrelated structures. J Biol Chem 272:4302–4309.

Faulkner DJ (1992) Chemical defenses of marine molluscs. In: Paul VJ (ed.) Ecological roles of marine natural products. Cornell Univ Press, Ithaca, NY, pp. 119–163.

Fenical W, Okuda RK, Bandurraga MM, Culver P, Jacobs RS (1981) Lophotoxin: a novel neuromuscular toxin from Pacific sea whip of the genus *Lophogorgia*. Science 212: 1512–1514.

Fogh E, Kem WR, Norton RS (1990) Solution structure of neurotoxin I from the sea anemone *Stichodactylus helianthus*. A nuclear magnetic resonance, distance geometry, and restrained molecular dynamics study. J Biol Chem 265:13016–13028.

Froscio SM, Humpage AR, Wickramasinghe W, Shaw G, Falconer IR (2008) Interaction of the cyanobacterial toxin cylindospermopsin with the eukaryotic protein sysnthesis system. Toxicon 51:191–198.

Fusetani N, Asai N, Matsunaga S (1994) Cyclostellettamines A-F, pyridine alkaloids which inhibit binding of methyl quinuclidinyl benzilate (QNB) to muscarinic acetylcholine receptors, from the marine sponge *Stelletta maxima*. Tetrahedron Lett 35:3967–3970.

Gerwick WH, Tan LT, Sitachitta N (2001) Nitrogen-containing metabolites from marine cyanobacteria. Alkaloids 57:75–184.

Ghadessy FJ, Chen D, Kini M, Chung MCM, Jeyaseelan K, Khoo HE, Yuen R (1996) Stonustoxin is a novel lethal factor from stonefish (*Synanceja horrida*) venom. J Biol Chem 271: 25575–25581.

Ghiaroni V, Sasaki M, Fuwa H, Rossini GP, Scalera G, Yasumoto T, Pietra P, Bigiani A (2005) Inhibition of voltage-gated potassium currents by gambierol in mouse taste cells. Toxicol Sci 85:657–665.

Ghiretti F (1959) Cephalotoxin: the crab-paralyzing agent of the posterior salivary glands of cephalopods. Nature 183:1192–1193.

Gill S, Murphy M, Clausen J, Richard D, Quilliam M, MacKinnon S, LaBlanc P, Mueller R, Pulido O (2003) Neural injury biomarkers of novel shellfish toxins, spriolides: a pilot study using immunochemical and transcriptional analysis. Neurotoxicology 24:593–604.

Gleibs S, Mebs D (1999) Distribution and sequestration of palytoxin in coral reef animals. Toxicon 37:1521–1527.

Gonoi T, Ashida K, Feller D, Schmidt J, Fujiwara M, Catterall WA (1986) Mechanism of action of a polypeptide neurotoxin from the coral *Goniopora* on sodium channels in mouse neuroblastoma cells. Mol Pharmacol 29:347–354.

Hagiwara H, Numata M, Konishi K, Oka Y (1965) Synthesis of nereistoxin and related compounds. Chem Pharm Bull 13:253–260.

Halstead BW (1965) Poisonous and venomous marine animals of the world (Vol. 1). US Government Printing Office, Wahington, DC.

Halstead BW (1967) Poisonous and venomous marine animals of the world (Vol. 2). US Government Printing Office, Wahington, DC.

Halstead BW (1970) Poisonous and venomous marine animals of the world (Vol. 3). US Government Printing Office, Wahington, DC.

Halstead BW (1984) Miscellaneous seafood toxicants. In: Ragelis EP (ed.) Seafood toxins. American Chemical Soceity, Washington, DC, pp. 37–51.

Hanson LA, Gusftafsson S, Rengefors K, Bomark L (2005) Cyanobacterial chemical warfare affects zooplankton community composition. Freshwater Biol 52:1290–1301.

Hashimoto Y (1979) Marine toxins and other bioactive marine metabolites. Japan Scientific Societies Press, Tokyo.

Hashimoto Y, Ashida K (1973) Screening of toxic corals and isolation of a toxic polypeptide from *Goniopora* spp. Publ Seto Mar Biol Lab 20:703–711.

Hashimoto Y, Oshima Y (1972) Separation of grammistins A, B and C from a soapfish *Pogonoperca punctata*. Toxicon 10:279–284.

Hayashi E, Isogai M, Kagawa Y, Takayanagi N, Yamada S (1984) Neosurugatoxin, a specific antagonist of nicotinic acetylcholine receptors. J Neurochem 42:1491–1494.

Hilgemann DW (2003) From a pump to a pore: how palytoxin opens the gates. Proc Natl Acad Sci USA 100:386–388.

Honma T, Iso T, Ishida M, Nagashima Y, Shiomi K (2003) Occurrence of type 3 sodium channel peptide toxins in two species of sea anemones (*Dofleinia armata* and *Entacmaea amsayi*). Toxicon 41:637–639.

Huerta V, Morera V, Guanche Y, Chinea G, Gonzalez LJ, Betancourt L, Martinez D, Alvarez C, Lanio ME, Besada V (2001) Primary structure of two cytolysin isoforms from *Stichodactyla helianthus* differing in their hemolytic activity. Toxicon 39:1253–1256.

Iguchi A, Iwanaga S, Nagai H (2007) Isolation and characterization of a novel protein toxin from fire coral. Biochem Biophys Res Commun 365:107–112.

Ito E, Satake M, Yasumoto T (2002) Pathological effects of lyngbyatoxin A upon mice. Toxicon 40:551–556.

Kaji T, Sugiyama N, Ishizaki S, Nagashima, Shiomi K (2006) Molecular cloning of grammistins, peptide toxins from the soapfish *Pogonoperca punctata*, by hemolytic screening of a cDNA library. Peptides 27:3069–3076.

Kalman K, Pennington MW, Lanigan MD, Nguyen A, Rauert H, Mahnir V, Paschetto K, Kem WR, Grissmer S, Gutman GA, Christian EP, Cahalan MD, Norton RS, Chandy KG (1998) ShK-Dap[22], a potent Kv1.3-specific immunosuppressive polypeptide. J Biol Chem 273: 32697–32707.

Kem WR (1971) A study of the occurrence of anabaseine in *Paranemertes* and other nemertines. Toxicon 9:23–32.

Kem WR (1976) Purification and characterization of a new family of polypeptide neurotoxins from the heteronemertine *Cerebratulus lacteus* (Leidy). J Biol Chem 251:4184–4192.

Kem WR (1988) Sea anemone toxin structure and action. In: Hessinger D, Lenhoff H (eds.) The biology of nematocysts. Academic Press, New York, pp. 375–405.

Kem WR (1994) Structure and membrane actions of a marine worm cytolysin, *Cerebratulus* toxin A-III. Toxicology 87:189–203.

Kem WR (2008) Alpha7 nicotinic acetylcholine receptor agonists as potential Alzheimer's drugs: their cognition enhancing, neuroprotective and anti-amyloid actions. In: Martinez A. (ed.) Medicinal chemistry of Alzheimer's disease. Research Signpost, pp. 135–159.

Kem WR, Blumenthal KM (1978) Purification and characterization of the cytolytic *Cerebratulus* A toxins. J Biol Chem 253:5752–5757.

Kem WR, Dunn BM (1988) Separation and characterization of four different amino acid sequence variants of a sea anemone (*Stichodactyla helianthus*) protein cytolysin. Toxicon 26:997–1008.

Kem WR, Scott JD (1980) Partial purification and characterization of a cytotoxic protein from squid (*Loligo pealei*) posterior salivary glands. Biol Bull 158:475.

Kem WR, Soti F (2001) *Amphiporus* alkaloid multiplicity implies functional diversity: initial studies on crustacean pyridyl receptors. Hydrobiology 456:221–231.

Kem WR, Abbott BC, Coates RM (1971) Isolation and structure of a hoplonemertine toxin. Toxicon 9:15–22.

Kem WR, Scott, KN, Duncan JH (1976) Hoplonemertine worms – a new source of pyridine neurotoxins. Experientia 32:684–686.

Kem WR, Parten B, Pennington MW, Dunn BM, Price D (1989) Isolation, characterization, and amino acid sequence of a polypeptide neurotoxin occurring in the sea anemone *Stichodactyla helianthus*. Biochemistry 28:3483–3489.

Kem WR, Mahnir VM, Papke R, Lingle C (1997) Anabaseine is a potent agonist upon muscle and neuronal α-bungarotoxin sensitive nicotinic receptors. J Pharmacol Exp Ther 283:979–992.

Kem WR, Pennington MW, Norton RS (1999) Sea anemone toxins as templates for the design of immunosuppressant drugs. Perspect Drug Discov Des 15–16:111–129.

Kem WR, Soti F, Rittschof D (2003) Inhibition of barnacle larval settlement and crustacean toxicity of some hoplonemertine pyridyl alkaloids. Biomol Eng 20:355–361.

Kem WR, Mahnir VM, Prokai L, Papke RM, Cao XF, LeFrancois S, Wildeboer K, Porter-Papke J, Prokai-Tatrai K, Soti F (2004). Hydroxy metabolites of the Alzheimer's drug candidate DMXBA (GTS-21): their interactions with brain nicotinic receptors and brain penetration. Mol Pharmacol 65:56–67.

Kem W, Soti F, Wildeboer K, LeFrancois S, MacDougall K, Wei DQ, Chou KC, Arias H (2006) The nemertine toxin anabaseine and its derivative DMXBA (GTS-21): chemical and pharmacological properties. Mar Drugs 4:255–273.

Khoo HE (2002) Bioactive proteins from stonefish venom. Clin Exp Pharmacol Physiol 29:802–806.

Kita M, Uemura D (2005) Iminium alkaloids from marine invertebrates: structure, biological activity, and biogenesis. Chem Lett 34:454–459.

Kita M, Ohishi N, Konishi K, Kondo M, Koyama T, Kitamura M, Yamada K, Uemura D (2007) Symbiodinolide, a novel polyol macrolide that activates N-type Ca^{2+} channel, from the symbiotic marine dinoflagellate *Symbiodinium* sp. Tetrahedron 63:6241–6251.

Kosuge Y, Tsuji K, Hirai K (1982) Isolation of neosurugatoxin from the Japanese ivory shell, *Babylonia japonica*. Chem Pharm Bull 30:3255–3259.

Kosuge Y, Tsuji K, Hirai K, Fukuyama T (1985) First evidence of toxin production by bacteria in marine organism. Chem Pharm Bull 33:3059–3061.

Lazarovici P, Primor N, Loew LM (1986) Purification and pore-forming activity of two hydrophobic polypeptides from the secreti on of the Red Sea Moses sole (*Pardachirus marmoratus*). J Biol Chem 261:16704–16713.

Lee JU, Kang DI, Zhu WL, Shin SY, Hahm KS, Kim Y (2007) Solution structures and biological functions of the antimicrobial peptide, arenicin-1, and its linear derivative. Pept Sci 88:208–216.

Llewellyn LE (2006) Saxitoxins, a toxic marine natural product that targets a multitide of receptors. Nat Prod Rep 23:200–222.

Maeda M, Kodama T, Tanaka T, Yoshizumi H, Nomoto K, Takemoto T, Fujita T (1985) Structures of insecticidal substances isolated from a red alga, *Chondria armata*. In: Proceedings of 27th symposium on the chemistry of natural products, pp. 616–623.

Mancheno JM, Martin-Benito J, Martinez-Ripoli, Gavilanes JG, Hermoso JA (2003) Crystal and electron microscopy structures of sticholysin II actinoporin reveal insights into the mechanism of membrane pore formation. Structure 11:1319–1328.

McClelland D, Evans RM, Abidin I, Sharma S, Choudry FZ, Jaspars M, Sepčić K, Scott RH (2003) Irreversible pore formation by polymeric alkylpyridinium salts (poly-APS) from the sponge *Reniera sarai*. Br J Pharmacol 139:1399–1408.

Minale L, Riccio R, Zollo F (1993) Steroidal oligosaccharides and polyhydroxysteroids from echinoderms. Fortschr Chem Org Naturst 62:75–308.

Miyazawa K, Noguchi T (2001) Distribution and origin of tetrodotoxin. J Toxicol Toxin Rev 20:11–33.

Miyazawa K, Jeon JK, Noguchi T, Ito K, Hashimoto K (1987) Distribution of tetrodotoxin in the tissues of the flatworm *Planocera multitentaculata* (Platyhelminthes). Toxicon 25:975–980.

Molgo J, Girard E, Benoit E (2007) Cyclic imines: an insight into this emerging group of bioactive marine toxins. Phycotoxins 319–335.

Moore RE (1985) Structure of palytoxin. Fortschr Chem Org Naturst 48:81–202.

Moriya T, Ishida Y, Nakamura H, Asari T, Murai A, Ohizumi Y (1998) Vasoconstriction induced by zooxanthellatoxin-B, a polyoxygenated long-chain product from a marine alga. Eur J Pharmacol 350:59–65.

Mos L (2001) Domoic acid: a fascinating marine toxin. Environ Toxicol Pharmacol 9:79–85.

Mosher HS, Fuhrman FA (1984) Occurrence and origin of tetrodotoxin. In: Ragelis EP (ed.) Seafood toxins. American Chemical Soceity, Washington, DC, pp. 333–344.

Müller WEG, Brümmer F, Batel R, Müller IM, Schröder HC (2003) Molecular biodiversity. Case study: Porifera (sponges). Naturwissenschaften 90:103–120.

Muramatsu I, Fujiwara M, Miura A, Narahashi T (1985) Effects of *Goniopora* toxin on crayfish giant axons. J Pharmacol Exp Ther 234:307–315.

Murata M, Yasumoto T (2000) The structure elucidation and biological activities of high molecular weight algal toxins: maitotoxin, prymnesins and zooxanthellatoxins. Nat Prod Rep 17:293–314.

Murayama K, Abbott NJ, Narahashi T, Shapiro BI (1972) Effects of allethrin and *Condylactis* toxin on the kinetics of sodium conductance of crayfish giant axon membranes. Comp Gen Pharmacol 3:391–400.

Naar JP, Flewelling LJ, Lenzi A, Abbott JP, Granholm A, Jacocks HM, Gannon D, Henry M, Pierce R, Baden DG, Wolny J, landsberg JH (2007) Brevetoxins, like ciguatoxins, are potent ichthyotoxic neurotoxins that accumulate in fish. Toxicon 50:707–723.

Nagai H, Yasumoto T, Hokama Y (1996) Aplysiatoxin and debromoaplysiatoxin as the causative agents of red alga *Gracilaria coronopifolia* poisoning in Hawaii. Toxicon 37:753–761.

Nagai H, Takuwa K, Nakao M, Ito E, Miyake M, Noda M, Nakajima T (2000) Novel proteinaceous toxins from the box jellyfish (sea wasp) *Carbdea rastoni*. Biochem Biophys Res Commun 275:582–588.

Nagai H, Takuwa-Kuroda K, Nakao M, Oshiro N, Iwanaga S, Nakajima T (2002) A novel protein toxin from the deadly box jellyfish (sea wasp, havu-kurage) *Chiropsalmus quadrigatus*. Biosci Biotechnol Biochem 66:97–102.

Nakagawa H, Tu A, Kimura A (1991) Purification and characterization of contractin A from the pedicellarial venom of sea urchin, *Toxopneustes pileolus*. Arch Biochem Biophys 284: 279–284.

Nicholson CM, Lewis RJ (2006) Ciguatoxins: cyclic polyether modulators of voltage-gated ion channel function. Mar Drugs 4:82–118.

Ngy L, Tada K, Yu CF, Takatani T, Arakawa O (2008) Occurrence of paralytic shellfish toxins in Cambodian Mekong pufferfish *Tetraodon turgidus*: selective toxin accumulation in the skin. Toxicon 51:280–288.

Norton TR, Shibata S, Kashiwagi M, Bentley J (1976) Isolation and characterization of the cardiotonic polypeptide Anthopleurin-A from the sea anemone *Anthopleurin xanthogrammica*. J Pharm Sci 65:1368–1374.

Okaichi T, Hashimoto T (1962) The structure of nereistoxin. Agric Biol Chem 26:224–227.

Onodera K, Nakamura H, Oba Y, Ohizumi Y, Ojika M (2005) Zooxanthellamide Cs: vasoconstrictive polyhydroxylated macrolides with the largest lactone ring size from a marine dinoflagellate of *Symbiodinium* sp. J Am Chem Soc 127:10406–10411.

Onuma Y, Satake M, Ukena T, Roux J, Chanteau S, Rasolofonirira, Ratsimaloto M, Naoki H, Yasumoto T (1999) Identification of putative palytoxin as the cause of clupeotoxism. Toxicon 37:55–65.

Ota E, Nagai H, Nagashima Y, Shiomi K (2006) Molecular cloning of two toxic phospholipasesA2 from the crown-of-thorns starfish *Acanthaster planci* venom. Comp Biochem Physiol B 143:54–60.

Paquette LA, Yotsu-Yamashita M (2007) Polycavernosides. Phycotoxins 275–296.

Pennington MW, Byrnes ME, Zaydenberg, Khaytin I, de Chastonay J, Krafte DS, Hill R, Mahnir VM, Volberg WA, Gorczyca W, Kem WR (1995) Chemical synthesis and characterization of ShK toxin: a potent potassium channel inhibitor from a sea anemone. Int J Pept Protein Res 46:354–358.

Pennington MW, Mahnir VM, Krafte DS, Zaydenberg I, Byrnes ME, Khaytin I, Crowley K., Kem WR (1996a) Identification of three separate binding sites on ShK toxin, a potent inhibitor of voltage-dependent potassium channels in human T-lymphocytes and rat brain. Biochem Biophys Res Commun 219:696–701.

Pennington MW, Mahnir VM, Khaytin I, Zaydenberg, I, Byrnes ME, Kem WR (1996b) An essential binding surface for ShK toxin interaction with rat brain potassium channels. Biochemistry 35:16407–16411.

Piel J (2004) Metabolites of symbiotic bacteria. Nat Prod Rep 21:519–538.

Qar J, Schweitz H, Schmid A, Lazdunski M (1986) A polypeptide toxin from the coral *Goniopora*. Purification and action on Ca^{2+} channels. FEBS Lett 202:331–336.

Riobo P, Paz B, Franco JM (2006) Analysis of palytoxin-like in *Ostreopsis* cultures by liquid chromatography with precolumn derivatization and fluorescence detection. Anal Chim Acta 566:217–223.

Ritson-Williams R, Yotsu-Yamashita M, Paul VJ (2006) Ecological functions of tetrodotoxin in a deadly polyclad flatworm. Proc Natl Acad Sci USA 103:3176–3179.

Ronzitti G, Hess P, Rehmann N, Rossini GP (2007) Azaspiracid-1 alters the E-cadherin pool in epithelial cells. Toxicol Sci 95:427–435.

Salgado VL, Kem WR (1992) Actions of three structurally distinct sea anemone toxins on crustacean and insect sodium channels. Toxicon 30:1365–1381.

Sanders JM, Pentikäinen OT, Settimo L, Pentikäinen U, Shoji M, Sasaki M, Sakai R, Johnson MS, Swanson GT (2006) Determination of binding site residues responsible for the subunit selectivity of novel marine-derived compounds on kinate receptors. Mol Pharmacol 69:1849–1860.

Sauviat MP, Vaucauteren J, Grimaud N, Nabil M, Petit JY, Biard JF (2006) Sensitivity of cardiac background inward recycling K+ outward current (IK1) to the alkaloids lepadiformines A, B, and C. J Nat Prod 69:558–562.

Savage IVE, Howden MEH (1977) Hapalotoxin, a second lethal toxin from the octopus *Hapalochlaena maculosa*. Toxicon 15:463–466.

Schmitz FJ, Hollenbeak KH, Campbell DC (1978) Marine natural products: halitoxin, toxic complex of several marine sponge of the genus Haliclona. J Org Chem 43:3916–3922.

Sepčić K (2000) Bioactive alkylpyridinium compounds from marine sponges. J Toxicol Toxin Rev 19:139–160.

Sepčić K, Batista U, Vacelet J, Maček P, Turk T (1997a) Biological activities of aqueous extracts from marine sponges and cytotoxic effects of 3-alkylpyridinium polymers from *Reniera salai*. Comp Biochem Physiol 117C:47–53.

Sepčić K, Guella G, Mancini I, Pietra F, Dalla Serra M, Menestrina G, Tubbs K, Maček P, Turk T (1997b) Characterization of anticholineesterase-active 3-alkylpyridinium polymers from the marine sponge *Reniera sarai* in aqueous solutions. J Nat Prod 60:991–996.

Sher D, Fishman Y, Zhang M, Lebendikers M, Gaathon A, Mancheno J-M, Zlotkin E (2005) Hydralysins, a new category of β-pore-forming toxins in cnidaria. J Biol Chem 280: 22847–22855.

Shiomi K, Takamiya M, Yamanaka H, Kikuchi, Konno K (1986) Hemolytic, lethal and edemaforming activities of the skin secretion from the oriental catfish (*Plotosus lineatus*). Toxicon 24:1015–1018.

Shiomi K, Yamamoto S, Yamanaka H, Kikuchi T (1988a) Purification and characterization of a lethal factor in venom from the Crown-of-Thorns starfish (*Acanthaster planci*). Toxicon 11:1077–1083.

Shiomi K, Yamamoto S, Yamanaka H, Kikuchi, Konno K (1988b) Liver damage by the crown-of-thorns (*Acanthaster planci*) lethal factor. Toxicon 28:469–475.

Shiomi K, Midorikawa S, Ishida M, Nagashima Y, Nagai H (2004) Plancitoxins, lethal factors from the crown-of-thorns starfish *Acanthaster planci*, are deoxyribonucleases II. Toxicon 44:499–506.

Sobel J, Painter J (2005) Illnesses caused by marine toxins. Food Saf 41:1290–1296.

Songdahl JH, Shapiro BI (1974) Purification and composition of a toxin from the posterior salivary glands of *Octopus dofleini*. Toxicon 12:109–115.

Stevens SM, Jr, Kem WR, Prokai L (2002) Investigation of cytolysin variants by peptide mapping: enhanced protein characterization using complementary ionization and mass spectrometric techniques. Rapid Commun Mass Spectrom 16:1–8.

Takei M, Nakagawa H (2006) A sea urchin lectin, SUL-1, from the toxopneustid sea urchin induces DC maturation from human monocyte and drives Th1 polarization in vitro. Toxicol Appl Pharmacol 213:27–36.

Thesing J, Muller A (1956) Synthese des Nicotellins. Angew Chem 68:577–578.

Thuesen EV, Kogure K (1989) Bacterial production of tetrodotoxin in four species of Chaetognatha. Biol Bull 176:191–194.

Thuesen EV, Kogure K, Hashimoto K, Nemoto T (1988) Poison arrowworms: a tetrodotoxin venom in the marine phylum Chaetognatha. J Exp Mar Biol Ecol 116:249–256.

Torigoe K, Murata M, Yasumoto T (1988) Prorocentrolide, a toxic nitrogenous macrocycle from a marine dinoflagelate. J Am Chem Soc 110:7876–7877.

Tsuneki H, You Y, Toyooka N, Sasaoka T, Nemoto H, Dani JA, Kimura I (2005) Marine alkaloids (-)-pictamine and (-)-lepadine B block neuronal nicotinic acetylcholine receptors. Biol Pharm Bull 28:611–614.

Tudor JE, Pallaghy PK, Pennington MW, Norton RS (1996) Solution structure of ShK toxin, a novel potassium channel inhibitor from a sea anemone. Nat Struct Biol 3:317–320.

Turk T, Frangez R, Sepčić K (2007) Mechanisms of toxicity of 3-alkylpyridinium polymers from marine sponge *Reniera sarai*. Mar Drugs 5:157–167.

Ueda A, Suzuki M, Honma T, Nagai H, Hanashima Y, Shiomi K (2006) Purification, properties and cDNA cloning of neoverrucotoxin (neoVTX), a hemolytic lethal factor from the stonefish *Synaceia verrucosa* venom. Biochim Biophys Acta 1760:1713–1722.

Uemura D (1991) Bioactive polyethers. In: Scheuer PJ (ed.) Bioorganic marine chemistry, Springer-Verlag, Berlin, Vol. 4, pp. 1–31.

Usami M, Satake M, Ishida S, Inoue A, Kan Y, Yasumoto T (1995) Palytoxin analogs from the dinoflagellate *Ostreopsis siamensis*. J Am Chem Soc 117:5389–5390.

Ushio-Sata N, Matsunaga S, Fusetani N, Honda K, Yasumuro K (1996) Penaramides, which inhibit binding of ω-conotoxin GVIA to N-type Ca²⁺ channels, from the marine sponge *Penares* aff. *incrustans*. Tetrahedron Lett 37:225–228.

Valcarcel CA, Serra MD, Potrich C, Bernhart I, Tejuca M, Martinez D, Pazos F, Lanio ME, Menestrina G (2001) Effects of lipid composition on membrane permeabilization by sticholysin I and II, two cytolysins of the sea anemone *Stichodactyla helianthus*. Biophys J 80:2761–2774.

van Apeldoorn ME, van Egmond HP, Speijers GJA, Bakker GJI (2007) Toxins of cyanobacteria. Mol Nutr Food Res 51:7–60.

Virariño N, Nicolau KC, Frederick MO, Vieytes MR, Botana LM (2007) Irreversible cytoskeletal disarrangement is indipendent of caspase activation during in vivo azaspiracid toxicity in human neuroblastoma cells. Biochem Pharmacol 74:327–335.

Warabi K, Nakao Y, Matsunaga S, Fukuyama T, Kan T, Yokoshima S, Fusetani N (2001) Dogger Bank itch revisited: isolation of (2-hydroxyethyl) dimethylsulfoxonium chloride as a cytotoxic constituent from the marine sponge *Theonella* aff. *mirabis*. Comp Biochem Physiol 128B:27–30.

Wattenberg EV (2007) Palytoxin: exploiting a novel skin tumor to explore signal transduction and carcinogenesis. Am J Physiol Cell Physiol 292:C24–C32.

Yamada, K, Kigoshi, H (1997) Bioactive compounds from the sea hares of two genera: *Aplysia* and *Dolabella*. Bull Chem Soc Jpn 70:1479–1489.

Yamanouchi T (1955) On the poisonous substance contained in holothurians. Publ Seto Mar Biol Lab 4:183–203.

Yasumoto T (2005) Chemistry, etiology, and food chain dynamics of marine toxins. Proc Japan Acad Ser B 81:43–51.

Yasumoto T, Murata M (1993) Marine toxins. Chem Rev 93:1897–1909.

Yasumoto T, Nagai H, Yasumura D, Michishita T, Endo A, Yotsu M, Kotaki Y (1986a) Interspecific distribution and possible origin of tetrodotoxin. Ann NY Acad Sci 479:44–51.

Yasumoto T, Yasumura D, Ohizumi Y, Takahashi M, Alcala AC, Alcala LC (1986b) Palytoxin in two species of xanthid crab from the Philippines. Agric Biol Chem 50:163–167.

Zhang M, Fishman Y, Sher D, Zlotkin E (2003) Hydralysin, a novel animal group-selective paralytic and cytolytic protein from a noncnidocystic origin in *Hydra*. Biochem 42:8939–8944.

Zucchi R, Ronca-Testoni S (1997) The sarcoplasmic reticulum Ca^{2+} channel/ryanodine receptor: modulation by endogenous effectors, drugs and disease states. Pharmacol Rev 49:1–51.

Conotoxins: Molecular and Therapeutic Targets

Richard J. Lewis

Abstract Marine molluscs known as cone snails produce beautiful shells and a complex array of over 50,000 venom peptides evolved for prey capture and defence. Many of these peptides selectively modulate ion channels and transporters, making them a valuable source of new ligands for studying the role these targets

R.J. Lewis (✉)

Institute for Molecular Biosciences, The University of Queensland, Brisbane 4072, Australia
e-mail: r.lewis@imb.uq.edu.au

N. Fusetani and W. Kem (eds.), *Marine Toxins as Research Tools*,
Progress in Molecular and Subcellular Biology, Marine Molecular Biotechnology 46,
DOI: 10.1007/978-3-540-87895-7, © Springer-Verlag Berlin Heidelberg 2009

play in normal and disease physiology. A number of conopeptides reduce pain in animal models, and several are now in pre-clinical and clinical development for the treatment of severe pain often associated with diseases such as cancer. Less than 1% of cone snail venom peptides are pharmacologically characterised.

1 Introduction

Cone snails are a large group (500+ species) of recently evolved, widely distributed marine molluscs of the family Conidae. They hunt prey using a highly developed envenomation apparatus that can paralyse prey within seconds ensuring its capture and reducing exposure to predation by larger fishes. Each species of cone snail produces a rich cocktail of mostly small, disulfide-bonded peptides evolved to rapidly immobilise their prey of either fish (piscivorous species), molluscs (molluscivorous species) or worms (vermivorous species). The venom peptides of cone snails are expressed in a venom duct and injected through a hollow-tipped and barbed harpoon-like structure (modified radula) into the soft tissue of animals, using a muscular proboscis (Fig. 1). Depending on their lethality to animals, individual venom peptides are referred to as either conotoxins (lethal) or conopeptides (nonlethal). Each venom contains a unique array of over 100 different peptides (Fig. 2). The defensive, attractant or prey tranquilizing effects of specific conopeptides are less clearly defined. This broadly evolved bioactivity provides a unique source of new research tools and potential therapeutic agents, with ω-MVIIA (Prialt or Ziconitide) already having reached the clinic (Lewis and Garcia 2003).

Conotoxins are genetically encoded as propeptides which following expression are cleaved by specialised venom endoproteases (e.g., Tex31) to produce the final mature venom peptide (Milne et al. 2003). Their small size (typically <5 kDa), relative ease of synthesis, structural stability and target specificity make them ideal pharmacological probes (Adams et al. 1999). Somewhat surprisingly, many of these classes of conotoxins act on pain targets, allowing the *specific* dissection of key ion channels and receptors underlying pain, and providing new ligands with potential as pain therapeutics (Lewis and Garcia 2003). It is estimated that in excess of 50,000 conopeptides have evolved, with only ~0.1% characterised pharmacologically. The use of high throughput and more recent multiplexed high content screens should accelerate target discovery, although many conotoxins are expected to be selective for the prey species over related mammalian targets and may be missed in most screens.

This chapter focuses on the families of conotoxins and conopeptides acting at calcium (ω-conotoxins), sodium (μ-, μO- and δ-conotoxins) and potassium channels (κ-conotoxins), the norepinephrine transporter (χ-conopeptides), nicotinic acetylcholine receptor (α-conotoxins), α_1-adrenoceptor (ρ-conopeptides), NMDA receptor (conantokins), vasopressin receptor (conopressins) and neurotensin receptor (contulakins). The therapeutic potential of selected members of these classes are highlighted in Table 1.

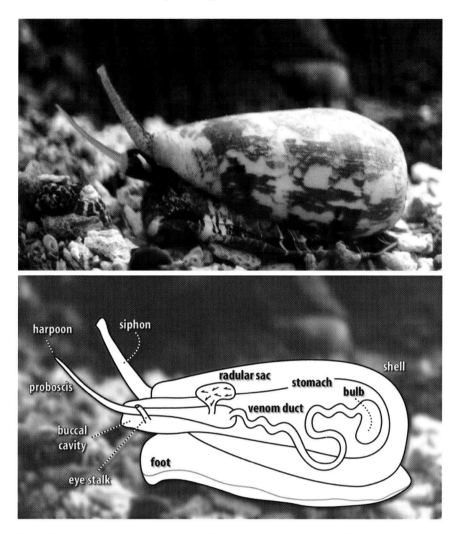

Fig. 1 *Conus striatus* in hunting mode with proboscis extended in readiness to harpoon and envenomate prey (fish). Lower panel shows the venom-containing venom duct, the radula sac where the hollow-tipped and barbed harpoons are made and stored prior to use, and the proboscis which holds the harpoon

2 Calcium Channel Inhibitors

2.1 *Calcium Channels*

ω-Conotoxins produced by fish hunting cone snails (Table 2) are amongst the most potent ichthyotoxins. Their selectivity for specific calcium channel subtypes found in nerves provide unique tools with which to identify and determine the

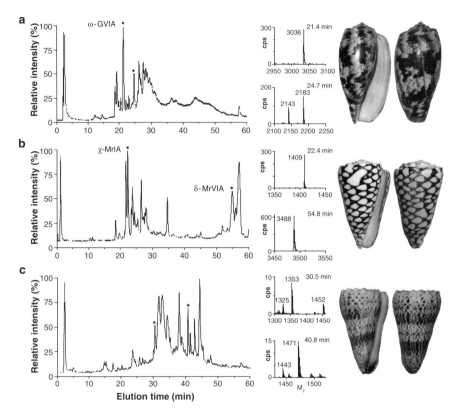

Fig. 2 Complexity of crude venom revealed by liquid chromatography/mass spectrometry (LC/MS) analyses of cone snail venoms. **a** The fish hunting (piscivorous) *C. geographus,* **b** the mollusc hunting (moluscivorous) *C. marmoreus* and **c** the worm hunting (vermivorous) *C. imperialis*. The insets show mass spectra of selected venom peptides. A shell of each species is also shown, indicating the typically wide aperture for fish hunting cones, intermediate aperture for mollusc hunting cones, and the narrow aperture for worm hunting cones. Electrospray mass spectra were acquired on a SCIEX QSTAR Pulsar QqTOF mass spectrometer over *m/z* 650–2600. Several known conotoxins (ω-GVIA, χ-MrIA and δ-MrVIA) were abundant, while other known conotoxins, including for example α-GI, μ-GIIIA, α-ImI and α-ImII, were present at trace levels or were undetectable in these batches of pooled venom. Spectra of three other marked peaks contained multiple uncharacterised conotoxins. It is estimated that the venom of each species contains in excess of 100 unique peptides. Reproduced from Lewis and Garcia (2003) with permission

physiological role of different neuronal calcium channels. Ca^{2+} influx into nerve terminals occurs through calcium channels that open on arrival of an action potential, initiating the neurotransmitter release process that allows signalling to other nerves and muscle. In recent years, much has been discovered about the nature of these voltage sensitive channels, which have been classified into six groups according to their electrophysiological and pharmacological properties, termed L-, N-, P-, Q-, T- and R-types (Catterall et al. 2005a). Given their diversity and overlapping roles in neurotransmitter release, subtype selective inhibitors are required to

Table 1 Amino acid sequence, mode of action for eleven conotoxin and conopeptide classes with clinical potential

Class	Action	Name	Sequence	Clinical potential
ω	$Ca_v2.2$ inhibitor	MVIIA	**CKGKGAKCSRLMYDCCTGSCRSGKC**[a]	pain (i.t., phase IV)
	$Ca_v2.2$ inhibitor	CVID	**CKSKGAKCSKLMYDCCSGSCSGTVGC**[a]	pain (i.t. phase II)[b]
μ	Na_v inhibitor	SIIIA	ZNCCNGGCSSKWCRDHARCC[a]	pain (preclinical lead)
μO	$Na_v1.8$ inhibitor	MrVIB	ACSKKWEYCIVPIIGFIYCCPGLICGPFVCV	pain (preclinical lead)
δ	Na_v enhancer	EVIA	DDCIKOYGFCSLPILKNGLCCSGACVGVCADL[a]	pain (preclinical lead)
κ	K_v inhibitor	PVIIA	CRIONQKCFQHLDDCCSRKCNNRFNKCV	cardiovascular (preclinical lead)
χ	NET inhibitor	Xen2174	Z[c]GVCCGYKLCHOC	pain (i.t., phase II)
α	nAChR inhibitor	Vc1.1	GCCSDPRCNYDHPEIC[a]	pain (i.v.. phase II)[b]
ρ	$α_1$-adrenoceptor inhibitor	TIA	FNW**RCCLIPACRRNHKKFC**[a]	cardiovascular (preclinical lead)
Conantokin	NMDAR antagonist	cona-G	GEγγLQγNQγLIRγKSN	pain and epilepsy (i.t., preclinical)[b]
Conopressin	Vasopressin agonist	cono-G	CFIRNCPKG[a]	cardiovascular (preclinical lead)
Contulakin	neurotensinR agonist	cont-G	ZSEEGGSNAT$_g$KKPYIL	pain (i.t. phase II)[b]

O is *trans*-4-hydroxyproline, γ γ-carboxyglutamic acid, B is 6-bromotryptophan, T_g is glycosylated threonine. Cysteines involved in disulfide bonds (bolded) that connect in discrete overlapping patterns depending on sequence (see Tables 2–7).

[a] Amidated C-termini

[b] Clinical development halted

[c] Z pyroglutamate

Table 2 Selected ω-conotoxins and N- versus P/Q-VGCC selectivity

CVID	CKSKGAKCSKLMYDCCSGSCSGTVGRC-[a]	10^6-fold
GVIA	CKSOGSSCSOTSYNCCR–SCNOYTKRCY[a]	$10^{4.5}$-fold
MVIIA	CKGKGAKCSRLMYDCCTGSCRS–GKC-[a]	10^4-fold
CVIB	CKGKGASCRKTMYDCCRGSCRS–GRC-[a]	$10^{0.2}$-fold
MVIIC	CKGKGAPCRKTMYDCCSGSCGRR–GKC-[a]	$10^{-1.2}$-fold

Disulfide connectivity for class shown above
[a]Amidated

determine their relative roles in normal physiological processes. In disease states, specific subtypes of a receptor or ion channel may be differentially regulated, providing opportunities to develop subtype selective ligands that specifically treat the disease.

2.2   ω-Conotoxins

N-type calcium channels (see Fig. 3) play a key role in pain transmission by controlling neurotransmitter release at spinal synapse and thus can act as a gate-keeper for responses to sensory pathway activation. Since opiates also inhibit N-type calcium channels at the spinal level (indirectly via G proteins), it is not surprising that ω-conotoxins specific for N-type channels might be analgesic. Sub-nanomolar doses of ω-conotoxin MVIIA or ω-CVID (Lewis et al. 2000) delivered directly to the spinal cord (intrathecally) produce analgesia for up to 24 h in rats (Malmberg and Yaksh 1995; Smith et al. 2002). Interestingly, the affinity of ω-conotoxins for N-type calcium channels is reduced upon co-expression with the auxilliary α2δ subunit protein (Mould et al. 2004; see Fig. 3), an effect that would reduce ω-conotoxin efficacy in pain states where this subunit is upregulated. ω-MVIIA (Prialt, Elan) recently gained FDA approval for the treatment of otherwise unmanageable severe chronic pain. However, development ω-CVID (AM336) by AMRAD has not progressed beyond a Phase I/IIa clinical trial in severe cancer pain sufferers, where it produced clear signs of efficacy. Unfortunately, both ω-conotoxins produced unwanted side effects at therapeutic doses. Avoidance of the P/Q-calcium channels is considered an important requirement for therapeutic ω-conotoxins, but on-target effects at inhibitory synapses and supraspinal sites also are likely to contribute to dose-limiting side effects.

The structure of CVID, the most structurally defined and selective N-type calcium channel inhibitor known, is shown in Fig. 4 (Lewis et al. 2000). The discovery of new ω-conotoxins with selectivity profiles that produce fewer side effects may lead to the development of better analgesics in this peptide class. Extensive structure–activity relationship studies have allowed the development of a pharmacophore model for ω-conotoxins (Nielsen et al. 2000), allowing the rational development of further specific N-type inhibitors, including macrocyclic peptides and

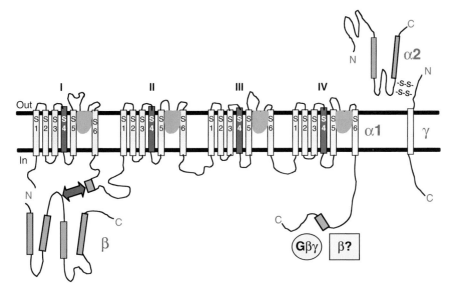

Fig. 3 Topology profile of a voltage-gated calcium channel (VGCC) membrane protein showing the pore forming α-subunit and the auxilliary β- and α2δ subunits and their likely sites of interaction. The ω-conotoxins are proposed to bind within or near the mouth of the VGCC to inhibit calcium ion influx, in contrast to the Gβγ subunit released from activated G-protein coupled receptors such as the μ-opioid receptor which act intraclealalrly to inhibit calcium influx

peptidomimetics (Schroeder et al. 2004). Based on the importance of N-type calcium channels in pain pathways revealed by ω-conotoxins, the Canadian biopharmaceutical company Neuromed Pharmaceuticals has developed a series of use-dependent small molecule inhibitors that are analgesic in rodents. Despite possessing functional selectivity, clinical studies to date on the small molecule NMED-160 have ceased, and follow-up leads are currently in development.

3   Sodium Channel Modulators

3.1   Voltage-gated Sodium Channels (VGSCs)

Like the structurally related calcium channels, voltage-gated sodium channels (VGSCs) play a key role in controlling neuronal excitability, but in this instance they are critical for initiation and transmission of action potentials along nerves. Based on their susceptibility to block by the puffer fish toxin tetrodotoxin (TTX), sodium channels can be divided into TTX-sensitive (TTX-S) and TTX-resistant (TTX-R) classes. Nine homologous α-subunit subtypes have been identified (Na$_v$1.1–1.9) (Catterall et al. 2005b). A number of these sodium channel subtypes

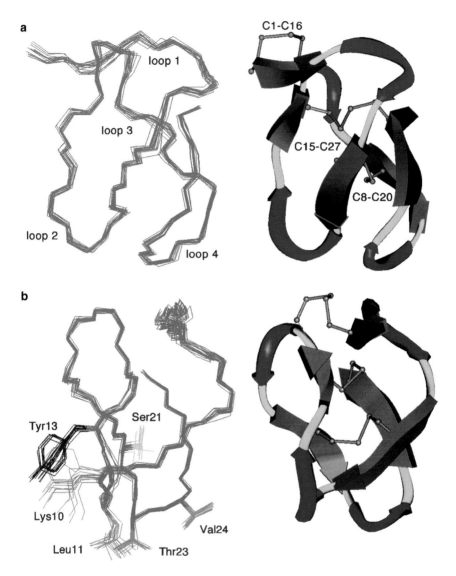

Fig. 4 The three-dimensional NMR structure of ω-CVID. **a** Shown are the 20 lowest energy structures (left) and the corresponding secondary structure (right). **b** Structures in A rotated 90°$_y$ showing position of side chains in the juxtapositioned loops 2 and 4 that are believed to contribute to the exquisite N-type selectivity of this ω-conotoxin. The hydroxyl of Tyr13 in loop 2 is the critical for high affinity interactions with the N-type calcium channel. Reproduced from Lewis et al. (2000) with permission

are implicated in clinical states such as pain (Wood and Boorman 2005; Lewis and Garcia 2003) as well as stroke and epilepsy. Given their critical role in the central and peripheral nervous system, it is not surprising that cone snails have evolved a number of different ways to target this ion channel class.

3.2 μ-Conotoxins

The first conotoxin in this class to be identified were μ-conotoxins GIIIA–C that target the skeletal muscle sodium channel $Na_v1.4$ (Table 3). More recently, μ-conotoxins active at neuronal subtypes in addition to $Na_v1.4$ have been identified, including PIIIA and TIIIA which inhibit $Na_v1.2$ (Shon et al. 1998a; Lewis et al. 2007), and KIIIA which inhibits $Na_v1.2$, 1.6, 1.7 (Zhang et al. 2007). However, despite subtype selective sodium channel inhibitors having considerable therapeutic potential, little progress has been made towards the development of μ-conotoxin inhibitors for specific therapeutically relevant VGSCs, including $Na_v1.3$, 1.6, 1.7 and 1.8. The structures of several μ-conotoxins including μ-PIIIA (Nielsen et al., 2002) (Lewis et al., 2007) are shown in Fig. 5. The NMR structures of the muscle selective GIIIA (Lancelin et al. 1991) may not reflect the structure that binds the VGSC based on comparisons with NMR structure of TIIIA (Lewis et al. 2007).

3.3 μO-Conotoxins

μO-conotoxins MrVIA and MrVIB from *Conus marmoreus* (Table 4) were originally identified as novel hydrophobic peptide blockers of *Aplysia* VGSCs (Fainzilber et al. 1995; McIntosh et al. 1995). More recently, μO-conotoxins were found to preferentially (tenfold selective) inhibit mammalian $Na_v1.8$ over other neuronal VGSC subtypes (Daly et al. 2004). Given the key role TTX-R sodium channels play in pain pathways (Wood and Boorman 2005), MrVIB was assessed for analgesic activity in animal models of pain. In these studies MrVIB delivered intrathecally was analgesic at doses that produced no local anaesthetic-like inhibitory effects on movement or coordination (Ekberg et al. 2006). In a separate study, synthetic MrVIB was again found to be analgesic when delivered subcutaneously to rats (Bulaj et al. 2006). The solution structure of MrVIB, which includes several well defined loops and a large and apparently flexible loop II (Daly et al. 2004), provides the first insights into the structural features contributing to this selective block

Table 3 Selected μ-conotoxins and sodium channel subtype preference

GIIIA	-RDCCTOOKKCKDRQCKOQ-RCCA[a]	1.4
GIIIB	-RDCCTOORKCKDRRCKOM-KCCA[a]	1.4>1.2
GIIIC	-RDCCTOOKKCKDRRCKOL-KCCA[a]	1.4
PIIIA	ZRLCCGFOKSCRSRQCKOH-RCC-[a]	1.4≈1.2
TIIIA	RHGCCKGOKGCSSRECROQ-HCC-[a]	1.4≈1.2
SmIIIA	-ZRCCNGRRGCSSRWCRDHSRCC-[a]	amphibian
SIIIA	-ZNCCNG--GCSSKWCRDHARCC-[a]	amphibian
KIIIA	---CCN----CSSKWCRDHSRCC-[a]	1.2≈1.4

Disulfide connectivity for class shown above
[a]Amidated

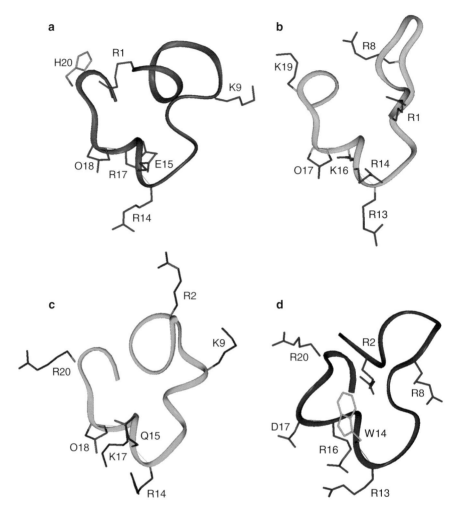

Fig. 5 The three-dimensional NMR structure of selected μ-conotoxins. Shown are the ribbon representation of μ-conotoxin **a** TIIIA red, **b** GIIIB cyan, **c** PIIIA green and **d** SmIIIA purple. Peptides are superimposed across the backbone residues 11–18 of TIIIA with the corresponding residues in PIIIA (11–18), GIIIA (10–17) and SmIIIA (10–17). Residues are numbered according to their individual primary sequence (see Table 3). Reproduced from Lewis et al. (2007) with permission

Table 4 Selected μO- and δ-conotoxin modulators of VGSCs

μO-MrVIA	--ACRKKWEYCIVPIIGFIYCCPGLICGPFVCV--
μO-MrVIB	--ACSKKWEYCIVPILGFVYCCPGLICGPFVCV--
δ-TxVIA	--WCKQSGEMC---NLLDQNCCDGY-CIVLVCT--[a]
δ-GmVIA	VKPCRKEGQLC---DPIFQNCCRGWNC-VLFCV--
δ-EVIA	-DDCIKOYGFCSLPILKNGLCCSGA-C-VGVCADL[a]
δ-PVIA	-EACYAOGTFC---GIKOGLCCSEF-CLPGVCFG--

Disulfide connectivity for class shown above
[a]Amidated

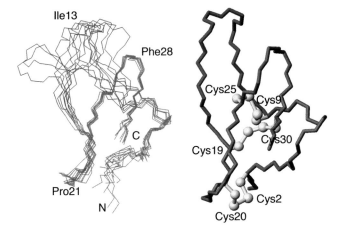

Fig. 6 The three-dimensional NMR structure of μO-conotoxin MrVIB. Shown on the left are the 20 lowest energy structures superimposed over residues 4–6 and 19–30 (Daly et al. 2004). Note the disordered structure of the large loop 2 that contrasts the more ordered loop 2 structure observed for ω-CVID (Fig. 3) and μ-conotoxins (Fig. 4). Shown on the right is the lowest energy structure and relative positions of the disulfide bonds

(Fig. 6). Unfortunately, a challenging oxidative folding step due to the hydrophobic nature of this class of conotoxins, has limited study of the structure-activity relationships of μO-conotoxins.

3.4 δ-Conotoxins

The hydrophobic "King Kong" peptide TxVIA from the molluscivorous *C. textile* was the first δ-conotoxin isolated due to the unusual toxic effects it produced upon injection into snails and lobster (Hillyard et al. 1989). The toxic effects of δ-conotoxins were subsequently attributed to a delay of VGSC inactivation in molluscan neurons (Fainzilber et al. 1991) through what appears to be a new site on the VGSC (Fainzilber et al. 1994). Interestingly, this site has been shown to overlap the μO-conotoxin binding site in rat brain (Ekberg et al. 2006), which is not surpising given their similar hydrophobicity and related structures. Since these early investigations, a range of new δ-conotoxins have been isolated with varying selectivity for different mammalian VGSCs (Table 4). δ-Conotoxin EvIA appears particularly interesting as it acts selectively at Nav1.2, 1.3 and 1.6 and spares Na$_v$1.4 and 1.5, making it the first δ-conotoxin identified that acts selectively on neuronal VGSCs of mammals (Barbier et al. 2004). The solution structures of TxVIA (Kohno et al. 2002) and EvIA (Volpon et al. 2004) again provide insights into the structural determinants important for their activity at VGSCs, although the specific residues involved remain to be identified.

4 Potassium Channel Inhibitors

4.1 κ-Conotoxins

Unlike the diversity of K⁺ channel inhibitors produced by scorpions and sea anemones, cone snails appear to have evolved relatively few peptides active at this physiologically important target. κ-PVIIA (Table 1) was identified as a key component of *C. purpurascens* that synergised with a sodium channel activator toxin to produce a potent lethal effect in fish (Terlau et al. 1996). The structure of PVIIA resembles other O-superfamily conotoxins characterised such as the ω-conotoxins (Scanlon et al. 1997). PVIIA interacts in the pore of the Shaker K⁺ channel (Shon et al. 1998b) at a position that confers voltage sensitivity (Scanlon et al. 1997) and state dependence (Terlau et al. 1999) confers to its block. Interestingly, PVIIA (CGX-1051) reduced infarct size in rat and dog models of myocardial ischaemia and reperfusion (Lubbers et al. 2005).

4.2 κA- and κM-Conotoxins

A second class of K⁺ channel blockers, the κA-conotoxins from piscivorous species, were first identified in the venom of *C. striatus*. Interestingly, the κA-SIVA sequence EKSLVPSVITTCCGYDOGTMCOOCRCTNSC* (Ser7 O-glycosylated) matched the mass of the major ichthytoxin (4083 Da) identified in the milked venom of *C. striatus* (Bingham et al. 1996). Shorter forms of the κA-conotoxins have recently been isolated (Teichert et al. 2007). κM-conotoxins have also been isolated from piscivorous species (Ferber et al. 2003). These latter K⁺ channel blockers have structures reminiscent of the μ-conotoxins and interact within the pore of the channel through a novel set of interactions (Al-Sabi et al. 2004). In addition to these inhibitors of K⁺ channels, κ-BtX was identified as a novel peptide in the venom of the vermivorous *C. betulineas* that selectively enhances large conductance calcium-activated K⁺ channel (BK) current in chromaffin cells (Fan et al. 2003).

5 Antagonists of Nicotinic Acetylcholine Receptors

5.1 α-Conotoxins

The nicotinic acetylcholine receptors (nAChRs) are a family of pentameric ligand gated cation channels that play a key signalling role at synapses and neuromuscular junctions where acetylcholine is released (Nicke et al. 2004). The α-conotoxins are a large and growing class of small peptides from cone snails that competitively inhibit specific subtpyes of the nAChR (Nicke et al. 2004). Muscle selective α-conotoxins (e.g. GI, Table 5) may present an alternative to the use of small molecule curare-mimetic muscle relaxants, which are used during surgery but have slower than ideal recovery periods. Potentially more interesting are the neuronally selective α-conotoxins

Table 5 Selected α-conotoxins and nAChR subtype preference

GI	---ECC--NPAC--GRHYSC[a]	Muscle
EI	RDOGCCYHPTCNMSNPQIC[a]	Muscle
ImI	---GCCSDPRC----AWRC[a]	α3β2 > α7
MII	---GCCSNPVCHLEHSNLC[a]	α3β2
EpI	---GCCSDPRCNMNNPDY$_s$C[a]	α7
PnIA	---GCCSLPPCAANNPDYC[a]	α3β2
GID	IRDBCCSNPACRVNNOHVC	α7 ≈ α3β2
Vc1a	---GCCSDORCNYDHPBIC[a]	α9/α10
TxIA	---GCCSRPPCIANNPDLC	α3β2

Disulfide connectivity for class shown above
[a]Amidated

that target nAChR subtypes comprising different combinations of α3–α10 and/or β2–β4 subunits expressed as either homomeric or heteromeric receptors depending on the neuronal cell type. Interestingly, α-conotoxin Vc1.1 (Sandall et al. 2003) and RgI (Vincler et al. 2006) have been identified as having potential analgesic properties following peripheral administration to rats. This result contrasts with the analgesic effects usually attributed to agonists of the nAChR acting centrally, and appears to be mediated by a specific subtype of the receptor (α9α10) possessing a previously unrecognised role in pain (Vincler et al. 2006), although the involvement of this target in its analgesic action has been questioned (Nevin et al. 2007). Metabolic Pharmaceuticals Ltd has discontinued clinical trials of Vc1.1 due to an anticipated lack of efficacy in humans when it was revealed experimentally that Vc1.1 had low affinity for the human form of the α9α10 receptor. Co-crystal structures with the acetylcholine binding protein (AChBP) are revealing precisely how α-conotoxins bind at this important therapeutic target. This is illustrated for α-TxIA, α-PnIA and α-ImI, showing distinct differences in their binding orientations at AChBP (Fig. 7) that provide a structural basis for the development of novel subtype specific inhibitors with clinical potential (Dutertre and Lewis 2006; Dutertre et al. 2006). In addition to the smaller α-, αA- and αS-conotoxins, certain vermivorus cone snail species utilise larger 11 kDa dimeric αD-conotoxins to target α7 and β2 containing receptors nAChRs (Loughnan et al. 2006).

6   Norepinephrine Transporter Inhibitors

6.1   χ-Conotoxins

The norepinephrine transporter (NET) plays a key role in reducing levels of neuronally released norepinephrine (NE = noradrenaline). The tricyclic antidepressants inhibit NET and appear to produce analgesia by enhancing the descending inhibitory pathway controlled by norepinephrine release. Unfortunately, this class of drugs also have antidepressant effects and significant off-target pharmacologies that limit their usefulness in the treatment of pain. χ-Conopeptides first isolated from *C. marmoreus* (Fig. 2) are highly specific, non-competitive inhibitors of norepinephrine uptake through the NET (Sharpe et al. 2001) that produce potent anal-

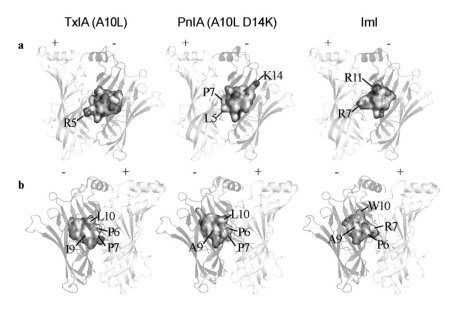

Fig. 7 Surface representation of the three α-conotoxins that have been co-crystallized with Ac-AChBP. The surface of the α-conotoxin facing the principal binding site is shown in (**a**), and the complementary binding site in (**b**). Ac-AChBP is shown in a transparent view for clarity. The principal subunit (+) of Ac-AChBP is shown in yellow, the complementary subunit (–) in blue. Reproduced from Dutertre et al. (2007) with permission

gesia in rats. As shown in Fig. 8, χ-MrIA has an unusual 1–4, 2–3 cysteine connectivity which produces a fold that is distinct from the α-conotoxins (see Table 5). An analogue of χ-conotoxin MrIA (Xen2174) has been developed with improved stability and duration of efficacy in animal models of pain (Nielsen et al. 2005). Xenome Ltd has completed its first evaluation of Xen2174 intrathecally in a Phase I/IIa safety trial in cancer patients suffering otherwise poorly managed pain. Initial results are promising both in terms of safety and efficacy. Interestingly, the binding site for χ-conopeptides on the NET partially overlaps the tricyclic anti-depressant binding site but not the NE binding site as illustrated in Fig. 9 (Paczkowski et al. 2007). Thus the structure of MrIA provides clues to the develop-ment of peptidomimetics and non-competitive small molecule inhibitors.

7 α_1-Adrenoceptor Antagonists

7.1   ρ-Conotoxins

A single ρ-conotoxin, TIA, has been isolated from cone snails to date (Table 1). This peptide has a cysteine spacing and structure reminiscent of the α-conotoxins, but its novel sequence allows it to selectively target mammalian α_{1A}-, α_{1B}- and

Fig. 8 χ-Conopeptide MrIA sequence, disulfide bond connectivity and NMR structure of the peptide in solution (Nilsson et al. 2005). An analogue of χ-MrIA, with the unstable asparagine replaced with a pyroglutamate (named Xen2174), is currently in the clinical in a Phase I/IIa trial in severe cancer pain patients

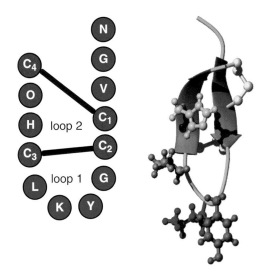

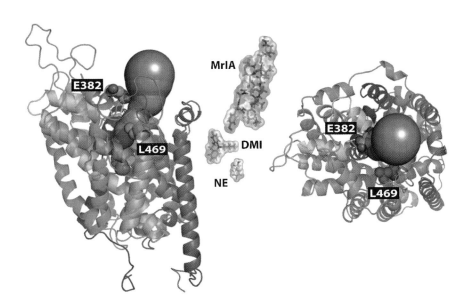

Fig. 9 Proposed binding positions of MrIA, the tricyclic desipramine (DMI) and NE at NET. The homology model of human NET in ribbon representation shows the presumed water-filled pathway (grey volume) linking the extracellular space to the substrate binding site. Residues with largest effect on MrIA binding (E382 and L469) are shown in CPK representation. MrIA, DMI and NE are shown in stick and transparent molecular surface rendition at the same scale as NET with vertical positions shown relative to the side view of NET (left panel). Right panel shows the top (90°x rotated) view of left panel. Overlap apparent between MrIA/DMI and DMI/NE but not MrIA/NE is consistent with the competitive MrIA/DMI and DMI/NE interactions and the noncompetitive MrIA/NE interaction. Reproduced from Paczkowski et al. (2007) with permission

α_{1D}-adrenoceptors (Sharpe et al. 2001). Interestingly, TIA noncompetitively inhibits α_{1B}-adrenoceptors but competitively inhibits α_{1A}- and α_{1D}-adrenoceptors (Sharpe et al. 2003; Chen et al. 2004). Given its unusual mode of action and specificity for the α_{1B} subtype, TIA has been used to determine the extent of α_{1B}-adrenoceptor involvement in the contractions of different tissues (Kamikihara et al. 2005; Lima et al. 2005).

8   NMDA Receptor Antagonists

8.1   *Conantokins*

Conantokins (Table 6) are specific inhibitors of the *N*-methyl-D-aspartate (NMDA) receptor (Layer et al. 2004). These peptides adopt slightly altered helical structures in solution in the presence of divalent metal ions such as Ca^{++} (Nielsen et al. 1999). Conantokins competitively inhibit glutamate activation, especially at NR2B containing subtypes, which explains their spermine dependence (Donevan and McCabe 2000), but also activate NMDA receptors containing NR2A containing subunits (Ragnarsson et al. 2006). Malmberg et al. (2003) showed that intrathecal conantokin G and T also have analgesic activity in pain models of tissue damage, nerve injury and inflammation in mice at doses that were ~20-fold lower than those affecting motor function. Conantokins also have therapeutic potential in epilepsy (Jimenez et al. 2002) and provide useful tools for monitoring changes in NR2B containing receptor density in brain, as observed in Alzheimer's diseased human cerebral cortex (Ragnarsson et al. 2002).

9   Vasopressin Receptor Modulators

9.1   *Conopressins*

A growing list of conopressins with sequences related to oxytoxin and vasopressin have been isolated from cone snail venoms (Table 7). Oxytocin and vasopressin act at the vasopressin receptor (a GPCR) and the related oxytocin receptors to regulate water balance, blood pressure and uterine smooth muscle contraction. The recently

Table 6   Selected conantokin inhibitors of the NMDA receptor

Con-G	GEγγLQγNQ-γLIRγKSN[a]
Con-T	GEγγYQKML-γNLRγAEVKKNA[a]
Con-R	GEγγVAKMAAγLARγNIAKGCKVNCYP[a]
Con-L	GEγγVAKMAAγLARγDAVN[a]

[a]Amidated

Table 7 Selected conopressin modulators of the vasopressin receptor

Conopressin-G	CFIRNCPKG[a]
Conopressin-S	CIIRNCPRG[a]
Conopressin-Vil	CLIQDCPγG[a]
Conopressin-T	CYIQNCLRV[a]
Vasopressin	CYFQNCPRG[a]
Oxytoxcin	CYIQNCPLG[a]

Disulfide connectivity for class shown above
[a]Amidated

discovered Con-T was found to be a selective V_{1a} antagonist with partial agonist activity at the OT receptor and no detectable activity at V_{1b} and V_2 receptors (Dutertre et al. 2008). Interestingly, this study found that replacing the Gly^9 in oxytoxin and arginine vasopressin with Val^9 found in Con-T converted these peptides from full agonists to full antagonists at the V_{1a} receptor, demonstrating the role of position 9 as an agonist/antagonist switch in these peptides. The role of conopressins in cone snail prey capture is unclear and their mode of action, receptor subtype specificity and structure are presently poorly defined. Therapeutic applications for this class of conotoxins, beyond potential in cardiovascular disorders and pre-term child birth, may relate to their central actions where effects on mood have been observed in animals.

10   Neurotensin Receptor Agonists

10.1   Contulakin-G

Cone snails produce a second endogenous peptide analogue, a glycosylated neurotensin analogue named contulakin-G (Craig et al. 1998). This peptide is a potent analgesic in a wide range of animal models of pain (Allen et al. 2007). Interestingly, contulakin-G is 100-fold less potent than neurotensin for neurotensin receptor I, but ~100-fold more potent as an analgesic. Based on its potency and wide therapeutic window, contulakin-G (CGX-1160) went into early stage clinical development by Cognetix Inc. for the treatment of pain. The development of the conantokins and contulakins is now on hold with the demise of this firm.

11   Conclusions

Given the vast array of conotoxins that remain pharmacologically uncharacterised, enormous opportunity remains to identify new research tools and potential therapeutics from amongst the highly evolved venoms of Conidae. The wider application of high-throughput and high content screens are expected to accelerate the discovery of new conotoxin pharmacologies, although the use of more traditional tissue and animal behavioural studies, pioneered in the laboratories of Baldomero Olivera and the late Robert Endean, still has a place in the discovery process. Given that cone snails have evolved to target worm, mollusc and fish targets, and non mammalian receptors, greater emphasis on screening across non-mammalian targets is expected to generate new discoveries. Likewise, expanding the investigation of Atlantic and Indian Ocean cone snail species is expected to yield new classes of biologically active conotoxins. Recent examples of novel peptides with undefined targets include the molluscan excitatory conopeptide conomap-Vt (Dutertre et al. 2006) and NMB-1, a novel inhibitor of mechanosensitive ionic currents and mechanical pain. For known classes of conotoxins, the application of peptide engineering approaches, guided by an understanding of the molecular details of receptor ligand interactions, are expected to generate conotoxins with optimised target specificity for specific disease states.

Acknowledgement Aspects of this work were supported by an NHMRC Program grant.

References

Adams D, Alewood P, Craik D, Drinkwater R, Lewis RJ (1999) Conotoxins and their potential pharmaceutical applications. Drug Discov Res 46:219–234.

Allen JW, Hofer K, McCumber D, Wagstaff JD, Layer RT, McCabe RT, Yaksh (2007) An assessment of the antinociceptive efficacy of intrathecal and epidural contulakin-G in rats and dogs. Anesth Analg 104:1505–1513.

Al-Sabi A, Lennartz D, Ferber M, Gulyas J, Rivier JE, Olivera BM, Carlomagno T, Terlau H (2004) κM-Conotoxin RIIIK, structural and functional novelty in a K+ channel antagonist. Biochemistry 43:8625–8635.

Barbier J, Lamthanh H, Le Gall F, Favreau P, Benoit E, Chen H, Gilles N, Ilan N, Heinemann SH, Gordon D, Menez A, Molgo J (2004) A δ-conotoxin from *Conus ermineus* venom inhibits inactivation in vertebrate neuronal Na+ channels but not in skeletal and cardiac muscles. J Biol Chem 279:4680–4685.

Bingham J-P, Jones A, Lewis RJ, Andrews PR, Alewood PF (1996) *Conus* venom peptides (conopeptides): inter-species, intra-species and within individual variation revealed by ion-spray mass spectrometry. In: Lazarovici P, Spiro M, Zlotkin E (eds) Biochemical aspects of marine pharmacology. Alaken, Inc., Fort Collins, CO, pp 13–27.

Bulaj G, Zhang MM, Green BR, Fiedler B, Layer RT, Wei S, Nielsen JS, Low SJ, Klein BD, Wagstaff JD, Chicoine L, Harty TP, Terlau H, Yoshikami D, Olivera BM (2006) Synthetic μO-conotoxin MrVIB blocks TTX-resistant sodium channel Na$_v$1.8 and has a long-lasting analgesic activity. Biochemistry 45:7404–7414.

Catterall WA, Goldin AL, Waxman SG (2005a) International Union of Pharmacology. XLVII. Nomenclature and structure-function relationships of voltage-gated sodium channels. Pharmacol Rev 57:397–409.

Catterall WA, Perez-Reyes E, Snutch TP, Striessnig J (2005b) International Union of Pharmacology. XLVIII. Nomenclature and structure-function relationships of voltage-gated calcium channels. Pharmacol Rev 57:411–425.

Chen Z, Rogge G, Hague C, Alewood D, Colless B, Lewis RJ, Minneman KP (2004) Subtype-selective noncompetitive or competitive inhibition of human α_1-adrenergic receptors by ρ-TIA. J Biol Chem 279:35326–35333.

Craig AG, Zafaralla G, Cruz LJ, Santos AD, Hillyard DR, Dykert J, Rivier JE, Gray WR, Imperial J, DelaCruz RG, Sporning A, Terlau H, West PJ, Yoshikami D, Olivera BM (1998) An O-glycosylated neuroexcitatory *Conus* peptide. Biochemistry 37:16019–16025.

Craig AG, Norberg T, Griffin D, Hoeger C, Akhtar M, Schmidt K, Low W, Dykert J, Richelson E, Navarro V, Mazella J, Watkins M, Hillyard D, Imperial J, Cruz LJ, Olivera BM (1999) Contulakin-G, an O-glycosylated invertebrate neurotensin. J Biol Chem 274: 13752–13729.

Daly NL, Ekberg JA, Thomas L, Adams DJ, Lewis RJ, Craik DJ (2004) Structures of μO-conotoxins from *Conus marmoreus*. Inhibitors of tetrodotoxin (TTX)-sensitive and TTX-resistant sodium channels in mammalian sensory neurons. J Biol Chem 279:25774–25782.

Donevan SD, McCabe RT (2000) Conantokin G is an NR2B-selective competitive antagonist of N-methyl-D-aspartate receptors. Mol Pharmacol 58:614–623.

Drew LJ, Rugiero F, Cesare P, Gale JE, Abrahamsen B, Bowden S, Heinzmann S, Robinson M, Brust A, Colless B, Lewis RJ, Wood JN (2007) High-threshold mechanosensitive ion channels blocked by a novel conopeptide mediate pressure-evoked pain. PLoS ONE 2:e515.

Dutertre S, Lewis RJ (2006) Toxin insights into nicotinic acetylcholine receptors. Biochem Pharmacol 72:661–670.

Dutertre S, Lumsden N, Alewood PF, Lewis RJ (2006) Isolation and characterisation of Conomap-Vt, a D-amino acid containing excitatory peptide from the venom of a vermivorous cone snail. FEBS Lett 580:3860–3866.

Dutertre S, Ulens C, Büttner R, Fish A, van Elk R, Kendel Y, Hopping G, Alewood PF, Schroeder C, Nicke A, Smit AB, Sixma TK, Lewis RJ (2007) AChBP-targeted α-conotoxin correlates distinct binding orientations with nAChR subtype selectivity. EMBO J 26:3858–3867.

Dutertre S, Croker D, Daly NL, Andersson A, Muttenthaler M, Lumdsen NG, Craik DJ, Alewood PF, Guillon G, Lewis RJ (2008) Conopressin-T from *Conus tulipa* reveals an antagonist switch in vasopressin-like peptides. J Biol Chem 83:7100–7108.

Ekberg J, Jayamanne A, Vaughan CW, Aslan S, Thomas L, Mould J, Drinkwater R, Baker MD, Abrahamsen B, Wood JN, Adams DJ, Christie MJ, Lewis RJ (2006) μO-conotoxin MrVIB selectively blocks Na$_v$1.8 sensory neuron specific sodium channels and chronic pain without motor deficits. Proc Natl Acad Sci USA 103:17030–17035.

Fainzilber M, Gordon D, Hasson A, Spira ME, Zlotkin E (1991) Mollusc-specific toxins from the venom of *Conus textile neovicarius*. Eur J Biochem 202:589–595.

Fainzilber M, Kofman O, Zlotkin E, Gordon D (1994) A new neurotoxin receptor site on sodium channels is identified by a conotoxin that affects sodium channel inactivation in molluscs and acts as an antagonist in rat brain. J Biol Chem 269:2574–2580.

Fainzilber M, van der Schors R, Lodder JC, Li KW, Geraerts WP, Kits KS (1995) New sodium channel-blocking conotoxins also affect calcium currents in *Lymnaea* neurons. Biochemistry 34:5364–5371.

Fan CX, Chen XK, Zhang C, Wang LX, Duan KL, He LL, Cao Y, Liu SY, Zhong MN, Ulens C, Tytgat J, Chen JS, Chi CW, Zhou Z (2003) A novel conotoxin from *Conus betulinus*, κ-BtX, unique in cysteine pattern and in function as a specific BK channel modulator. J Biol Chem 278:12624–12633.

Ferber M, Sporning A, Jeserich G, DeLaCruz R, Watkins M, Olivera BM, Terlau H (2003) A novel *Conus* peptide ligand for K$^+$ channels. J Biol Chem 278:2177–2183.

Hillyard DR, Olivera BM, Woodward S, Corpuz GP, Gray WR, Ramilo CA, Cruz LJ (1989) A molluscivorous *Conus* toxin: conserved frameworks in conotoxins. Biochemistry 28:358–361.

Jimenez EC, Donevan S, Walker C, Zhou LM, Nielsen J, Cruz LJ, Armstrong H, White HS, Olivera BM (2002) Conantokin-L, a new NMDA receptor antagonist: determinants for anti-convulsant potency. Epilepsy Res 51:73–80.

Kamikihara SY, Mueller A, Lima V, Silva AR, da Costa IB, Buratini J Jr, Pupo AS (2005) Differential distribution of functional α$_1$-adrenergic receptor subtypes along the rat tail artery. J Pharmacol Exp Ther 314:753–761.

Kohno T, Sasaki T, Kobayashi K, Fainzilber M, Sato K (2002) Three-dimensional solution structure of the sodium channel agonist/antagonist δ-conotoxin TxVIA. J Biol Chem 277:36387–36391.

Lancelin JM, Kohda D, Tate S, Yanagawa Y, Abe T, Satake M, Inagaki F (1991) Tertiary structure of conotoxin GIIIA in aqueous solution. Biochemistry 30:6908–6916.

Layer RT, Wagstaff JD, White HS (2004) Conantokins: peptide antagonists of NMDA receptors. Curr Med Chem 11:3073–3084.

Lewis RJ, Garcia ML (2003) Therapeutic potential of venom peptides. Nat Rev Drug Discov 2:790–802.

Lewis RJ, Nielsen KJ, Craik DJ, Loughnan ML, Adams DA, Sharpe IA, Luchian T, Adams DJ, Bond T, Thomas L, Jones A, Matheson JL, Drinkwater R, Andrews PR, Alewood PF (2000) Novel ω-conotoxins from *Conus catus* discriminate among neuronal calcium channel subtypes. J Biol Chem 275:35335–35344.

Lewis RJ, Schroeder CI, Ekberg J, Nielsen KJ, Loughnan M, Thomas L, Adams DA, Drinkwater R, Adams DJ, Alewood PF (2007) Isolation and structure-activity of μ-conotoxin TIIIA, a potent inhibitor of TTX-sensitive voltage-gated sodium channels. Mol Pharmacol 71:676–685.

Loughnan M, Nicke A, Jones A, Schroeder CI, Nevin ST, Adams DJ, Alewood DJ, Lewis RJ (2006) Identification of a novel class of nicotinic receptor antagonists: dimeric conotoxins VxXIIA, VxXIIB and VxXIIC from *Conus vexillum*. J Biol Chem 281, 24745–24755.

Lima V, Mueller A, Kamikihara SY, Raymundi V, Alewood D, Lewis RJ, Chen Z, Minneman KP, Pupo AS (2005) Differential antagonism by conotoxin ρ-TIA of contractions mediated by distinct α_1-adrenoceptor subtypes in rat vas deferens, spleen and aorta. Eur J Pharmacol 508:183–192.

Lubbers NL, Campbell TJ, Polakowski JS, Bulaj G, Layer RT, Moore J, Gross GJ, Cox BF (2005) Postischemic administration of CGX-1051, a peptide from cone snail venom, reduces infarct size in both rat and dog models of myocardial ischemia and reperfusion. J Cardiovasc Pharmacol 46:141–146.

Malmberg AB, Yaksh TL (1995) Effect of continuous intrathecal infusion of ω-conopeptides, N-type calcium-channel blockers, on behavior and antinociception in the formalin and hot-plate tests in rats. Pain 60:83–90.

Malmberg AB, Gilbert H, McCabe RT, Basbaum AI (2003) Powerful antinociceptive effects of the cone snail venom-derived subtype-selective NMDA receptor antagonists conantokins G and T. Pain 101:109–116.

McIntosh JM, Hasson A, Spira ME, Gray WR, Li W, Marsh M, Hillyard DR, Olivera BM (1995) A new family of conotoxins that blocks voltage-gated sodium channels. J Biol Chem 270:16796–16802.

Milne TJ, Abbenante G, Tyndall JD, Halliday J, Lewis RJ (2003) Isolation and characterization of a cone snail protease with homology to CRISP proteins of the pathogenesis-related protein superfamily. J Biol Chem 278:31105–31110.

Mould J, Yasuda T, Schroeder CI, Beedle AM, Clinton J, Doering CJ, Zamponi GW, Adams DJ, Lewis RJ (2004) The $\alpha 2\delta$ auxiliary subunit reduces affinity of ω-conotoxins for recombinant N-type calcium channels J Biol Chem 279:34705–34714.

Nevin ST, Clark RJ, Klimis H, Christie MJ, Craik DJ, Adams DJ (2007) Are $\alpha 9\alpha 10$ nicotinic acetylcholine receptors a pain target for α-conotoxins? Mol Pharmacol 72:1406–1410.

Nicke A, Wonnacott S, Lewis RJ (2004) α-Conotoxins as tools for the elucidation of structure and function of neuronal nicotinic acetylcholine receptor subtypes. Eur J Biochem 271:2305–2319.

Nielsen CK, Lewis RJ, Alewood D, Drinkwater R, Palant E, Patterson M, Yaksh TL, McCumber D, Smith MT (2005) Anti-allodynic efficacy of the χ-conopeptide, Xen2174, in rats with neuropathic pain. Pain 118:112–124.

Nielsen K, Schroeder T, Lewis R (2000) Structure-activity relationships of ω-conotoxins at N-type voltage-sensitive calcium channels. J Mol Recognit 13:55–70.

Nielsen KJ, Skjaerbaek N, Dooley M, Adams DA, Mortensen M, Dodd PR, Craik DJ, Alewood PF, Lewis RJ (1999) Structure-activity studies of conantokins as human *N*-methyl-D-aspartate receptor modulators. J Med Chem 42:415–426.

Nielsen KJ, Watson M, Adams DJ, Hammarstrom AK, Gage PW, Hill JM, Craik DJ, Thomas L, Adams D, Alewood PF, Lewis RJ (2002) Solution structure of μ-conotoxin PIIIA, a preferential inhibitor of persistent tetrodotoxin-sensitive sodium channels. J Biol Chem 277: 27247–27255.

Nilsson KP, Lovelace ES, Caesar CE, Tynngard N, Alewood PF, Johansson HM, Sharpe IA, Lewis RJ, Daly NL, Craik DJ (2005) Solution structure of χ-conopeptide MrIA, a modulator of the human norepinephrine transporter. Biopolymers 80:815–823.

Paczkowski FA, Sharpe IA, Dutertre S, Lewis RJ (2007) χ-Conopeptide and tricyclic antidepressant interactions at the norepinephrine transporter define a new transporter model. J Biol Chem 282:17837–17844.

Ragnarsson L, Mortensen M, Dodd PR, Lewis RJ (2002) Spermine modulation of the glutamate (NMDA) receptor is differentially responsive to conantokins in normal and Alzheimer's disease human cerebral cortex. J Neurochem 81:765–79.

Ragnarsson L, Yasuda T, Lewis RJ, Dodd PR, Adams DJ (2006) NMDA receptor subunit-dependent modulation by conantokin-G and Ala7-conantokin-G. J Neurochem 96:283–291.

Sandall DW, Satkunanathan N, Keays DA, Polidano MA, Liping X, Pham V, Down JG, Khalil Z, Livett BG, Gayler KR (2003) A novel α-conotoxin identified by gene sequencing is active in suppressing the vascular response to selective stimulation of sensory nerves *in vivo*. Biochemistry 42:6904–6911.

Scanlon MJ, Naranjo D, Thomas L, Alewood PF, Lewis RJ, Craik DJ (1997) Solution structure and proposed binding mechanism of a novel potassium channel toxin κ-conotoxin PVIIA. Structure 5:1585–1597.

Schroeder CI, Smythe ML, Lewis RJ (2004) Development of small molecules that mimic the binding of ω-conotoxins at the N-type voltage-gated calcium channel. Mol Diversity 8:127–134.

Sharpe IA, Gehrmann J, Loughnan ML, Thomas L, Adams DA, Atkins A, Palant E, Craik DJ, Adams DF, Alewood PF, Lewis RJ (2001) Two new classes of conopeptides inhibit the α1-adrenoceptor and noradrenaline transporter. Nat Neurosci 4:902–907.

Sharpe IA, Thomas L, Loughnan M, Motin L, Palant E, Croker DE, Alewood D, Chen S, Graham RM, Alewood PF, Adams DJ, Lewis RJ (2003) Allosteric α$_1$-adrenoreceptor antagonism by the conopeptide ρ-TIA. J Biol Chem 278:34451–34457.

Shon KJ, Olivera BM, Watkins M, Jacobsen RB, Gray WR, Floresca CZ, Cruz LJ, Hillyard DR, Brink A, Terlau H, Yoshikami D (1998a) μ-Conotoxin PIIIA, a new peptide for discriminating among tetrodotoxin-sensitive Na channel subtypes. J Neurosci 18:4473–81.

Shon KJ, Stocker M, Terlau H, Stuhmer W, Jacobsen R, Walker C, Grilley M, Watkins M, Hillyard DR, Gray WR, Olivera BM (1998b) κ-Conotoxin PVIIA is a peptide inhibiting the shaker K$^+$ channel. J Biol Chem 273:33–38.

Smith M., Cabot PJ, Ross FB, Robertson AD, Lewis RJ (2002) The novel N-type calcium channel blocker, AM336, produces potent dose-dependent antinociception after intrathecal dosing in rats and inhibits substance P release in rat spinal cord slices. Pain 96:119–127.

Teichert RW, Jacobsen R, Terlau H, Yoshikami D, Olivera BM (2007) Discovery and characterization of the short κA-conotoxins: a novel subfamily of excitatory conotoxins. Toxicon 49:318–328.

Terlau H, Shon KJ, Grilley M, Stocker M, Stuhmer W, Olivera BM (1996) Strategy for rapid immobilization of prey by a fish-hunting marine snail. Nature 381:148–151.

Terlau H, Boccaccio A, Olivera BM, Conti F (1999) The block of Shaker K$^+$ channels by κ-conotoxin PVIIA is state dependent. J Gen Physiol 114:125–140.

Vincler M, Wittenauer S, Parker R, Ellison M, Olivera BM, McIntosh JM (2006) Molecular mechanism for analgesia involving specific antagonism of α9α10 nicotinic acetylcholine receptors. Proc Natl Acad Sci USA 103:17880–17884.

Volpon L, Lamthanh H, Barbier J, Gilles N, Molgo J, Menez A, Lancelin JM (2004) NMR solution structures of δ-conotoxin EVIA from *Conus ermineus* that selectively acts on vertebrate neuronal Na$^+$ channels. J Biol Chem 279:21356–21366.

Wood JN, Boorman J (2005) Voltage-gated sodium channel blockers; target validation and therapeutic potential. Curr Topics Med Chem 5:529–537.

Zhang MM, Green BR, Catlin P, Fiedler B, Azam L, Chadwick A, Terlau H, McArthur JR, French RJ, Gulyas J, Rivier JE, Smith BJ, Norton RS, Olivera BM, Yoshikami D, Bulaj G (2007) Structure/function characterization of μ-conotoxin KIIIA, an analgesic, nearly irreversible blocker of neuronal mammalian sodium channels. J Biol Chem 282:30699–30706.

Sodium Channel Inhibiting Marine Toxins

Lyndon E. Llewellyn

Abstract Saxitoxin (STX), tetrodotoxin (TTX) and their many chemical relatives are part of our daily lives. From killing people who eat seafood containing these toxins, to being valuable research tools unveiling the invisible structures of their pharmacological receptor, their global impact is beyond measure. The pharmacological receptor for these toxins is the voltage-gated sodium channel which transports Na ions between the exterior to the interior of cells. The two structurally divergent families of STX and TTX analogues bind at the same location on these Na channels

L.E. Llewellyn (✉)
Australian Institute of Marine Science, Townsville MC, QLD 4810, Australia
e-mail: L.Llewellyn@aims.gov.au

N. Fusetani and W. Kem (eds.), *Marine Toxins as Research Tools*,
Progress in Molecular and Subcellular Biology, Marine Molecular Biotechnology 46,
DOI: 10.1007/978-3-540-87895-7, © Springer-Verlag Berlin Heidelberg 2009

to stop the flow of ions. This can affect nerves, muscles and biological senses of most animals. It is through these and other toxins that we have developed much of our fundamental understanding of the Na channel and its part in generating action potentials in excitable cells.

1   Introduction

Building on the energetic properties of different elements and their charged ionic forms, in particular sodium (Na), calcium (Ca), potassium (K) and chloride (Cl), life has evolved because of an ability to move, grow, sense and reproduce which all rely on cellular electric potential. As is the nature of evolution and the pressures of natural selection, some organisms synthesize chemicals to interfere with transmembrane potential differences enabling the toxin-producers to capture food, defend against attack and out-compete other organisms for their place in the world. Rapid passage of Na ions into excitable cells such as in nerves and muscles is essential for the generation of cellular action potentials. This is facilitated by the transport of Na ions across cell membranes through sophisticated proteins that can sense changes in cell membrane electrical potential differences. These so-called voltage gated sodium channels (Na channels) are the focus of this chapter along with the marine subset of natural products which target them and prohibit the flow of sodium through the channel preventing action potential generation.

1.1   The Sodium Ion

Sodium is an alkali metal that is one of the most common elements in the world. It is the cationic form of this element within the dominant oceanic salt, a critical component of soil and rocks and one of the main physiological ions in many living organisms and in all animals. In solution Na is a univalent cation surrounded by a "shell" of water molecules. For a Na ion to pass through the sodium channel this aqueous shell must be penetrated by polar side chains of the amino acids that line the channel's pore to enable ion ligation and selection, until the ion makes contact with the aqueous internal surface of the cell membrane. This is all energetically complex and demanding, as the ion and water molecules have a high attraction for each other due to the above mentioned charge interactions.

   Under normal circumstances the concentration of Na ions in extracellular physiological fluid is ~140–150 mM and 10- to 30-fold less inside the cell. K and Ca occur at concentrations of about 5 mM and 1–2 mM, respectively, in normal extracellular human physiological fluids, and at ~140 mM and 100 nM inside cells, respectively. For every millimolar ion concentration in physiological fluid, this translates to 6×10^5 Na, K and Ca ions per μm^3 of water. Bearing in mind that a μm^3 equals $10^{12} Å^3$ and that Na, K and Ca have ionic radii of 1–1.3 Å, these ions occupy only a small fraction of the solution that bathes the extra- and intra-cellular surfaces of the membranes of the electrically excitable cell. These are important figures to bear in mind when later considering the Na channel's ability to attract Na ions to the channel's mouth, quickly extract the ions from their watery enclosure, rapidly move the ions, while all the time selecting them over K and Ca ions.

1.2   Generation and Propagation of Cellular Action Potentials

Commercially-generated electricity is commonly transmitted along metallic wires to power our homes and industries. While considered by convention as the flow of positive charges to regions of negative charge, commercially generated electricity is actually transmitted by the movement of electrons in the opposite direction. In contrast, action potential generation in cells of excitable tissue is caused by complex, but elegant series of processes allowing the rapid flow of Na and then K ions across the membrane, thereby driving the membrane's internal potential from negative to positive and then back to negative at the end of the action potential. An electrical potential difference normally exists across a cell membrane that is determined by the relative permeabilities of the major ions and their concentration gradients across the membrane. Rapid and selective movements of the different ions across the cell membrane generates a self-propagating positive electrical charge along (and across) the cell membrane. This process is enabled by integral membrane proteins which traverse the cell membrane, bridging the intra- and extracellular fluids. They undergo structural conformations in response to changes in the electrical potential that causes the sequential opening and closing of their channels, leading to their being named voltage-gated ion channels.

Changes in electrical potential can remain localized or be transferred along a cell membrane. This transfer along a cell membrane is achieved by channels sensing a change in electrical potential responding by opening and altering potential of the membrane adjacent to the original membrane alteration. The channel's response to the voltage change stimulus is to not only open; slower additional conformational changes are initiated that eventually lead to the channel's inactivation (Aldrich et al. 1983). Both the rate of channel opening and inactivation therefore control a channel's open time. For example, rapid opening and slow inactivation leads to longer open times and increased conductance due to a longer period of ion flow with slow opening and rapid inactivation having the opposite effect. These conformational changes are dependent upon structure. In Na channels, the time constants for channel activation are often sub-millisecond with channel inactivation time constants ranging from sub-millisecond to over 10 ms (Catterall et al. 2005). After inactivation, and membrane repolarization, the channels return to their resting condition to be responsive to the next electrical stimulus. In this way, a localized depolarization can lead to an action potential that can rapidly propagate to the end of the axon and there act as a stimulus for neurotransmitter secretion and synaptic transmission.

Cellular action potentials are critical to nerves, muscles, sensing, flagellary and ciliary movement to name but a few key functions. For example:

- nerves carry messages throughout animal bodies;
- muscles contract and relax in response to bioelectrical impulse;
- vision, touch, smell and hearing all rely upon bioelectric signaling processes;
- cilia and flagella flex in response to changes in concentrations of ions along their length; and
- plant guard cells open and close in response to environmental conditions.

Thus, bioelectric signaling is an ancient and fundamental property critical to the evolution of life. Na channels had not been found in any of these ancient unicellular organisms until recently; a simplified Na channel was found in the bacterium *Bacillus halodurans* (Ren et al. 2001) but it was more related in amino acid sequence and toxin and drug sensitivity to voltage activated Ca channels. It should also be borne in mind that voltage gated ion channels are only one group of ion channels; others open and close as a result of binding ligands such as neurotransmitters, ATP, hydrogen ions, to name a few examples (Hamill 2006; Saier 2000).

One of the most unusual bioelectric signaling systems are electric organs such as are found in electric eels and electric rays; one species of the latter (*Electrophorus electricus*) occupies a special place in the history of the study of voltage gated ion channels. It was from this species' electric organ that a Na channel was first purified. Partial amino acid sequence was then obtained from the purified protein that then allowed the cloning and first unveiling of the complete amino acid sequence of a voltage-gated Na channel (Noda et al. 1984).

2   Sodium Channels and Their Evolutionary Origin

2.1   *Discovery of Sodium Channels*

The importance of sodium for the nerve impulse was well established before the existence of ion channels was realized and the giant axons of crustaceans and especially the squid had a pivotal role in this realization. These giant axons allowed placement of current-passing as well as voltage-measuring electrodes along a segment of the axon, making possible the "clamping" of the membrane's potential at various potentials. This allowed the accurate measurement of Na and K generated currents across the membrane, which were shown to occur through separate pathways (Hodgkin 1958). The use of radioisotopes of these cations also enabled quantification of the passage of Na, K and additionally Ca ions across the cell membrane. These and subsequent investigations, often using toxins as probes, led to the hypothesis that protein mediated passive diffusion was occurring, and the ion channel concept was given strong support. We now know that there are not only families of channels which specialize in transporting Na, K and Ca, but that there are channel isoforms within each of these ion channel families, including Na channels (Catterall et al. 2003, 2005).

2.2   Structure

The Na channel comprises several subunits with the largest subunit, the α-subunit (~260 kDa), containing the pore through which the ions travel. The α-subunit contains four internal amino acid sequence repeats or domains, with each domain containing six transmembrane α-helices. The four connected domains assemble in the plane of the membrane to constitute the pore-forming structure, with the fourth domain interacting with the first as well as third domain (Terlau and Stuhmer 1998). At present there are nine isoforms of the Na channel known in mammals, all having a separate gene coding their expressed sequence (Catterall et al. 2005). The different amino acid sequences between these isoforms generates measurable differences in channel properties such as their sensitivity to different toxins and drugs. Three of these Na channel isoforms possess markedly decreased sensitivity to the channel inhibiting toxin tetrodotoxin and these three channel families are more related to each by amino acid sequence similarity than to other toxin-sensitive families (Catterall et al. 2005). Depending upon the tissue and Na channel isoform, multiple smaller proteins, the β-subunits, may attach to the α-subunit and affect channel properties such as the speed of ion conduction and channel activation and inactivation rates (Cannon et al. 1993; Makita et al. 1996). The Na channel is closely related in amino acid sequence and the motifs and patterns therein to the other major voltage dependent cation channels, the Ca and K channels (Guy and Durell 1995). Despite these similarities, the Na channel has the extraordinary ability to select Na cations over K and Ca, and additionally enable the passage of many billions of ions per second (Hille 2001).

The most recent model of the entire molecular structure of the Na^+ channel (Fig. 1) is again from the electric organ of the electric eel *Electrophorus electricus*. Ultra-high resolution cryo-electron microscopy using single particle image analysis (Sato et al. 2001) was used to develop a structure at 19 Å resolution. This model consists of a bell shaped molecule honeycombed by channels radiating like the diagonals of a cube with their intersection in the centre being the ion pore. If this model is correct, the channel actually has four entry points converging onto a central cavity which then radiate out to four exit pathways into the interior of the cell. The visualized channel is 135 Å high and 100 Å and 65 Å wide at the intracellular base and the exterior surface, respectively. The central cavity is ~40 Å long and has a 35 Å diameter, large enough to accommodate a range of pore-blocking toxins.

2.3   Function with Regards Ion Transport

The sodium channel allows the passive transport of billions of Na ions per second across a cell membrane which is generally considered about 75–100 Å thick. As alluded to above, this remarkable property is made even more impressive by the fact that to enable this massive ionic transportation, the channel must also rapidly concentrate Na ions from the watery milieu at its surface to feed this ionic surge through the channel and at the same time as the Na ions are being selected over

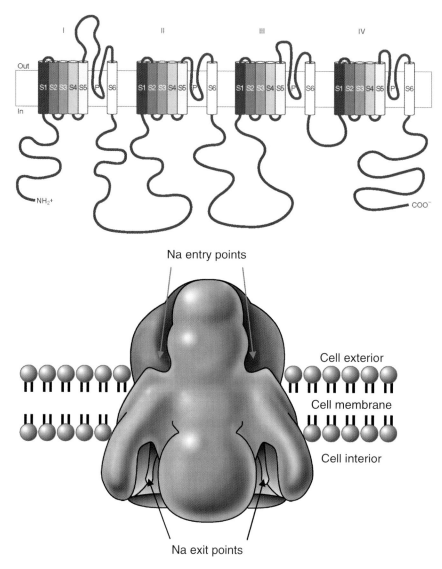

Fig. 1 Cartoon depicting the structure of the Na channel. The uppermost figure depicts the four domains of the channel containing six transmembrane helices with the connecting amino acid sequence between helices 5 and 6 coming together to line the pore through which the ions travel. The lower cartoon shows the side-on view of the whole channel from the electric organ of the electric eel after (Catterall 2001). The structure was derived at 19 Å resolution using helium-cooled cryo-electron microscopy and single-particle image analysis (Sato et al. 2001). The structure observed using this method was of a bell shape with multiple external apertures proposed to be inlets into a network of passageways through the channel connected by a central pore. This network then leads to a ring of four exits into the cellular interior

other cations they are being dehydrated to allow their passage through such a small pore. The physiological media at the orifices of the pore are even more complicated because of the presence of diverse organic molecules ranging from small molecules to large proteins, as well as other inorganic ions not directly involved in generating the action potential.

2.4 Taxonomic Distribution and Evolution

Na channels are widely distributed within the animal kingdom, but not as widely as K and Ca channels. The Na channel has been detected in vertebrates and most invertebrate phyla including mollusks (Dyer et al. 1997), jellyfish (Spafford et al. 1998) and flatworms (Jeziorski et al. 1997). The currently known taxonomic distribution of Na channels probably reflects search effort and as more and more genomes are sequenced, the true extent of their occurrence in organisms will become more apparent.

As mentioned earlier, Na channels shares significant amino acid sequence homologies with the voltage dependent Ca and K channels (Guy and Durell 1995). Like the Na channel, the Ca channel also contains a quadruplicate of homologous but different domains. Na and Ca channels are so closely related that relatively simple *in vitro* mutations of their sequences can transfer their cation selectivity properties between the two channel families (Heinemann et al. 1992). Unlike the Na and Ca channel families, the K channel has no replication within its amino acid sequence and is homologous to the single domains of tetrameric Na and Ca channels. To produce functional channels, K channel single domains must not only fold correctly but also combine with three other homologous (but frequently different) domains to form a functioning ion channel. By contrast, the Na and Ca channel α-subunits need merely to fold together into the correct formation within the cell membrane to function. The tetrameric domain Na channel is evolutionarily more recent than the monomeric K channel domains – it is thought that it's tetrameric structure arose from gene duplications of monomers such as are used to form the K channel (Anderson and Greenberg 2001; Goldin 2002; Pohorille et al. 2005).

3 The Sodium Channel Inhibitors Saxitoxin and Tetrodotoxin

Saxitoxin (STX) derives its name from the bivalve *Saxidomus giganteus* while tetrodotoxin (TTX) derives its name from the taxonomic name of the fish family in which it was first discovered, namely the Family Tetraodontidae. Both have a close association with humanity, causing fatalities through seafood poisoning but also because of their use as biomedical tools to reveal sub-microscopic properties of the proteins which bind them as well as the cells and organs that possess the toxin receptors.

3.1   Their Histories

The occurrence of a paralyzing poison amongst filter-feeding animals like clams and oysters, some of which are eaten by people, led to a concerted effort to identify the biological agent that causes this public health problem, dubbed paralytic shell-fish poisoning (PSP). The filter feeding sand crab *Emerita analoga* was one of the first animals other than bivalves in which this type of toxicity was discovered, prior to its being realized that it came from a dinoflagellate (Sommer 1932).

Pufferfish poison (later referred to as TTX, see below) has long been known as part of the Asian dining culture through the "fugu" (Japanese name for pufferfish) tradition. The poison was also classified in the Chinese pharmacopeia, Pen-T'so Chin (The Book of Herbs of The Herbal, 2838–2698 BC) as a drug having tonic effects but that could be toxic depending on dosage (Kao 1966). The discovery of the use of ground up dried pufferfish livers in the making of "zombie" powder in Haiti led to the suggestion that TTX could cause the appearance of death through its paralytic properties without killing the victim. Its reputed role in the zombifica-tion process is however, highly controversial (Kao and Yasumoto 1990; Yasumoto and Kao 1986).

Both TTX and STX are highly lethal. Intravenous injection of doses in the low nmol/kg range of either STX or TTX are all that is needed to kill some animals (Chang et al. 1997; Flachsenberger 1986; Llewellyn 2006; Matsumura 1995b). While less potent if administered orally, only small (a few micrograms per kilo-gram) of either toxin are sufficient to kill mammals (Llewellyn 2006; Xu et al. 2005). Fortunately, this high potency is not generally observed for many analogues of these toxins.

3.2   Structure of Saxitoxins and Tetrodotoxins

STX (Fig. 2; $C_{10}H_{17}N_7O_4$; MW = 299) is a trialkyl tetrahydropurine (Schantz et al. 1975) that is extremely stable in biological fluids and physiological solutions unless exposed for a lengthy period to alkaline pHs (Rogers and Rapoport 1980; Stafford and Hines 1995). It possesses two pKa's of 8.2 and 11.3 which belong to the 7,8,9 and 1,2,3 guanidinium groups, respectively (Rogers and Rapoport 1980; Schantz et al. 1966; Shimizu et al. 1981). These pKa's confer upon STX a polar nature enabling its ready solubility in water and lower alcohols (Schantz et al. 1957). At physiological pH, the 1,2,3-guanidino carries a positive charge, whereas the 7,8,9-guanidino group is only partially deprotonated (Shimizu et al. 1981; Strichartz 1984). The tricyclic core of STX is rigid in solution (Bordner et al. 1975; Schantz et al. 1975) and deviates significantly from a planar surface (Niccolai et al. 1980). However, the carbamoyl side chain is flexible, displaying considerable con-formational heterogeneity in solution (Fig. 2).

The structure for crystalline TTX ($C_{11}H_{17}N_3O_8$; MW = 319) was elucidated in 1964 (Goto et al. 1965; Tsuda et al. 1964; Woodward 1964) (Fig. 2). It is an orthoester

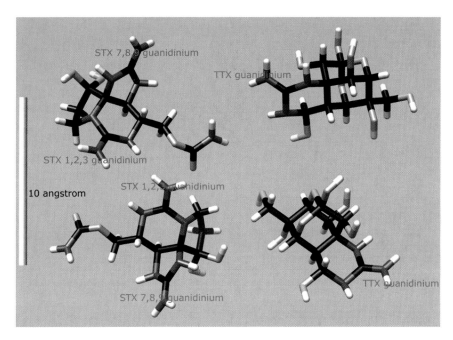

Fig. 2 Structures of STX (left) and TTX (right) with carbon, nitrogen, oxygen and hydrogen atoms being black, dark grey, light grey and white, respectively. The upper and lower toxin images in different orientations are simply provided to assist the reader to better appreciate the three-dimensional structures of the toxins. The vertical white line is a scale bar 10 Å long. The gua-nidino groups in both STX and TTX are labeled with those for the two guanidino groups of STX also identifying which atoms are involved. Images were produced using the UCSF Chimera pack-age from the Resource for Biocomputing, Visualization, and Informatics at the University of California, San Francisco (Pettersen et al. 2004)

with a guanidinium group and a polyoxygenated ring system. Insoluble in neutral organic solvents and pure water, it readily dissolves in slightly acidic solutions and will degrade at alkaline pH's. The TTX structure is somewhat prismatic with its core being tetrahedral but like STX it also possesses a guanidinium group exposed to solution which has a pKa of 8.8 and is essential to its toxicity (Goto et al. 1965; Tsuda et al. 1964; Woodward 1964). It is also amphoteric due to the cationic nature of its guanidinium moiety and the presence of an anionic acid hydroxyl group off C-10.

3.3 Analogues of Saxitoxin and Tetrodotoxin

STX has many chemical relatives that vary in the nature of their carbamoyl which ranges from being non-existent to terminating in sulfate, benzoyl and methyl moieties. This is exemplified by the two analogues depicted in Fig. 3. Chemical

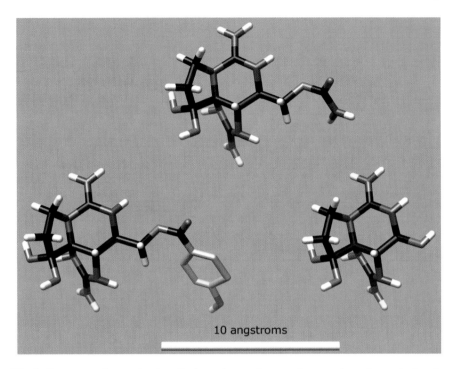

Fig. 3 Structures of two natural saxitoxin analogues from marine organisms demonstrating the variation observed in structure within the saxitoxin family. The uppermost molecule is saxitoxin itself with the two analogues shown below having had their carbamoyl chain modified. The leftmost analogue is a hydroxybenzoate saxitoxin (Negri et al. 2003) while the carbamoyl chain has been completely removed in the right-hand analogue, decarbamoylSTX (Koehn et al. 1981). In all cases, the 1,2,3 guanidinium group is pointing upwards and the elements are colored as in Fig. 1

variations also occur at N-1 which can be hydroxylated. The C-11 sometimes bears sulfate groups and the C-12 hydroxyls can be replaced with protons. Only one analogue of STX has been discovered in a terrestrial organism. Zetekitoxin was isolated from the skin of a rare South American frog and differs dramatically from STX in having a chain bearing a pentacyclic ring connecting carbons 6 and 11. The carbamoyl side chain is absent with a similar side chain now extending from N-7 (Yotsu-Yamashita et al. 2004).

In marine organisms TTX has a number of structural analogues (Nakamura and Yasumoto 1985; Noguchi et al. 1991) but their chemical diversity is limited compared to the complexity found to date for marine saxitoxins. Unlike STX, TTX has many terrestrial relatives with it being produced in newts (Wakely et al. 1966) and frogs (Kim et al. 1975); the structural diversity in this biome (Pires et al. 2003, 2005; Yotsu-Yamashita et al. 1992; Yotsu et al. 1990) is comparable with that observed for the saxitoxins in the ocean. Chiriquitoxin is an unusual TTX analogue from the frog *Atelopus chiriquiensis* which remains equipotent to

TTX even though a glycine residue replaces a methylene hydrogen of the C-11 hydroxymethyl function (Yang and Kao 1992).

3.4 Taxonomic Distribution and Biosynthesis

Both marine and freshwater microscopic organisms produce STX and TTX. In the ocean, dinoflagellates are the dominant microscopic producer of the saxitoxins whereas in freshwater, cyanobacteria produce STX. *Alexandrium minutum*, a dinoflagellate known to synthesize saxitoxins, may also produce TTX (Kodama et al. 1996). Bacteria may be involved in the production of both toxins (Do et al. 1990; Gallacher et al. 1997; Matsui et al. 1989; Noguchi et al. 1987; Yotsu et al. 1987) but this is the subject of considerable debate (Baker et al. 2003; Gallacher and Smith 1999; Matsumura 1995a).

In higher organisms, both invertebrates and vertebrates can harbor STX and TTX. When the above mentioned microalgae bloom, filter-feeding organisms that consume them often become toxic. Scallops, oysters, clams and some filter feeding crustaceans are renowned for bioaccumulating saxitoxins, so during algal blooms they may become a public health hazard. Non-filter feeding mollusks such as carnivorous gastropods (Chen and Chou 1998; Ito et al. 2004; Kanno et al. 1976; Yasumoto and Kotaki 1977) and cephalopods (Robertson et al. 2004) may also accommodate PSTs in their flesh. An interesting example of a non-filter-feeding mollusk harboring either STX or TTX is the Australian blue-ringed octopus whose venom contains TTX (Sheumack et al. 1978). This toxin may originate from an endosymbiotic micro-organism (Hwang et al. 1989), which implies that the octopus essentially farms the microbes for its toxin. Like the blue-ringed octopus, some polyclad worms (Ritson-Williams et al. 2006) and chaetognaths (Thuesen et al. 1988) also contain TTX and may have coopted this toxin as a tool for capturing prey.

Several crustaceans, primarily but not solely those that belong to a family (Xanthidae) of crabs, regularly carry STX and TTX in their tissue (Arakawa et al. 1995; Llewellyn and Endean 1989; Noguchi et al. 1969; Noguchi et al. 1983; Yasumoto et al. 1981). These crabs are not filter feeders and an unequivocal identification of the toxin source for their toxins remains unidentified. One hypothesis is that they accumulate toxins from grazing on the red alga *Jania* which can contain minor amounts of the STX analogues called gonyautoxins (Kotaki et al. 1983)

Regarding vertebrates, fish and amphibia are the main carriers of TTX and STX. Mackerel containing STX in their viscera have caused whale deaths (Geraci et al. 1989), but the most famous toxic fish would have to be pufferfish. TTX and pufferfish are almost synonymous because of the "fugu" tradition but pufferfish can also contain STX (Kodama et al. 1983; Nakamura et al. 1984; Nakashima et al. 2004; Oliveira et al. 2006). Frogs, newts and salamanders have proved a rich source of TTX and a wide array of analogues (Brodie 1982; Daly et al. 1994; Geffeney et al. 2002, 2005; Kim et al. 1975; Mebs et al. 1995; Pires et al. 2005; Williams et al. 2004; Yotsu-Yamashita et al. 2007; Yotsu et al. 1990) and the ecological interplay between these amphibia and their predators has become a useful chemical ecological model (Geffeney et al. 2002, 2005; Williams et al. 2004).

4   Interaction Between the Na Channel and the Toxins

Saxitoxins and tetrodotoxins block the Na channel from the outside of the cell and are ineffective from the channel's interior (Narahashi et al. 1966; Uehara and Moczydlowski 1986). The prototypes of the tetrodotoxin and saxitoxin families, TTX and STX, possess equivalent affinities for the same binding site on toxin sensitive Na channels (Colquhon et al. 1974; Guo et al. 1987; Ritchie 1975; Weigele and Barchi 1978).

4.1   Effects of pH

Both STX and TTX require a protonated guanidinium group to allow efficient binding to the Na channel with the 7,8,9 guanidinium of STX being the most critical to that toxin's bioactivity. pH therefore affects their ability to be attracted to their binding site on the Na channel by altering the ratio of charged to uncharged forms. In the saxitoxins, pH also affects the charge state of the sulfate groups in sulfated analogues and the N-1 hydroxy in the neosaxitoxins. The 7,8,9 guanidinium in STX has a pKa near 8 and so deprotonates with increased alkalinity within the normal pH range (Rogers and Rapoport 1980; Schantz et al. 1966; Shimizu et al. 1981). The 1,2,3 guanidinium has a pKa of 11.3 (Rogers and Rapoport 1980; Schantz et al. 1966; Shimizu et al. 1981) and so this guanidinium is always in the charged state unless it is exposed to a highly alkaline solution which would degrade the toxin. The guanidinium on TTX has a pKa of 8.8 (Goto et al. 1965; Tsuda et al. 1964; Woodward 1964) and so at physiological pH, the concentration of TTX molecules with a charged guanidinium would be nearly the same as the total TTX concentration. But pH also affects the ability of the Na channel to bind these toxins, as shown at the nodes of Ranvier in myelinated frog nerve (Ulbricht and Wagner 1975) and with garfish olfactory nerves (Henderson et al. 1973, 1974) where acidic pH decreased TTX and STX potency. Together these experiments pointed to a carboxylic acid being part of the toxin binding sites (Hille 1975), an observation confirmed by later experiments (see next section).

4.2   Effects of Na Channel Modification

Because the Na channel is a protein it is amenable to modification with protein modification reagents. One of these (trimethyloxonium ion) revealed that acidic amino acids were important in the binding of TTX and STX. Pretreatment with this reagent abolished the channel's ability to bind the toxin, whereas pre-exposure of the channel to the toxin protected it from protein modification (Barchi and Weigele 1979; Doyle et al. 1993; Spalding 1980; Worley et al. 1986). Later, after Na channel genes were sequenced and successfully expressed in recombinant systems, mutagenesis studies pinpointed a series of carboxylate sidechain-containing amino acids in homologous positions within each of the four domains of the channel involved in binding the toxin (Terlau et al. 1991) and were hypothesized to form two predominantly negatively charged rings near the ion pore. The role of these carboxylates was demonstrated by their mutation to a related amino acid lacking a negatively charged sidechain (e.g., aspartic acid to asparagine). Figure 4 shows one model of how STX and TTX integrates itself into the pore of the Na channel to stop ion flow. This model shows the close coordination needed by the amino acid side chains to ligate the toxins and how these multiple interactions cumulatively generate the high affinity binding between the toxins and the channel.

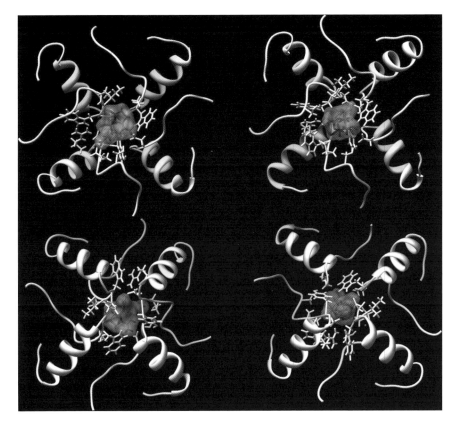

Fig. 4 The upper pair of images show the view looking inwards of the modeled structure of the pore of a rat skeletal muscle Na channel including the hypothetical location of bound STX and TTX molecules, respectively (Tikhonov and Zhorov 2005). Coordinates used are those provided by the cited authors in their supporting material. The lower pair of images are the same as the uppermost pair but rotated to show the view from the intracellular side of the channel pore. Molecular graphics images were produced using the UCSF Chimera package from the Resource for Biocomputing, Visualization, and Informatics at the University of California, San Francisco (Pettersen et al. 2004)

4.3 Temperature

Increasing temperature decreases affinity of both STX and TTX for the toxin sensitive isoforms of the Na channel (Barchi and Weigele 1979; Bay and Strichartz 1980; Weigele and Barchi 1978). Warming 20°C can reduce affinity of the toxins 10- to 20-fold. Temperature may affect the potency of TTX and STX circulating in the blood of certain ectothermic animals which come into contact with these guanidinium toxins. Besides prey, predators which ingest toxic prey may be more sensitive at lower temperatures (Yotsu-Yamashita et al. 2001).

4.4 Toxin Resistance

Resistance to STX and TTX may arise from the possession of insensitive Na channels. Mammals express Na channel isoforms to which STX can bind with only micromolar affinity such as those in mammalian sensory neurons (Akopian et al. 1996), cardiac muscle (Cribbs et al. 1990), skeletal muscle in early stages of development (White et al. 1991) and experimentally denervated skeletal muscle (Kallen et al. 1990). But in the main, most mammalian nerve and striated muscle Na channels are sensitive to STX and TTX (Rogart 1986). Differences in Na channel amino acid sequences explain this diversity in toxin sensitivity, and conversion of a single amino acid can make a toxin insensitive Na channel more easily able to be blocked by tetrodotoxins and saxitoxins (Satin et al. 1992; Soong and Venkatesh 2006). Mutations can have significant effects on populations of organisms that are exposed to these toxins in the environment. A single amino acid mutation within the Na channel of the bivalve *Mya arenaria* was found to reduce the shellfish' Na channel sensitivity by several orders of magnitude, an effect which translated to changes in the population structure of these mollusks with regards their sensitivity to toxic microalgae (Bricelj et al. 2005). Similarly, garter snakes which actively prey upon tetrodotoxic newts have evolved Na channels which are resistant to the toxin enabling them to predate upon an animal not previously available as a prey item (Geffeney et al. 2005). In those animals that have co-opted these toxins for prey capture, they too have resistant nerves. For example, TTX does not affect the nerves of the blue-ringed octopus (Flachsenberger and Kerr 1985). This makes sense because it would be of little value for the octopus to capture prey by envenomating it with TTX and then succumbing to the toxin's effect upon consuming the poisoned prey. Likewise, it would not be surprising if the simple nervous systems of the polyclad worm and chaetognaths described earlier were resistant to TTX (Ritson-Williams et al. 2006; Thuesen et al. 1988). Like the STX resistant mollusks, the crabs which routinely carry STX and TTX in their flesh also possess nerves resistant to these particular toxins as mentioned earlier (Daigo et al. 1988). Closely related species not known to carry either of these two toxin families possess nerves that cannot tolerate the toxins.

An alternative means by which resistance may be conferred is physical separation of the toxin away from its site of action. TTX and STX need to act upon the extracellular opening of the Na channel, and if the toxins are retained intracellularly or prevented from reaching the cell membrane's exterior surface, then they will never be able to exert their pharmacological action. Some animals possess proteins in their blood which can sequester tetrodotoxins or saxitoxins which would then render them inactive. Saxiphilin was first discovered in the North American bullfrog *Rana catesbeiana* (Mahar et al. 1991; Moczydlowski et al. 1988) which binds STX with sub-nanomolar affinity but is unable to bind TTX. It later proved to be a member of the transferrin family of proteins (Li and Moczydlowski 1991; Li et al. 1993; Morabito and Moczydlowski 1994). Hydrophilic proteins similar to bullfrog saxiphilin which exhibit pH dependent STX binding and are unable to bind tetrodotoxin exist in a variety of ectothermic vertebrates and invertebrates (Llewellyn 1997; Llewellyn et al. 1997). Of the vertebrates, saxiphilin is readily detected in numerous fish, amphibians and reptiles, and in animals from diverse habitats within every continent and most climactic zones (Llewellyn et al. 1997). Importantly, some of the species known to possess saxiphilins are some of the xanthid crabs (Llewellyn 1997) and it was speculated that a complex interplay between these two toxin defense systems (i.e. resistant nerves and possessing of saxiphilin) may underpin these animal's ability to actively accumulate the toxins for an as yet unknown evolutionary advantage while being able to resist the toxin's effects.

The pufferfish *Fugu pardalis* possesses yet another STX binding protein but unlike saxiphilin it can also bind TTX and its sequence was unrelated to the Na channel or bullfrog saxiphilin (Yotsu-Yamashita et al. 2001; Yotsu-Yamashita et al. 2002). Similar TTX binding proteins have been observed in horseshoe crabs (Ho et al. 1994) and malacostracan crustacea (Nagashima et al. 2002). The role of these soluble TTX and STX receptors is not yet understood, but in the normal course of events it could be expected that they would inhibit the ability of freely circulating guanidinium toxins in an animal's blood from reaching sodium channels. Once the toxin was bound it would no longer be able to exert its toxic effect because they would now be unable to bind to the Na channel.

4.5   Other Marine Toxins that Inhibit Sodium Channel Function

At this point it is worth digressing to learn about several other marine toxins that could be considered the pharmacological antitheses of STX and TTX. The ciguatoxins and brevetoxins are polyether toxins significantly larger than either STX, TTX or any of their known analogues (Fig. 5a, b). Like STX and TTX, they both act upon the Na channel (Dechraoui et al. 1999, 2006; Gawley et al. 1992; Lombet et al. 1987), but at a different location within its structure. Rather than inhibit Na flow, their action shifts the voltage at which the Na channel responds to changes in potential difference across the cell membrane (Atchison et al. 1986; Huang et al. 1984; Jeglitsch et al. 1998; Sheridan and Adler 1989). This stimulates the opening of a few Na channels sufficient to depolar-

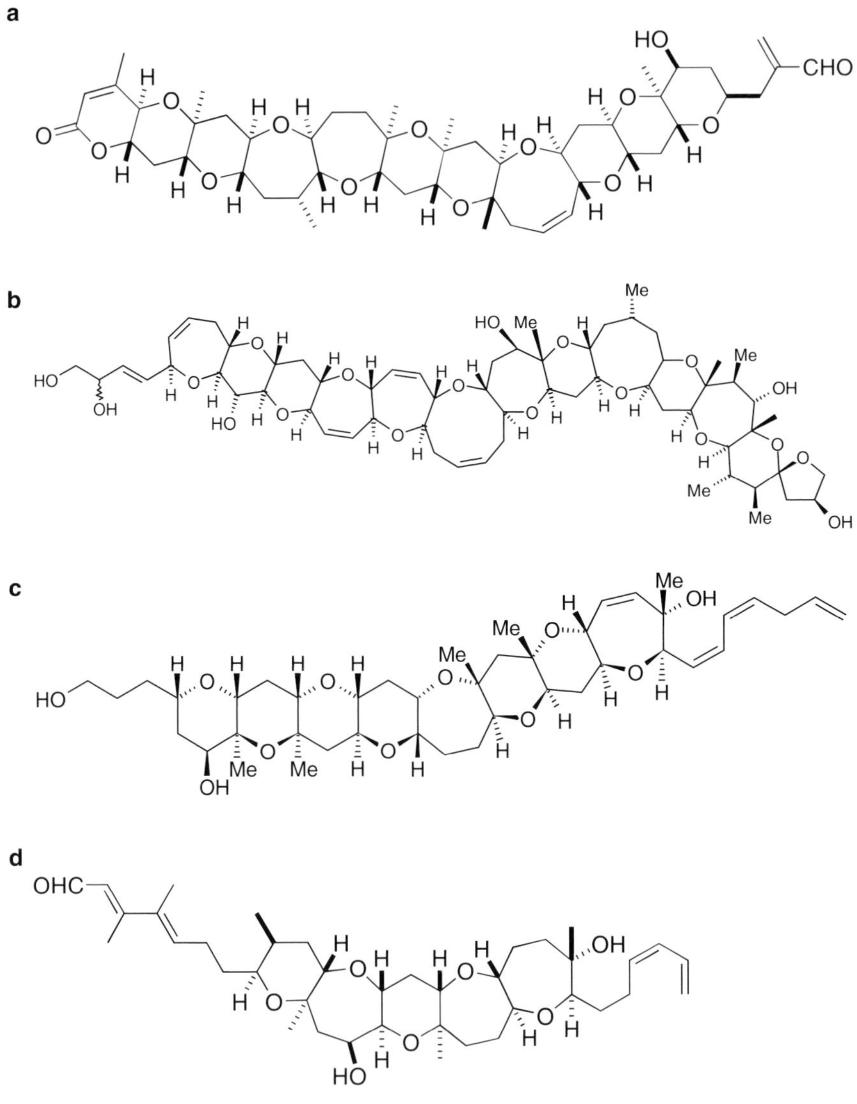

a

b

c

d

Fig. 5 Structures of several examples of brevetoxins and ciguatoxins. (**a**) Brevetoxin B (Lin et al. 1981), (**b**) P-ciguatoxin 1 (Murata et al. 1989), (**c**) gambierol (Satake et al. 1993) and (**d**) brevenal (Bourdelais et al. 2004, 2005)

ize the resting membrane potential enough to cause the inactivation of the remaining Na channels. There is some evidence that the inactivation step is also slowed by these polyether toxins (Gawley et al. 1992; Jeglitsch et al. 1998). Overall, unlike STX and TTX, ciguatoxins and brevetoxins increase Na flow into the cell. These physiologically antagonistic effects between the excitatory polyether toxins and the Na blocking toxins have been used to develop bioassays for the detection of the guanidinium toxins in seafood (Manger et al. 2003). But the ciguatoxins and brevetoxins are public health threats in their own right. Brevetoxins are accumulated by filter-feeding shellfish and can cause the syndrome neurotoxic shellfish poisoning in those persons that eat brevetoxin contaminated mollusks. Ciguatoxin is accumulated through the oceanic food chain with ciguatoxic fish causing debilitating and long lasting effects in humans (Gatti et al. 2008; Lewis 2001).

Brevetoxins and ciguatoxins bind to sensitive channels at nanomolar or even lower concentrations (Dechraoui et al. 1999; Poli et al. 1986). Only recently have insensitive Na channel subtypes been discovered (Dechraoui et al. 2006) with rat heart Na channels being significantly less sensitive to the brevetoxins and ciguatoxins. This is reminiscent of the insensitivity of mammalian cardiac Na channels to both STX and TTX. Potency towards sensitive Na channels translates to strong lethality towards test mammals. Several of the brevetoxins possess LD_{50}'s to rats between $60\,\mu g/kg$ and $200\,\mu g/kg$ via either the intravenous or intraperitoneal route (Baden and Mende 1982; Templeton et al. 1989). Ciguatoxins can be orders of magnitude more lethal via the same administration routes than brevetoxins to some test mammals (Lewis et al. 1991). The toxins' large size provides many potential points of contact between the toxins and the Na channel, all of which combine energetically yield the very high binding affinities observed. This potential to interact with large regions of the channel is exemplified by the brevetoxins whose binding involves the first and fourth of the four homologous domains within the Na channel's α-subunit at a site which incorporates regions of transmembrane helix 6 in domain I and transmembrane helix 5 of domain IV (Poli et al. 1986; Trainer et al. 1991, 1994). Their large size also makes it possible for the toxins to potentially stretch through much of the channel's axis that runs perpendicular to the cell membrane (Gawley et al. 1992; Jeglitsch et al. 1998; Rein et al. 1994a, b).

Also linking these groups of toxins together is that they are all produced by dinoflagellates. The brevetoxins are produced by the dinoflagellate *Karenia brevis* (formerly *Ptychodiscus brevis* and *Gymnodinium breve*) from which they derive their name. The dinoflagellate originally implicated in the production of ciguatoxin was *Gambierdiscus toxicus* (Yasumoto et al. 1977), but in recent years related species have also been found to be ciguatoxic, namely *Gambierdiscus polynesiensis*, *G. pacificus*, *G. australis* (Chinain et al. 1999) and *G. yasumotoi* (Holmes 1998).

Both brevetoxins and ciguatoxins are lipophilic and contain many organic rings of which most are modified to contain ethers. The structures of these natural products are extraordinarily complex and their chemical synthesis in the laboratory are considered highlights in this field (Hirama et al. 2001; Nicolaou et al. 1998). Their structures provide ample opportunity for flexibility, enabling the toxins to adopt many conformations facilitating their binding to protein receptors. This flexibility can be readily seen from the structure of a fragment of one of the ciguatoxins (Fig. 6) comprising the first three ring structures, obtained when they are conjugated to an antibody (Tsumoto et al. 2008). Not only can the toxins bend

Fig. 6 Two different orientations of the three-dimensional structure of the ABC rings of the ciguatoxin molecule as ligated by an antibody. Coordinates are taken from Protein Database Bank entry 2e27 (Tsumoto et al. 2008) by removing the protein structure coordinates. The toxin is visualized as before using UCSF Chimera (Pettersen et al. 2004). Carbon and oxygen atoms are black and light grey and hydrogens are not shown

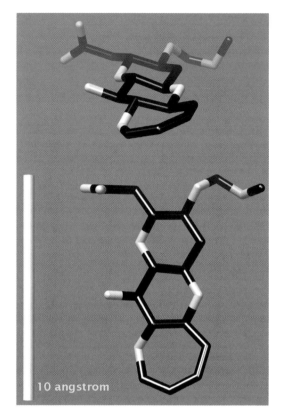

at bonds, likened to the rungs of a ladder, that link the rings, they can also bend and twist around the axis that runs the length of the molecule.

A number of smaller polyether natural products related to the ciguatoxins and brevetoxins are now being discovered. These chemicals, namely gambierol and brevenal (Fig. 5c, d) are not only interesting in that they expand the chemical diversity within these particular polyether families but also because they have diversified our knowledge of the biological activity of these chemicals. Brevenals retain their ability to bind to the Na channel and antagonize the toxicity caused by their larger cousins (Bourdelais et al. 2004, 2005). Their structures may point us in the direction of producing drugs that can treat humans affected by these toxins and other maladies caused by Na channel dysfunction. Like ciguatoxin, gambierol was isolated from *Gambierdiscus toxicus* (Satake et al. 1993) and it also elicits pathological effects in test animals (Ito et al. 2003) reminiscent of its polyether relatives. Despite these similarities, it is significantly smaller than the ciguatoxins and brevetoxins but not as small as the brevenals. The increasing structural complexity of the polyether marine toxin family, linked with the growing knowledge of their biological activity will facilitate further structural probing of the Na channel, already empowered by the saxitoxins and tetrodotoxins.

5   Biomedical Potential of These Toxins

Suppression of Na channel activity and thereby of nervous function provides an obvious means of temporarily deadening nerves to the insult of pain and other stimuli. Their potential use as anesthetics has been obvious for some time (Adams et al. 1976; Takman 1975) but a challenge that still remains is administering an effective but sub-lethal dose. A means by which this systemic toxicity can be avoided is by using it in regions where the effects remain localized. For example, application of TTX to the eye can produce significant anesthesia that may enable ocular surgical procedures without the toxin being able to exert its toxic effects elsewhere (Schwartz et al. 1998). Similarly, the use of encapsulation techniques which control the release of the toxin and keep the toxin localized within the target tissue may enable non-toxic anesthesia (Martinov and Njå 2005). Administering the toxins with other drugs may improve the likelihood of obtaining safe anesthesia; co-administration of STX with local anesthetics has proved synergistic (Barnet et al. 2004). Target tissue containing Na channels insensitive to STX and TTX also need to be anesthetized. Combination of STX with capsaicin (a TRP-1 receptor targeting natural product) proved effective as an anesthetic in a peripheral nerve model in rats (Kohane et al. 1999).

Much attention has been paid to the use of TTX as a local anesthetic but in recent times its utility in treating cancer induced pain resistant to other analgesic medications has also been demonstrated (Hagen et al. 2007). Further, there have been hints that the action of these toxins on the Na channel may be a fruitful avenue for treatment of metastatic cancers. Expression of Na channels is significantly increased in metastatic human breast cancer cells and this increased channel activity enhanced cellular motility and invasion. Addition of TTX inhibited Na channel activity and subsequently in vitro cellular metastasis (Fraser et al. 2005).

A previous examination of the saxitoxins as drug candidates using the "Lipinski rule of five" (Lipinski et al. 1997) highlighted the fact that the high number of violations of this rule would severely hinder their progression as oral drug candidates in most pharmaceutical drug discovery programs (Llewellyn 2006). The "rule of five" is a series of simple chemical descriptors common to over 90% of orally available drugs successfully developed by the pharmaceutical industry. Yet, the saxitoxins as well as the tetrodotoxins are highly effective orally available chemicals; their systemic uptake from toxic shellfish or pufferfish meals being rapid with signs of poisoning occurring within minutes of ingestion and death may occur within hours. More recently, a detailed chemoinformatic study (Llewellyn 2007) identified a group of chemical descriptors that when combined with multivariate statistics could be used to predict the known toxicity of saxitoxins (Fig. 7). Chemical descriptors

Fig. 7 Correlation between
reported toxicity values for
saxitoxins along with the
predicted toxicity using multivariate pattern analysis and
a selection of mathematical
descriptors of chemical properties. Panel A shows a fit
derived without excluding
any subsets of the dataset for
cross validation while panel
B shows quality of the fit that
can be obtained using leave-
one-out cross validation
(Llewellyn 2007)

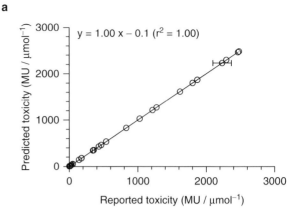

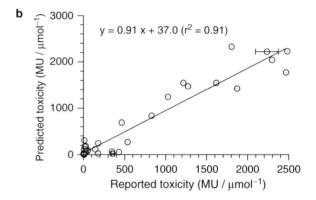

describe molecular properties or convert chemical structures and formulae into a single representative mathematical value (Todeschini and Consonni 2000). Descriptors may be simple such as molecular weight or numbers of heteroatoms, or complex such as estimates of polar surface area or molecular volumes. Other descriptors are derived after mathematically converting three-dimensional struc-tures to a single number representative of, or is an estimate of, a particular chemical property such as surface charge distribution, electronegativity, electrotopology, to name but a few. When a series of structural analogues are compared, the heteroge-neity within their associated descriptors can reflect their structural diversity. Multivariate statistics and other mathematical approaches can then be used to asso-ciate subsets of these descriptors combined in an almost infinite number of possi-bilities with biological activity data. In this study, leave-one-out cross-validation correlation coefficient (q^2) values greater than 0.9 were achieved when combining only six descriptors (selected from an original pool of over 3,000 descriptors). Like the correlation coefficient r^2, the closer q^2 approaches 1 the better the predictivity of the model. The six descriptors were JGI4, E3s, E3e, Mor30p, X2Av and DISPp, with JGI4 being a topological charge descriptor; E3s and E3e being Weighted Holistic Invariant Molecular descriptors which capture information about a molecular structure regarding size, shape, symmetry and atom distributions; Mor30p is one of the 3D-MorSE (Three-Dimensional Molecule Representation of Structures Based on Electron Diffraction) descriptors which are derived from electron diffraction equations; X2Av is a valence connectivity index calculated using the number of atomic valence electrons; and DISPp is a geometrical descriptor that relates theoretical displacement of the centre of a molecule's charge from the centre of its mass. Virtually all of these descriptors are variants that incorporate weighting by estimates of atomic polarizability and electronic states. The fact that there might be a simple and common theme in the chemical properties of the saxitoxins that enables prediction of their bioavailability and toxicity may contribute to further understanding of what makes one chemical more bioavailable than others and guide efforts in this aspect of drug development.

6   Future Directions

How identical are the binding sites on the Na channel for the saxitoxins and the tetrodotoxins? They are believed to be almost identical, but the reality is that despite their similarities, TTX and STX are structurally very different toxins. For example, the saxitoxins have two guandinium groups, of which the 7,8,9 guanidinium is the most critical for bioactivity, while the tetrodotoxins have only one. If you examine Fig. 8 which shows the STX and TTX structures aligned by these guanidinium groups, it is obvious that their shapes have some fundamental differences.

In recent years, it has also been shown that STX binds to many other cell membrane receptors apart form the Na channel such as L-type Ca channels and K channels with channel function being significantly affected at STX concentrations of $10\,\mu$M or less (Su et al. 2004; Wang et al. 2003). Are there other channels that the saxitoxins can bind to? In the above mentioned studies, TTX had minimal effect on the Ca channels, but a more thorough exploration may reveal that the tetrodotoxins may also be more promiscuous channel inhibitors than we currently believe. The actions of STX and TTX on these two families of non-Na cationic channels will elucidate further why these particular structural motifs have such a propensity to bind to a wide array of proteins with very high affinities.

Beyond the obvious, what is different about the terrestrial and aquatic biospheres that have guided the evolution of these two toxin families? There exists a fundamental difference between the STX and TTX family in that there are few terrestrial saxitoxins but many terrestrial tetrodotoxins and that the marine saxitoxins show greater structural heterogeneity than the marine tetrodoxins. These two toxin families then remain excellent models for investigating biosynthesis of toxins and their evolution (Daly 2004).

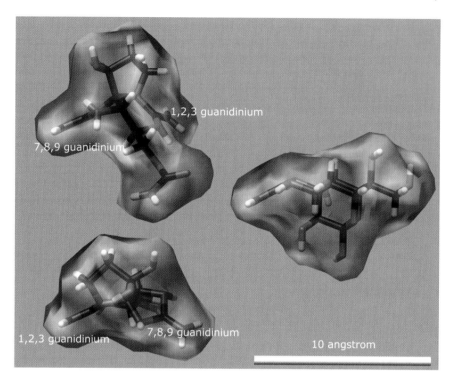

Fig. 8 Structures of STX (left) and TTX (right) as shown in Fig. 1 but now showing the solvent excluded molecular surface. The structures are aligned by their guanidinium moieties with the moiety used for alignment being on the left side of each molecule. The STX structure is duplicated to show the two possible orientations because of it possessing two guanidinium groups but it should be borne in mind that the 7,8,9 guanidinium of STX is the most critical for its bioactivity. The molecular volumes of STX and TTX are $240\,\text{Å}^3$ and $250\,\text{Å}^3$, respectively while their molecular surfaces are $241\,\text{Å}^2$ and $230\,\text{Å}^2$, respectively

The discovery of new STX analogues has accelerated recently partly due to the realization that standard analytical techniques and sample preparation eliminated them from many extracts (Negri et al. 2003, 2007). These toxins are potent Na channel binding toxins (Llewellyn et al. 2004). Are there other cryptic STX and TTX analogues waiting to be discovered? Furthermore, are there compounds yet to be discovered, or that have been discovered but yet to be tested for Na channel inhibiting activity like the saxitoxins and tetrodotoxins, with enough structural similarity to STX and TTX that they too might act as probes of the sodium channel? A search of publicly available chemical databases such as Pubchem (National Centre for Biotechnology Information, US National Library of Medicine, NIH, USA) and Marinlit (ver 14.3, University of Canterbury, Canterbury, New Zealand) quickly identified several marine natural products with structures reminiscent of STX and TTX (Fig. 9).

STX, TTX and their chemical cousins have advanced our knowledge in some of the most fundamental and unanticipated ways. Electrical signaling is critical to life

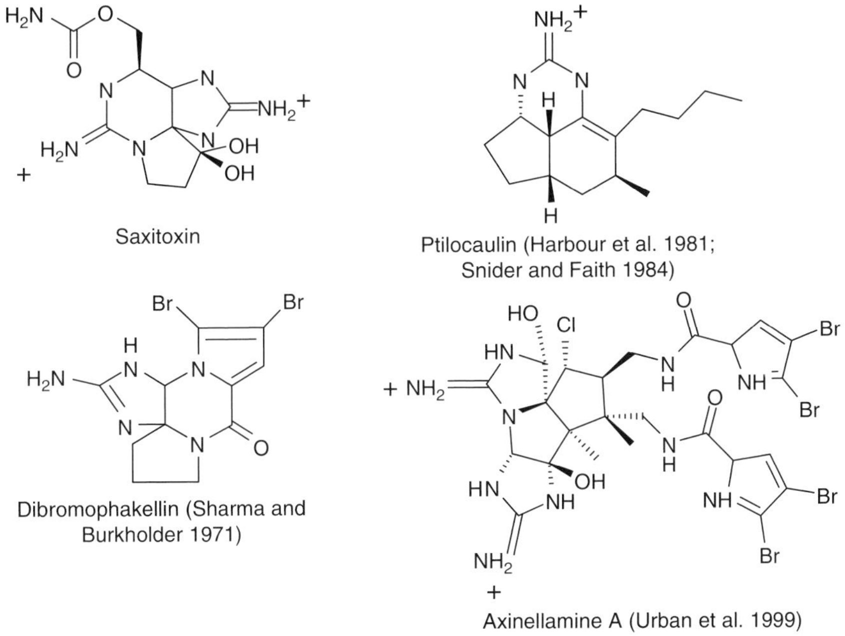

Fig. 9 Several examples of structures of other marine natural products bearing some resemblance to saxitoxin but not yet tested on sodium channels

and the Na channel (and Ca channels in some cases) is responsible for generating and propagating most action potentials. These toxins have enabled us to reveal the existence of the Na channel, count its concentrations at certain membrane sites, explore its structure and unravel many of its properties. Why did these two toxins evolve to target the same site on the Na channel? What is special and unique about this region of the channel that it has attracted the evolutionary "attention" of a multitude of organisms that have manufactured two quite different toxins, namely the saxitoxins and tetrodotoxins? We will probably understand more when technology improves to the point that it removes the barriers between our minds and eyes and the sub-microscopic world and directly reveals the detailed protein structure of their binding sites. When we are able to do so, STX and TTX will undoubtedly be there to aid our quest for knowledge and in so doing, we may demystify some of the evolutionary and biological mysteries that still surround these two toxins.

References

Adams HJ, Blair MR, Jr., Takman BH (1976) The local anesthetic activity of saxitoxin alone and with vasoconstrictor and local anesthetic agents. Arch Int Pharmacodyn Ther 224:275–282.

Akopian AN, Sivilotti L, Wood JN (1996) A tetrodotoxin-resistant voltage-gated sodium channel expressed by sensory neurons. Nature 379:257–262.

Aldrich RW, Corey DP, Stevens CF (1983) A reinterpretation of mammalian sodium channel gating based on single channel recording. Nature 306:436–441.

Anderson PA, Greenberg RM (2001) Phylogeny of ion channels: Clues to structure and function. Comp Biochem Physiol 129B:17–28.

Arakawa O, Nishio S, Noguchi T, Shida Y, Onque Y (1995) A new saxitoxin analogue from a xanthid crab atergatis floridus. Toxicon 33:1577–1584.

Atchison WD, Luke VS, Narahashi T, Vogel SM (1986) Nerve membrane sodium channels as the target site of brevetoxins at neuromuscular junctions. Br J Pharmacol 89:731–738.

Baden DG, Mende TJ (1982) Toxicity of two toxins from the Florida red tide marine dinoflagellate, *Ptychodiscus brevis*. Toxicon 20:457–461.

Baker TR, Doucette GJ, Powell CL, Boyer GL, Plumley FG (2003) GTX$_4$ imposters: Characterization of fluorescent compounds synthesized by *Pseudomonas stutzeri* SF/PS and *Pseudomonas/alteromonas* PTB-1, symbionts of saxitoxin-producing A*lexandrium* spp. Toxicon 41:339–347.

Barchi RL, Weigele JB (1979) Characteristics of saxitoxin binding to the sodium channel of sarcolemma isolated from rat skeletal muscle. J Physiol (Lond) 295:383–396.

Barnet CS, Tse JY, Kohane DS (2004) Site 1 sodium channel blockers prolong the duration of sciatic nerve blockade from tricyclic antidepressants. Pain 110:432–438.

Bay CM, Strichartz GR (1980) Saxitoxin binding to sodium channels of rat skeletal muscles. J Physiol (Lond) 300:89–103.

Bordner J, Thiessen WE, Bates HA, Rapoport H (1975) The structure of a crystalline derivative of saxitoxin. The structure of saxitoxin. J Am Chem Soc 97:6008–6012.

Bourdelais AJ, Campbell S, Jacocks H, Naar J, Wright JL, Carsi J, Baden DG (2004) Brevenal is a natural inhibitor of brevetoxin action in sodium channel receptor binding assays. Cell Mol Neurobiol 24:553–563.

Bourdelais AJ, Jacocks HM, Wright JL, Bigwarfe PM, Jr., Baden DG (2005) A new polyether ladder compound produced by the dinoflagellate *Karenia brevis*. J Nat Prod 68:2–6.

Bricelj VM, Connell L, Konoki K, Macquarrie SP, Scheuer T, Catterall WA, Trainer VL (2005) Sodium channel mutation leading to saxitoxin resistance in clams increases risk of PSP. Nature 434:763–767.

Brodie ED, Jr. (1982) Toxic salamanders. JAMA 247:1408.

Cannon SC, McClatchey AI, Gusella JF (1993) Modification of the Na$^+$ current conducted by the rat skeletal muscle alpha subunit by coexpression with a human brain beta subunit. Pfluegers Arch 423:155–157.

Catterall WA (2001) A 3D view of sodium channels. Nature 409:988–991.

Catterall WA, Goldin AL, Waxman SG (2003) International Union of Pharmacology. XXXIX. Compendium of voltage-gated ion channels: Sodium channels. Pharmacol Rev 55:575–578.

Catterall WA, Goldin AL, Waxman SG (2005) International Union of Pharmacology. XIVII. Nomenclature and structure-function relationships of voltage-gated sodium channels. Pharmacol Rev 57:397–409.

Chang FC, Spriggs DL, Benton BJ, Keller SA, Capacio BR (1997) 4-aminopyridine reverses saxitoxin (STX)- and tetrodotoxin (TTX)-induced cardiorespiratory depression in chronically instrumented guinea pigs. Fundam Appl Toxicol 38:75–88.

Chen CY, Chou HN (1998) Transmission of the paralytic shellfish poisoning toxins, from dinoflagellate to gastropod. Toxicon 36:515–522.

Chinain M, Faust MA, Pauillac S (1999) Morphology and molecular analyses of three toxic species of *Gambierdiscus* (Dinophyceae): *G. pacificus*, sp nov., *G. australes*, sp nov., and *G. polynesiensis*, sp nov. J Phycol 35:1282–1296.

Colquhon D, Rang HP, Ritchie JM (1974) The binding of tetrodotoxin and alpha-bungarotoxin to normal and denervated mammalian muscle. J Physiol (Lond) 240:199–226.

Cribbs LL, Satin J, Fozzard HA, Rogart RB (1990) Functional expression of the rat heart I Na$^+$ channel isoform. Demonstration of properties characteristic of native cardiac Na$^+$ channels. FEBS Lett 275:195–200.

Daigo K, Noguchi T, Miwa A, Kawai N, Hashimoto K (1988) Resistance of nerves from certain toxic crabs to paralytic shellfish poison and tetrodotoxin. Toxicon 26:485–490.

Daly JW (2004) Marine toxins and nonmarine toxins: Convergence or symbiotic organisms? J Nat Prod 67:1211–1215.

Daly JW, Gusovsky F, Myers CW, Yotsu-Yamashita M, Yasumoto T (1994) First occurrence of tetrodotoxin in a dendrobatid frog (*Colostethus inguinalis*), with further reports for the bufonid genus *Atelopus*. Toxicon 32:279–285.

Dechraoui MY, Naar J, Pauillac S, Legrand AM (1999) Ciguatoxins and brevetoxins, neurotoxic polyether compounds active on sodium channels. Toxicon 37:125–143.

Dechraoui MY, Wacksman JJ, Ramsdell JS (2006) Species selective resistance of cardiac muscle voltage gated sodium channels: Characterization of brevetoxin and ciguatoxin binding sites in rats and fish. Toxicon 48:702–712.

Do HK, Kogure K, Simidu U (1990) Identification of deep-sea-sediment bacteria which produce tetrodotoxin. Appl Environ Microbiol 56:1162–1163.

Doyle DD, Guo Y, Lustig SL, Satin J, Rogart RB, Fozzard HA (1993) Divalent cation competition with [^{3}H]saxitoxin binding to tetrodotoxin-resistant and -sensitive sodium channels. A two-site structural model of ion/toxin interaction. J Gen Physiol 101:153–182.

Dyer JR, Johnston WL, Castellucci VF, Dunn RJ (1997) Cloning and tissue distribution of the *Aplysia* Na$^+$ channel alpha-subunit cDNA. DNA Cell Biol 16:347–356.

Flachsenberger WA (1986) Respiratory failure and lethal hypotension due to blue-ringed octopus and tetrodotoxin envenomation observed and counteracted in animal models. J Toxicol Clin Toxicol 24:485–502.

Flachsenberger W, Kerr DI (1985) Lack of effect of tetrodotoxin and of an extract from the posterior salivary gland of the blue-ringed octopus following injection into the octopus and following application to its brachial nerve. Toxicon 23:997–999.

Fraser SP, Diss JK, Chioni AM, Mycielska ME, Pan H, Yamaci RF, Pani F, Siwy Z, Krasowska M, Grzywna Z, Brackenbury WJ, Theodorou D, Koyuturk M, Kaya H, Battaloglu E, De Bella MT, Slade MJ, Tolhurst R, Palmieri C, Jiang J, Latchman DS, Coombes RC, Djamgoz MB (2005) Voltage-gated sodium channel expression and potentiation of human breast cancer metastasis. Clin Cancer Res 11:5381–5389.

Gallacher S, Smith EA (1999) Bacteria and paralytic shellfish toxins. Protist 150:245–255.

Gallacher S, Flynn KJ, Franco JM, Brueggemann EE, Hines HB (1997) Evidence for production of paralytic shellfish toxins by bacteria associated with *Alexandrium* spp. (Dinophyta) in culture. Appl Environ Microbiol 63:239–245.

Gatti C, Oelher E, Legrand AM (2008) Severe seafood poisoning in French Polynesia: A retrospective analysis of 129 medical files. Toxicon 51:746–753.

Gawley RE, Rein KS, Kinoshita M, Baden DG (1992) Binding of brevetoxins and ciguatoxin to the voltage-sensitive sodium channel and conformational analysis of brevetoxin B. Toxicon 30:780–785.

Geffeney S, Brodie ED, Jr., Ruben PC, Brodie ED, III (2002) Mechanisms of adaptation in a predator-prey arms race: TTX-resistant sodium channels. Science 297:1336–1339.

Geffeney SL, Fujimoto E, Brodie ED, III, Brodie ED, Jr., Ruben PC (2005) Evolutionary diversification of TTX-resistant sodium channels in a predator-prey interaction. Nature 434:759–763.

Geraci JR, Anderson DM, Timperi RJ, St. Aubin DJ, Early GA, Prescott JH, Mayo CA (1989) Humpback whales (*Megaptera novaeangliae)* fatally poisoned by dinoflagellate toxin. Can J Fish Aquat Sci 46:1895–1898.

Goldin AL (2002) Evolution of voltage-gated Na$^+$ channels. J Exp Biol 205:575–584.

Goto T, Kishi Y, Takahashi S, Hirata Y (1965) Tetrodotoxin. Tetrahedron 21:2059–2088.

Guo XT, Uehara A, Ravindran A, Bryant SH, Hall S, Moczydlowski E (1987) Kinetic basis for insensitivity to tetrodotoxin and saxitoxin in sodium channels of canine heart and denervated rat skeletal muscle. Biochemistry 26:7546–7556.

Guy HR, Durell SR (1995) Structural models of Na$^+$, Ca^{2+}, and K$^+$ channels. Soc Gen Physiol Ser 50:1–16.

Hagen NA, Fisher KM, Lapointe B, du Souich P, Chary S, Moulin D, Sellers E, Ngoc AH (2007) An open-label, multi-dose efficacy and safety study of intramuscular tetrodotoxin in patients with severe cancer-related pain. J Pain Symptom Manage 34:171–182.

Hamill OP (2006) Twenty odd years of stretch-sensitive channels. Pflugers Arch 453:333–351.

Harbour GC, Tymiak AA, Rinehart KL, Jr., Shaw PD, Hughes R, Jr., Mizsak SA, Coats JH, Zurenko GE, Li LH, Kuentzel SL (1981) Ptilocaulin and isoptilocaulin, antimicrobial and cytotoxic cyclic guanidines from the caribbean sponge *Ptilocaulis* aff. *P. spiculifer* (Lamarck, 1814). J Am Chem Soc 103:5604–5606.

Heinemann SH, Terlau H, Stuhmer W, Imoto K, Numa S (1992) Calcium channel characteristics conferred on the sodium channel by single mutations. Nature 356:441–443.

Henderson R, Ritchie JM, Strichartz GR (1973) The binding of labelled saxitoxin to the sodium channels in nerve membranes. J Physiol (Lond) 235:783–804.

Henderson R, Ritchie JM, Strichartz GR (1974) Evidence that tetrodotoxin and saxitoxin act at a metal cation binding site in the sodium channels of nerve membrane. Proc Natl Acad Sci USA 71:3936–3940.

Hille B (1975) An essential ionized acid group in sodium channels. Fed Proc 34:1318–1321.

Hille B (2001) Ion channels of excitable membranes. Sinauer Associates, Sunderland, MA.

Hirama M, Oishi T, Uehara H, Inoue M, Maruyama M, Oguri H, Satake M (2001) Total synthesis of ciguatoxin CTX3c. Science 294:1904–1907.

Ho B, Yeo DS, Ding JL (1994) A tetrodotoxin neutralizing system in the haemolymph of the horseshoe crab, *Carcinoscorpius rotundicauda*. Toxicon 32:755–762.

Hodgkin AL (1958) Ionic movements and electrical activity in giant nerve fibres. Proc R Soc B 148:1–37.

Holmes MJ (1998) *Gambierdiscus yasumotoi* sp. Nov. (Dinophyceae), a toxic benthic dinoflagellate from southeastern Asia. J Phycol 34:661–668.

Huang JM, Wu CH, Baden DG (1984) Depolarizing action of a red-tide dinoflagellate brevetoxin on axonal membranes. J Pharmacol Exp Ther 229:615–621.

Hwang DF, Arakawa O, Saito T, Noguchi T, Simidu U, Tsukamoto K, Shida Y, Hashimoto K (1989) Tetrodotoxin-producing bacteria from the blue-ringed octopus *Octopus maculosus*. Mar Biol 100:327–332.

Ito E, Suzuki-Toyota F, Toshimori K, Fuwa H, Tachibana K, Satake M, Sasaki M (2003) Pathological effects on mice by gambierol, possibly one of the ciguatera toxins. Toxicon 42: 733–740.

Ito K, Asakawa M, Beppu R, Takayama H, Miyazawa K (2004) PSP-toxicification of the carnivorous gastropod *Rapana venosa* inhabiting the estuary of Nikoh River, Hiroshima Bay, Hiroshima Prefecture, Japan. Mar Pollut Bull 48:1116–1121.

Jeglitsch G, Rein K, Baden DG, Adams DJ (1998) Brevetoxin-3 (PbTx-3) and its derivatives modulate single tetrodotoxin-sensitive sodium channels in rat sensory neurons. J Pharmacol Exp Ther 284:516–525.

Jeziorski MC, Greenberg RM, Anderson PA (1997) Cloning of a putative voltage-gated sodium channel from the turbellarian flatworm *Bdelloura candida*. Parasitology 115:289–296.

Kallen RG, Sheng ZH, Yang J, Chen LQ, Rogart RB, Barchi RL (1990) Primary structure and expression of a sodium channel characteristic of denervated and immature rat skeletal muscle. Neuron 4:233–242.

Kanno K, Kotaki Y, Yasumoto T (1976) Distribution of toxins in molluscs associated with coral reefs. Nippon Suisan Gakkaishi 42:1395–1398.

Kao CY (1966) Tetrodotoxin, saxitoxin, and their significance in the study of excitation phenomena. Pharmacol Rev 18:997–1049.

Kao CY, Yasumoto T (1990) Tetrodotoxin in "zombie powder". Toxicon 28:129–132.

Kim YH, Brown GB, Mosher FA (1975) Tetrodotoxin: Occurrence in atelopid frogs of Costa Rica. Science 189:151–152.

Kodama M, Ogata T, Noguchi T, Maruyama J, Hashimoto K (1983) Occurrence of saxitoxin and other toxins in the liver of the pufferfish *Takifugu pardalis*. Toxicon 21:897–900.

Kodama M, Sato S, Sakamoto S, Ogata T (1996) Occurrence of tetrodotoxin in *Alexandrium tamarense*, a causative dinoflagellate of paralytic shellfish poisoning. Toxicon 34:1101–1105.

Koehn FE, Ghazarossian VE, Schantz EJ, Schnoes HK, Strong FM (1981) Derivatives of saxitoxin. Bioorg Chem 10:412–428.

Kohane DS, Kuang Y, Lu NT, Langer R, Strichartz GR, Berde CB (1999) Vanilloid receptor agonists potentiate the *in vivo* local anesthetic activity of percutaneously injected site 1 sodium channel blockers. Anesthesiology 90:524–534.

Kotaki Y, Tajiri M, Oshima Y, Yasumoto T (1983) Identification of a calcareous red alga as the primary source of paralytic shellfish toxins in coral reef crabs and gastropods. Nippon Suisan Gakkaishi 49:283–286.

Lewis RJ (2001) The changing face of ciguatera. Toxicon 39:97–106.

Lewis RJ, Sellin M, Poli MA, Norton RS, MacLeod JK, Sheil MM (1991) Purification and characterization of ciguatoxins from moray eel (*Lycodontis javanicus*, Muraenidae). Toxicon 29:1115–1127.

Li Y, Moczydlowski E (1991) Purification and partial sequencing of saxiphilin, a saxitoxin-binding protein from the bullfrog, reveals homology to transferrin. J Biol Chem 266:15481–15487.

Li Y, Llewellyn L, Moczydlowski E (1993) Biochemical and immunochemical comparison of saxiphilin and transferrin, two structurally related plasma proteins from *Rana catesbeiana*. Mol Pharmacol 44:742–748.

Lin Y, Risk M, Ray SM, van Engen D, Clardy J, Golik J, James JC, Nakanishi K (1981) Isolation and structure of brevetoxin B from "red tide" dinoflagellate *Ptychodiscus brevis* (*Gymnodinium breve*). J Am Chem Soc 103:6773–6776.

Lipinski CA, Lombardo F, Dominy BW, Feeney PJ (1997) Experimental and computational approaches to estimate solubility and permeability in drug discovery and development settings. Adv Drug Deliv Rev 23:3–25.

Llewellyn LE (1997) Haemolymph protein in xanthid crabs: Its selective binding of saxitoxin and possible role in toxin bioaccumulation. Mar Biol 128:599–606.

Llewellyn LE (2006) Saxitoxin, a toxic marine natural product that targets a multitude of receptors. Nat Prod Rep 23:200–222.

Llewellyn LE (2007) Predictive toxinology: An initial foray using calculated molecular descriptors to describe toxicity using the saxitoxins as a model. Toxicon 50:901–913.

Llewellyn LE, Endean R (1989) Toxicity and paralytic shellfish toxin profiles of the xanthid crabs, *Lophozozymus pictor* and *Zosimus aeneus*, collected from some Australian coral reefs. Toxicon 27:596–600.

Llewellyn LE, Bell PM, Moczydlowski EG (1997) Phylogenetic survey of soluble saxitoxin-binding activity in pursuit of the function and molecular evolution of saxiphilin, a relative of transferrin. Proc Roy Soc B 264:891–902.

Llewellyn L, Negri A, Quilliam M (2004) High affinity for the rat brain sodium channel of newly discovered hydroxybenzoate saxitoxin analogues from the dinoflagellate *Gymnodinium catenatum*. Toxicon 43:101–104.

Lombet A, Bidard JN, Lazdunski M (1987) Ciguatoxin and brevetoxins share a common receptor site on the neuronal voltage-dependent Na$^+$ channel. FEBS Lett 219:355–359.

Mahar J, Lukacs GL, Li Y, Hall S, Moczydlowski E (1991) Pharmacological and biochemical properties of saxiphilin, a soluble saxitoxin-binding protein from the bullfrog (*Rana catesbeiana*). Toxicon 29:53–71.

Makita N, Bennett PB, George AL (1996) Molecular determinants of beta 1 subunit-induced gating modulation in voltage-dependent Na$^+$ channels. J Neurosci 16:7117–7127.

Manger RL, Leja LS, Lee SY, Hungerford JM, Kirkpatrick MA, Yasumoto T, Wekell MM (2003) Detection of paralytic shellfish poison by rapid cell bioassay: Antagonism of voltage-gated sodium channel active toxins in vitro. J AOAC Int 86:540–543.

Martinov V, Njå A (2005) A microcapsule technique for long-term conduction block of the sciatic nerve by tetrodotoxin. J Neurosci Methods 141:199–205.

Matsui T, Taketsugu S, Kodama K, Ishii A, Yamamori K, Shimizu C (1989) Production of tetrodotoxin by the intestinal bacteria of a puffer fish *Takifugu niphobles*. Bull Jpn Soc Sci Fish 55:2199–2203.

Matsumura K (1995a) Reexamination of tetrodotoxin production by bacteria. Appl Environ Microbiol 61:3468–3470.

Matsumura K (1995b) In vivo neutralization of tetrodotoxin by a monoclonal antibody. Toxicon 33:1239–1241.

Mebs D, Yotsu-Yamashita M, Yasumoto T, Lotters S, Schluter A (1995) Further report of the occurrence of tetrodotoxin in *Atelopus* species (family: Bufonidae). Toxicon 33:246–249.

Moczydlowski E, Mahar J, Ravindran A (1988) Multiple saxitoxin-binding sites in bullfrog muscle: Tetrodotoxin-sensitive sodium channels and tetrodotoxin-insensitive sites of unknown function. Mol Pharmacol 33:202–211.

Morabito MA, Moczydlowski E (1994) Molecular cloning of bullfrog saxiphilin: A unique relative of the transferrin family that binds saxitoxin. Proc Natl Acad Sci USA 91:2478–2482.

Murata M, Legrand AM, Ishibashi Y, Yasumoto T (1989) Structures of ciguatoxin and its congener. J Am Chem Soc 111:8929–8931.

Nagashima Y, Yamamoto K, Shimakura K, Shiomi K (2002) A tetrodotoxin-binding protein in the hemolymph of shore crab *Hemigrapsus sanguineus*: purification and properties. Toxicon 40:753–760.

Nakamura M, Yasumoto T (1985) Tetrodotoxin derivatives in puffer fish. Toxicon 23:271–276.

Nakamura M, Oshima Y, Yasumoto T (1984) Occurrence of saxitoxin in puffer fish. Toxicon 22:381–385.

Nakashima K, Arakawa O, Taniyama S, Nonaka M, Takatani T, Yamamori K, Fuchi Y, Noguchi T (2004) Occurrence of saxitoxins as a major toxin in the ovary of a marine puffer *Arothron firmamentum*. Toxicon 43:207–212.

Narahashi T, Anderson N, Moore J (1966) Tetrodotoxin does not block excitation from inside the nerve membrane. Science 153:765–767.

Negri A, Stirling D, Quilliam M, Blackburn S, Bolch C, Burton I, Eaglesham G, Thomas K, Walter J, Willis R (2003) Three novel hydroxybenzoate saxitoxin analogues isolated from the dinoflagellate *Gymnodinium catenatum*. Chem Res Toxicol 16:1029–1033.

Negri AP, Bolch CJS, Geier SC, Green DH, Park TG, Blackburn SI (2007) Widespread presence of hydrophobic paralytic shellfish toxins in *Gymnodinium catenatum*. Harmful Algae 6: 774–780.

Niccolai N, Schnoes HK, Gibbons WA (1980) Study of the stereochemistry, relaxation mechanisms, and internal motions of natural products utilizing proton relaxation parameters: Solution and crystal structure of saxitoxin. J Am Chem Soc 102:1513–1517.

Nicolaou KC, Yang Z, Shi G, Gunzner JL, Agrios KA, Gartner P (1998) Total synthesis of brevetoxin A. Nature 392:264–269.

Noda M, Shimizu S, Tanabe T, Takai T, Kayano T, Ikeda T, Takahashi H, Nakayama H, Kanaoka Y, Minamino N, Kangawa K, Matsuo H, Raftery MA, Hirose T, Inayama S, Hayashida H, Miyata T, Numa S (1984) Primary structure of *Electrophorus electricus* sodium channel deduced from cDNA sequence. Nature 312:121–127.

Noguchi T, Konosu S, Hashimoto Y (1969) Identity of the crab toxin with saxitoxin. Toxicon 7:325–326.

Noguchi T, Uzu A, Koyama K, Maruyama J, Nagashima Y, Hashimoto K (1983) Occurrence of tetrodotoxin as the major toxin in a xanthid crab *Atergatis floridus*. Nippon Suisan Gakkaishi 49:1881–1892.

Noguchi T, Hwang DF, Arakawa O, Sugita H, Deguchi Y, Shida Y, Hashimoto K (1987) *Vibrio alginolyticus*, a tetrodotoxin-producing bacterium, in the intestines of the fish *Fugu vermicularis vermicularis*. Mar Biol 94:625–630.

Noguchi T, Ali AE, Arakawa O, Miyazawa K, Kanoh S, Shida Y, Nishio S, Hashimoto K (1991) Tetrodonic acid-like substance; a possible precursor of tetrodotoxin. Toxicon 29:845–885.

Oliveira JS, Fernandes SC, Schwartz CA, Bloch CJ, Melo JA, Rodrigues Pires OJ, de Freitas JC (2006) Toxicity and toxin identification in *Colomesus asellus*, an Amazonian (Brazil) freshwater puffer fish. Toxicon 48:55–63.

Pettersen EF, Goddard TD, Huang CC, Couch GS, Greenblatt DM, Meng EC, Ferrin TE (2004) UCSF Chimera – A visualization system for exploratory research and analysis. J Comput Chem 25:1605–1612.

Pires OR, Sebben A, Schwartz EF, Bloch C, Morales RA, Schwartz CA (2003) The occurrence of 11-oxotetrodotoxin, a rare tetrodotoxin analogue, in the brachycephalidae frog *Brachycephalus ephippium*. Toxicon 42:563–566.

Pires OR, Jr., Sebben A, Schwartz EF, Morales RA, Bloch C, Jr., Schwartz CA (2005) Further report of the occurrence of tetrodotoxin and new analogues in the Anuran family Brachycephalidae. Toxicon 45:73–79.

Pohorille A, Schweighofer K, Wilson MA (2005) The origin and early evolution of membrane channels. Astrobiology 5:1–17.

Poli MA, Mende TJ, Baden DG (1986) Brevetoxins, unique activators of voltage-sensitive sodium channels, bind to specific sites in rat brain synaptosomes. Mol Pharmacol 30:129–135.

Rein K, Baden DG, Gawley RE (1994a) Conformational analysis of the sodium channel modulator, brevetoxin A, comparison with brevetoxin B conformations, and a hypothesis about the common pharmacophore of the site 5 toxins. J Org Chem 59:2102–2106.

Rein K, Lynn B, Gawley R, Baden DG (1994b) Brevetoxin B: Chemical modifications, synaptosome binding, toxicity, and an unexpected conformational effect. J Org Chem 59:2107–2113.

Ren D, Navarro B, Xu H, Yue L, Shi Q, Clapham D (2001) A prokaryotic voltage-gated sodium channel. Science 294:2372–2375.

Ritchie JM (1975) Binding of tetrodotoxin and saxitoxin to sodium channels. Phil Trans R Soc B 270:319–336.

Ritson-Williams R, Yotsu-Yamashita M, Paul VJ (2006) Ecological functions of tetrodotoxin in a deadly polyclad flatworm. Proc Natl Acad Sci USA 103:3176–3179.

Robertson A, Stirling D, Robillot C, Llewellyn L, Negri A (2004) First report of saxitoxin in octopi. Toxicon 44:765–771.

Rogart RB (1986) High-STX-affinity vs. low-STX-affinity Na$^+$ channel subtypes in nerve, heart, and skeletal muscle. Ann NY Acad Sci 479:402–430.

Rogers RS, Rapoport H (1980) The pK$_a$'s of saxitoxin. J Am Chem Soc 102:7335–7339.

Saier MH (2000) Families of proteins forming transmembrane channels. J Membr Biol 175:165–180.

Satake M, Murata M, Yasumoto T (1993) Gambierol: A new toxic polyether compound isolated from the marine dinoflagellate *Gambierdiscus toxicus*. J Am Chem Soc 115:361–362.

Satin J, Kyle JW, Chen M, Bell P, Cribbs LL, Fozzard HA, Rogart RB (1992) A mutant of TTX-resistant cardiac sodium channels with TTX-sensitive properties. Science 256:1202–1205.

Sato C, Ueno Y, Asai K, Takahashi K, Sato M, Engel A, Fujiyoshi Y (2001) The voltage-sensitive sodium channel is a bell-shaped molecule with several cavities. Nature 409:1047–1051.

Schantz EJ, Mold JD, Stanger DW, Shavel J, Riel FJ, Bowden JP, Lynch JM, Wyler RS, Riegel B, Sommer H (1957) Paralytic shellfish poison. VI. A procedure for the isolation and purification of the poison from toxic clam and mussel tissues. J Am Chem Soc 79:5230–5235.

Schantz EJ, Lynch JM, Vayvada G, Matsumoto K, Rapoport H (1966) The purification and characterization of the poison produced by *Gonyaulax catanella* in axenic culture. Biochemistry 5:1191–1195.

Schantz EJ, Ghazarossian VE, Schnoes HK, Strong FM, Springer JP, Pezzanite JO, Clardy J (1975) The structure of saxitoxin. J Am Chem Soc 97:1238–1239.

Schwartz D, Fields H, Duncan K, Duncan J, Jones M (1998) Experimental study of tetrodotoxin, a long-acting topical anesthetic. Am J Ophthalmol 125:481–487.

Sharma GM, Burkholder PR (1971) Structure of dibromophakellin, a new bromine-containing alkaloid from the marine sponge *Phakellia flabellata*. J Chem Soc D 151–152, **DOI:** 10.1039/C29710000151.

Sheridan RE, Adler M (1989) The actions of a red tide toxin from *Ptychodiscus brevis* on single sodium channels in mammalian neuroblastoma cells. FEBS Lett 247:448–452.

Sheumack DD, Howden ME, Spence I, Quinn RJ (1978) Maculotoxin: A neurotoxin from the venom glands of the octopus *Hapalochlaena maculosa* identified as tetrodotoxin. Science 199:188–189.

Shimizu Y, Hsu CP, Genenah AA (1981) Structure of saxitoxin in solutions and stereochemistry of dihydrosaxitoxins. J Am Chem Soc 103:605–609.

Snider BB, Faith WC (1984) Total synthesis of (+)- and (−)-ptilocaulin. J Am Chem Soc 106: 1443–1445.

Sommer H (1932) The occurrence of paralytic shellfish poison in the common sand crab. Science 76:574–575.

Soong T, Venkatesh B (2006) Adaptive evolution of tetrodotoxin resistance in animals. Trends Genet 22:621–626.

Spafford JD, Spencer AN, Gallin WJ (1998) A putative voltage-gated sodium channel alpha subunit (PPSCN1) from the hydrozoan jellyfish, *Polyorchis penicillatus*: Structural comparisons and evolutionary considerations. Biochem Biophys Res Commun 244:772–780.

Spalding BC (1980) Properties of toxin-resistant sodium channels produced by chemical modification in frog skeletal muscle. J Physiol (Lond) 305:485–500.

Stafford RG, Hines HB (1995) Urinary elimination of saxitoxin after intravenous injection. Toxicon 33:1501–1510.

Strichartz G (1984) Structural determinants of the affinity of saxitoxin for neuronal sodium channels. Electrophysiological studies on frog peripheral nerve. J Gen Physiol 84:281–305.

Su Z, Sheets M, Ishida H, Li F, Barry WH (2004) Saxitoxin blocks L-type I_{Ca}. J Pharmacol Exp Ther 308:324–329.

Takman BH (1975) The chemistry of local anaesthetic agents: Classification of blocking agents. Br J Anaesth 47(Suppl):183–190.

Templeton CB, Poli MA, Solow R (1989) Prophylactic and therapeutic use of an anti-brevetoxin (PBTx-2) antibody in conscious rats. Toxicon 27:1389–1395.

Terlau H, Stuhmer W (1998) Structure and function of voltage-gated ion channels. Naturwissenschaften 85:437–444.

Terlau H, Heinemann SH, Stuhmer W, Pusch M, Conti F, Imoto K, Numa S (1991) Mapping the site of block by tetrodotoxin and saxitoxin of sodium channel II. FEBS Lett 293:93–96.

Thuesen EV, Kogure K, Hashimoto K, Nemoto T (1988) Poison arrowworms: A tetrodotoxin venom in the marine phylum Chaetognatha. J Exp Mar Biol Ecol 116:249–256.

Tikhonov DB, Zhorov BS (2005) Modeling P-loops domain of sodium channel: Homology with potassium channels and interaction with ligands. Biophys J 88:184–197.

Todeschini R, Consonni V (2000) Handbook of molecular descriptors. Wiley-VCH, Weinheim, Germany.

Trainer VL, Thomsen WJ, Catterall WA, Baden DG (1991) Photoaffinity labeling of the brevetoxin receptor on sodium channels in rat brain synaptosomes. Mol Pharmacol 40: 988–994.

Trainer VL, Baden DG, Catterall WA (1994) Identification of peptide components of the brevetoxin receptor site of rat brain sodium channels. J Biol Chem 269:19904–19909.

Tsuda K, Ikuma S, Kawamura M, Tachikawa R, Sakai K (1964) Tetrodotoxin. VII. On the structure of tetrodotoxin and its derivatives. Chem Pharm Bull 12:1357–1374.

Tsumoto K, Yokota A, Tanaka Y, Ui M, Tsumuraya T, Fujii I, Kumagai I, Nagumo Y, Oguri H, Inoue M, Hirama M (2008) Critical contribution of aromatic rings to specific recognition of polyether rings: The case of ciguatoxin CTX3c-ABC and its specific antibody 1c49. J Biol Chem 283:12259–12266.

Uehara A, Moczydlowski E (1986) Blocking mechanisms of batrachotoxin-activated Na channels in artificial bilayers. Membr Biochem 6:111–147.

Ulbricht W, Wagner HH (1975) The reaction between tetrodotoxin and membrane sites at the node of ranvier: Its kinetics and dependence on pH. Phil Trans R Soc B 270:353–363.

Urban S, de Almeida Leone P, Carroll AR, Fechner GA, Smith J, Hooper JN, Quinn RJ (1999) Axinellamines A-D, novel imidazo-azolo-imidazole alkaloids from the Australian marine sponge *Axinella* sp. J Org Chem 64:731–735.

Wakely JF, Fuhrman GJ, Fuhrman FA, Fischer HG, Mosher HS (1966) The occurrence of tetrodotoxin (tarichotoxin) in amphibia and the distribution of the toxin in the organs of newts (*Taricha*). Toxicon 3:195–203.

Wang J, Salata JJ, Bennett PB (2003) Saxitoxin is a gating modifier of hERG K^+ channels. J Gen Physiol 121:583–598.

Weigele JB, Barchi RL (1978) Analysis of saxitoxin binding in isolated rat synaptosomes using a rapid filtration assay. FEBS Lett 91:310–314.

White MM, Chen LQ, Kleinfield R, Kallen RG, Barchi RL (1991) SKM2, a Na$^+$ channel cDNA clone from denervated skeletal muscle, encodes a tetrodotoxin-insensitive Na$^+$ channel. Mol Pharmacol 39:604–608.

Williams BL, Brodie ED, Jr., Brodie ED, III (2004) A resistant predator and its toxic prey: Persistence of newt toxin leads to poisonous (not venomous) snakes. J Chem Ecol 30:1901–1919.

Woodward RB (1964) The structure of tetrodotoxin. Pure Appl Chem 9:49–74.

Worley JF, III, French RJ, Krueger BK (1986) Trimethyloxonium modification of single batrachotoxin-activated sodium channels in planar bilayers. Changes in unit conductance and in block by saxitoxin and calcium. J Gen Physiol 87:327–349.

Xu QH, Zhao XN, Wei CH, Rong KT (2005) Immunologic protection of anti-tetrodotoxin vaccines against lethal activities of oral tetrodotoxin challenge in mice. Int Immunopharmacol 5:1213–1224.

Yang L, Kao CY (1992) Actions of chiriquitoxin on frog skeletal muscle fibers and implications for the tetrodotoxin/saxitoxin receptor. J Gen Physiol 100:609–622.

Yasumoto T, Kao CY (1986) Tetrodotoxin and the Haitian zombie. Toxicon 24:747–749.

Yasumoto T, Kotaki Y (1977) Occurrence of a saxitoxin in a green turban shell. Nippon Suisan Gakkaishi 43:207–211.

Yasumoto T, Nakajima I, Bagnis R, Adachi R (1977) Finding of a dinoflagellate as a likely culprit of ciguatera. Bull Jpn Soc Sci Fish 43:1021–1026.

Yasumoto T, Oshima Y, Konta T (1981) Analysis of paralytic shellfish toxins of xanthid crabs in Okinawa. Nippon Suisan Gakkaishi 47:957–959.

Yotsu M, Yamazaki T, Meguro Y, Endo A, Murata M, Naoki H, Yasumoto T (1987) Production of tetrodotoxin and its derivatives by *Pseudomonas* sp. isolated from the skin of a pufferfish. Toxicon 25:225–228.

Yotsu M, Iorizzi M, Yasumoto T (1990) Distribution of tetrodotoxin, 6-epitetrodotoxin, and 11-deoxytetrodotoxin in newts. Toxicon 28:238–241.

Yotsu-Yamashita M, Mebs D, Yasumoto T (1992) Tetrodotoxin and its analogues in extracts from the toad *Atelopus oxyrhynchus* (family: Bufonidae). Toxicon 30:1489–1492.

Yotsu-Yamashita M, Sugimoto A, Terakawa T, Shoji Y, Miyazawa T, Yasumoto T (2001) Purification, characterization, and cDNA cloning of a novel soluble saxitoxin and tetrodotoxin binding protein from plasma of the puffer fish, *Fugu pardalis*. Eur J Biochem 268:5937–5946.

Yotsu-Yamashita M, Shoji Y, Terakawa T, Yamada S, Miyazawa T, Yasumoto T (2002) Mutual binding inhibition of tetrodotoxin and saxitoxin to their binding protein from the plasma of the puffer fish, *Fugu pardalis*. Biosci Biotechnol Biochem 66:2520–2524.

Yotsu-Yamashita M, Kim YH, Dudley SC, Jr., Choudhary G, Pfahnl A, Oshima Y, Daly JW (2004) The structure of zetekitoxin AB, a saxitoxin analog from the Panamanian golden frog *Atelopus zeteki*: A potent sodium-channel blocker. Proc Natl Acad Sci USA 101:4346–4351.

Yotsu-Yamashita M, Mebs D, Kwet A, Schneider M (2007) Tetrodotoxin and its analogue 6-epitetrodotoxin in newts (*Triturus* spp.; Urodela, Salamandridae) from southern Germany. Toxicon 50:306–309.

Sea Anemone Toxins Affecting Potassium Channels

Sylvie Diochot and Michel Lazdunski

Abstract The great diversity of K⁺ channels and their wide distribution in many tissues are associated with important functions in cardiac and neuronal excitability that are now better understood thanks to the discovery of animal toxins. During the past few decades, sea anemones have provided a variety of toxins acting on voltage-sensitive sodium and, more recently, potassium channels. Currently there are three major structural groups of sea anemone K⁺ channel (SAK) toxins that have been characterized. Radioligand binding and electrophysiological experiments revealed that each group contains peptides displaying selective activities for different subfamilies of K⁺ channels. Short (35–37

S. Diochot
Institut de Pharmacologie Moléculaire et Cellulaire, Centre National de la Recherche
Scientifique, Université de Nice-Sophia-Antipolis, 660 Route des Lucioles 06560 Valbonne,
France

M. Lazdunski (✉)
Institut de Pharmacologie Moléculaire et Cellulaire, CNRS, Sophia Antipolis,
F-06560 Valbonne, France
e-mail: lazdunski@ipmc.cnrs.fr

N. Fusetani and W. Kem (eds.), *Marine Toxins as Research Tools*,
Progress in Molecular and Subcellular Biology, Marine Molecular Biotechnology 46,
DOI: 10.1007/978-3-540-87895-7, © Springer-Verlag Berlin Heidelberg 2009

amino acids) peptides in the group I display pore blocking effects on Kv1 channels. Molecular interactions of SAK-I toxins, important for activity and binding on Kv1 channels, implicate a spot of three conserved amino acid residues (Ser, Lys, Tyr) surrounded by other less conserved residues. Long (58–59 amino acids) SAK-II peptides display both enzymatic and K+ channel inhibitory activities. Medium size (42–43 amino acid) SAK-III peptides are gating modifiers which interact either with cardiac HERG or Kv3 channels by altering their voltage-dependent properties. SAK-III toxins bind to the S3C region in the outer vestibule of Kv channels. Sea anemones have proven to be a rich source of pharmacological tools, and some of the SAK toxins are now useful drugs for the diagnosis and treatment of autoimmune diseases.

1 Introduction

1.1 The First Isolated Sea Anemone Toxins Affect Voltage-Gated Sodium Channels

Sea anemones belong to the phylum Cnidaria (Older name is Coelenterata). They possess specialized stinging organelles (cnidocysts) for capturing prey and self-protection. Cnidocysts contain potent paralysing toxins which have been isolated during the last 35 years from numerous sea anemone species (Beress et al. 1975; Kelso and Blumenthal 1998; Norton 1991; Romey et al. 1976; Schweitz et al. 1981). They are polypeptides of molecular weights between 3,000 Da and 6,500 Da that are cross-linked by several disulfide bridges. The first toxins that were characterized act on voltage-dependent Na+ channels (Nav channels). At low concentrations (<40 μg/kg) these toxins, after in vivo intravenous, intracisternal (mammals) or intramuscular (crustaceans) injections, induce severe toxic syndromes including paralysis, general hyperexcitability, cardiac disorders, convulsions and death (Alsen et al. 1978; Schweitz 1984). These toxins interact with a large variety of excitable cells including neurons, cardiac and skeletal muscle cells (Abita et al. 1977; Bergman et al. 1976; Kelso et al. 1996; Renaud et al. 1986; Romey et al. 1976; Shibata et al. 1976). In cardiac tissues they induce positive inotropic effects, arrhythmias or cell fibrillation (Hanck and Sheets 1995; Khera et al. 1995; Reimer et al. 1985; Renaud et al. 1986; Shibata et al. 1976). In neuronal cells they produce a massive release of neurotransmitters from nerve terminals (Abita et al. 1977). Most of the excitatory and paralyzing effects of the sea anemone toxins acting on Nav channels in excitable membranes are due to a prolongation of action potential duration. By binding to a specific receptor (site 3) they alter the gating of Nav channels and delay the Na+ current inactivation without altering activation kinetics (Barhanin et al. 1981; Cestele and Catterall 2000; Norton 1991; Romey et al. 1976; Vincent et al. 1980). Structure-activity relationship studies using mutants of toxins, and the comparison of their relative toxicity and Nav channel activities have been largely investigated and described in more recent publications (Benzinger et al. 1998; Gallagher and Blumenthal 1994; Loret et al. 1994; Norton 1991; Rogers et al. 1996).

1.2   Potassium Channel Toxins

In the last two decades venomous animals provided pharmacological tools that can block a variety of Ca^{2+} activated (e.g. apamin, iberiotoxin, charybdotoxin, scylla-toxin), voltage-dependent (Kv) (e.g. dendrotoxins (DTX), kaliotoxin, conotoxins, hanatoxins, phrixotoxins), and inward-rectifier (tertiapin) K^+ channels. These tox-ins have been isolated from scorpion, snake, cone-shell, spider, bee or sea anemone venoms (Bidard et al. 1987; Diochot et al. 1999; Grissmer et al. 1994; Halliwell et al. 1986; Jin and Lu 1998; Kauferstein et al. 2003; Laraba-Djebari et al. 1994; Miller 1995; Shon et al. 1998; Swartz and MacKinnon 1995). They have been crucial tools for determining the involvement of particular Kv channels in patho-physiological pathways. For example, DTX-I, MCD-peptide, kaliotoxin and ShK peptide, isolated from snake, bee, scorpion and sea anemone venoms respectively, were of primary importance in characterizing the function of Kv1.1 channels in epilepsy and the contribution of Kv1.3 channels in inflammatory processes (Beeton et al. 2001, 2003; Mourre et al. 1997). Due to their high specificity and affinity for K^+ channels, these toxins have facilitated the purification of K^+ channels, determi-nation of their subunit stoichiometry and sub-cellular localization and tissue distri-bution (Aiyar et al. 1995; Legros et al. 2000; MacKinnon 1991; Mourre et al. 1986; Rehm and Lazdunski 1988).

1.3   Sea Anemone Potassium Channel Toxins

Only 12 sea anemone toxins have been purified and characterized for their blocking action on K^+ channels. They are able to act in synergy with depolarizing sodium neurotoxins to induce neuronal, muscular and cardiovascular hyperexcitability. The first K^+ channel blocker to be described was BgK, a peptide purified from *Bunodosoma granulifera* (Aneiros et al. 1993). Actually, the structure of a peptide designated as a Blood Depressing Substance, BDS-I isolated from *Anemonia sul-cata,* had been solved by proton nuclear magnetic resonance (^{1}H NMR) spectros-copy in 1989, but its K^+ channel blocking activity was described 10 years later (Diochot et al. 1998; Driscoll et al. 1989a, b). The sea anemone potassium channel (SAK) toxins facilitate the evoked release of acetylcholine at neuromuscular junc-tions by selectively blocking Kv channels. Three major groups of SAK inhibitors have been described so far based on their structural properties. Each group includes peptides displaying different pharmacological properties. Group I toxins are short (35–37 amino acids) peptides, characterized by two short α-helices; these toxins block sub-families of Kv1 channels. Group II includes longer peptides (58–59 amino acids) with a two-stranded β-sheet and two α-helices. These peptides are homologous to bovine pancreatic trypsin inhibitor (BPTI); they may also display protease inhibitory activity in addition to their capacity to block with less potency some of the Kv1 currents (Schweitz et al. 1995). In group III are peptides of 42–43 amino acids, folded within a triple-stranded antiparallel β-sheet but without α-helix; they act on Kv3 or HERG subfamilies.

There is a remarkable diversity of K$^+$ channels that are endowed with various physiological functions. Sea anemones have evolved to produce structurally different peptides that target some of the Kv channel subfamilies and use different binding mechanisms and modes of action. These SAK toxins are generally not the most abundant peptides in the venom, but they act in synergism with other peptides such as Na$^+$ channel toxins or anticholinesterases to create a very toxic arsenal.

2   K$^+$ Channel Structures

Potassium channels are the most diverse class of ion channels and are expressed in a large variety of tissues, in excitable and non-excitable cells. They are key regulators of neuronal excitability by setting the resting membrane potential and controlling the shape, frequency and repolarization phase of action potentials (Shieh et al. 2000). The cloning of about 80 potassium channel genes has allowed a classification into three structural groups characterized by the number of transmembrane segments (TMS) and by the number of pore (P)domains: (1) the 6 TMS, 1 P domain K$^+$ channels which are either voltage-dependent or calcium-dependent, (2) the 2 TMS, 1 P domain K$^+$ channels which generate inward rectifier currents, and (3) the 4 TMS, 2 P domain K$^+$ channels, also called 2P channels (Coetzee et al. 1999; Jan and Jan 1997; Lesage and Lazdunski 2000; Patel et al. 2001; Pongs 1992). A functional K$^+$ channel needs 4P domains to be active, which means four 6 TMS/1P or 2 TMS/1P subunits or two 2 P subunits. The functional diversity of potassium channels is *further* increased by the existence of splice variants and the formation of heteromeric channels (Coetzee et al. 1999).

Kv channels allow potassium efflux from cells in response to membrane voltage depolarization. Their open probability increases with membrane depolarization, and is essentially zero at hyperpolarized membrane voltage values. Structurally, a functional Kv channel is a tetramer which can be formed by the association of the same (homotetramer) or two different (heterotetramer) α-subunits (MacKinnon 1991; Rehm and Lazdunski 1988). Each Kv α-subunit has six transmembrane segments (S1–S6) linked by extracellular and intracellular loops; both N and C-termini are intracellular (Fig. 1). The structure of Kv channels is mainly defined as a central pore allowing K$^+$ ion selectivity and conduction, a gate which controls the ion flow, and a voltage sensor that detects changes in membrane voltage. The recent crystallization of two prokaryotic potassium channels [KcsA which only display two transmembrane segments, and KvAP that contains the whole voltage sensing domain (S4)] has provided structural information about the central pore, and the molecular basis for K$^+$ selectivity and gating motions (Doyle et al. 1998; Jiang et al. 2003b; Ruta et al. 2003). According to the current models, S1–S4 adopt α-helical secondary structures, while S3 is composed of two α-helices, termed S3N and S3C (or S3a and S3b in some publications) separated by a short non-helical stretch

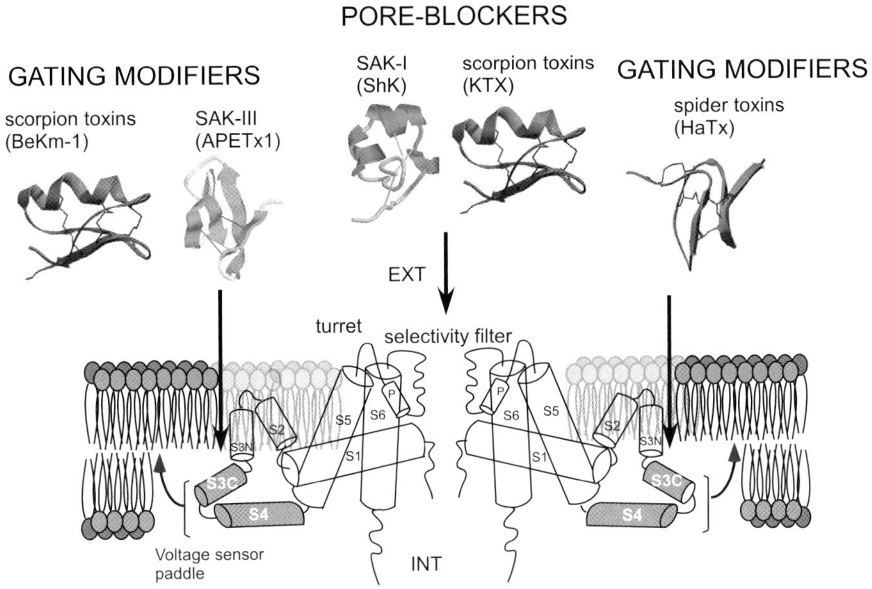

PORE-BLOCKERS

GATING MODIFIERS

SAK-I
(ShK)

scorpion toxins
(KTX)

GATING MODIFIERS

scorpion toxins
(BeKm-1)

SAK-III
(APETx1)

spider toxins
(HaTx)

EXT

turret selectivity filter

Voltage sensor
paddle

INT

Fig. 1 Schematic representation, in a transversal section, of the structure of a Kv channel inserted within a lipid bilayer membrane, according to the model described by the Swartz laboratory (Lee et al. 2003). S1 to S6 are transmembrane helical segments. In the pore lies the turret region, the selectivity filter and P is the helix of the pore. S3C and S4 form the voltage sensor paddle predicted to move towards the extracellular solution. Binding sites of sea anemones (SAK-I type represented by ShK and SAK-III type represented by APETx1), scorpion pore blockers (KTX type) or gating modifiers (BeKm1 type) and spider gating modifiers (hanatoxin type) are identified by arrows

including a conserved proline residue. S3C is at least around ten residues long, is partially exposed to the extracellular compartment and represents the target for gating modifier toxins (Jiang et al. 2003a, b).

The P domain which contains the most conserved region of all K$^+$ channels connects the S5–S6 membrane spanning α-helices. The P domain comprises the turret region, the pore helix and the selectivity filter (Fig. 1). The second important element in K$^+$ channels is the "gate," in the intracellular mouth of the ion pore, which includes negatively charged residues that determine the selectivity for K$^+$ ions. Open or closed conformations of crystallized K$^+$ channels support the idea that its gating involves changes in the dimensions of the internal pore (Doyle et al. 1998; Jiang et al. 2002). The third crucial structural element is the voltage sensor, represented by the S4 segment, which is rich in positively charged basic residues (Papazian et al. 1991) and which has been modeled using a crystal structure of the archaebacterial potassium channel KvAP (Jiang et al. 2003a). Upon membrane depolarization, charges in the voltage sensor contributed by four Arg residues display a net movement outwards to open the pore (Bezanilla 2000; Jiang et al. 2003b; Swartz 2004). The S4 segment and the carboxy terminus helix of S3 (S3C) are tightly coupled, forming an helix-turn-helix motif known as the "voltage sensor paddle" (Jiang et al. 2003b; Swartz 2004).

3 Purification and Synthesis of Sea Anemone K Channel Toxins

Sea anemones can be collected in large amounts at depths between 1 m and 5 m. Toxins are either extracted from electrically stimulated nematocysts, from mucus (a jelly like substance) secreted by the animal under stress, or from the whole body after homogenization and centrifugation steps at low temperature (Beress et al. 1975). Fats are extracted with chloroform from the supernatant obtained by centrifugation. Toxin extracts concentrated by freeze-drying can be dissolved in ammonium acetate and purified next on Sephadex G-50 gel columns, followed by successive ion-exchange and reversed-phase HPLC chromatographies (Aneiros et al. 1993; Beress et al. 1975; Bruhn et al. 2001; Castaneda et al., 1995; Schweitz et al. 1995). Sea anemone K$^+$ channel toxins are extremely basic peptides, and have very stable and compact globular structures cross-linked by several disulfide bridges.

BgK, the first potassium channel effector isolated from a marine organism, like other SAK peptides, is a minor component of the sea anemone *Bunodosoma granulifera* extract (0.05% of the mucus) (Aneiros et al. 1993). Thus, the purification of such low abundant SAK toxins could be hard and long owing to the difficulty to collect sea anemones in larger amounts. This is why, to allow a more extensive use of these very useful pharmacological tools, especially for *in vivo* studies which use large amounts of material, a number of the sea anemone K$^+$ channel inhibitors like ShK or BgK, have been chemically synthesized. Solid-phase synthesis allows the obtaining of "mg" of linear peptides which then require to be folded to their biologically active forms (Alessandri-Haber et al. 1999; Cotton et al. 1997; Pennington et al. 1995). The synthetic peptides display the same activity on K$^+$ channels as the native toxins. Toxins can also be produced by recombinant expression. A cDNA of HmK, a 35 aminoacid peptide isolated from *Heteractis magnifica*, was cloned and expressed in *E. coli* to produce a recombinant HmK peptide sharing the same biochemical properties and activity as the native one (Gendeh et al. 1997a). Another sea anemone peptide, BgK, was produced as a soluble cytoplasmic protein in *E. coli* displaying functional properties similar to those of native BgK (Braud et al. 2004).

4 Sea Anemone K$^+$ Channel Toxin Structures

The first group of Sea Anemone Kv channel blockers (SAK-I) includes short peptides with 35–37 amino acid residues, and three disulfide bonds. Six peptides (ShK, BgK, AsKs AeK, HmK and AETX K) belong to this group, and their sequences are presented in Fig. 2a; ShK, BgK and HmK disulfide bonds are paired as C1–C6,

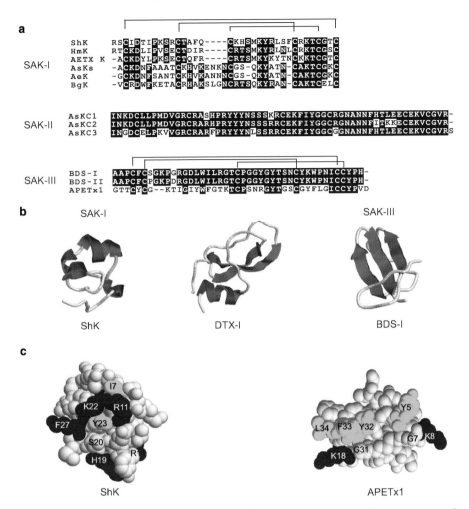

Fig. 2 Structures of SAK peptides. **a** Multiple peptide sequence alignment of the three groups of known sea-anemone toxins, which block Kv channels subfamilies. Amino acid identities (black boxes) and homologies (grey boxes) are shown. **b** The sea anemone K⁺ channel inhibitors belong to three different structural groups. The ribbon structures were drawn with the RASMOL program, using coordinates from the Protein Data Bank. In the SAK-I group, peptides with two short α-helices and one helical turn are represented by ShK (PDB code 1ROO). The SAK-II type includes peptides with a two-stranded β-sheet and two α-helices, represented here by BPTI (PDB code 1BPI). In the SAK-III group are peptides showing a triple-stranded antiparallel β-sheet without α-helix, like BDS-I (PDB code 1BDS). Lines correspond to disulfide bridges. **c** Model representing the active surface of BgK, ShK and APETx1. Coloured amino acids represent key residues for the binding and function of BgK (Gilquin et al. 2002; Racape et al. 2002), ShK (Rauer et al. 1999), and APETx1 (Chagot et al. 2005a; Zhang et al. 2007). Hydrophobic, basic and aliphatic residues are highlighted in green, blue, and yellow, respectively

C2–C4, C3–C5 (Fig. 2b) (Cotton et al. 1997; Dauplais et al. 1997; Gendeh et al. 1997b; Tudor et al. 1996). Two-dimensional ^{1}H NMR studies have shown that the SAK-I toxins secondary structure is mainly characterized by two short α-helices without β-sheet and belongs to the αα-type of fold (Dauplais et al. 1997; Mouhat et al. 2004; Tudor et al. 1996). These toxins do not share structure homologies in sequence and disulfide pairings with scorpion (e.g. charybdotoxin, agitoxin) or snake toxins (DTX) that have a similar activity and share the same binding site on Kv channels (Dauplais et al. 1997; Gross et al. 1994; Schweitz et al. 1995; Tudor et al. 1996; Tytgat et al. 1995).

The second group (SAK-II) includes longer peptides, the three kalicludines, AsKC1, AsKC2 and AsKC3 isolated from *A. sulcata*, containing 58–59 amino acids and three disulfide bonds (Fig. 2b). They are Kunitz-type protease inhibitors and present sequence homologies with BPTI (bovine pancreatic trypsin inhibitor) and also with dendrotoxin-I (DTX-I) (a Kv1 channel inhibitor) and calcicludine (a voltage-gated calcium channel blocker); both of these toxins are isolated from snake venoms (Harvey et al. 1994; Schweitz et al. 1995).

The third group (SAK-III) of toxins belongs to the β-defensin fold family which includes peptides of various origins, including human (produced by neutrophils and epithelial cells), snake, *Platypus* and sea anemone. These display antimicrobial, or, analgesic or myonecrotic activities (Torres and Kuchel 2004). SAK-III peptides include the BDS (Blood Depressing Substances) toxins, two very similar 43 amino acid peptides isolated from *A. sulcata* extracts in 1985 (Béress et al. 1985) and originally characterized for their antihypertensive and antiviral activities. BDS-I was structurally characterized in 1989 (Driscoll et al. 1989a, b) and BDS-I and BDS-II activities on Kv channels were described 13 years later (Diochot et al. 1998). APETx1, a 42 amino acid peptide isolated from *Anthopleura elegantissima* also belongs to the family of SAK-III peptides (Diochot et al. 2003). Both BDS-I and APETx1 display three disulfide bridges, paired in the C1–C5, C2–C4, C3–C6 connection which stabilizes the overall fold of the "all beta structure"(Torres and Kuchel 2004). Their secondary structure is a triple-stranded antiparallel β-sheet, without α-helix (Fig. 2b) (Driscoll et al. 1989a, b) also shared by Nav channel toxins isolated from sea anemones of the genera *Anemonia* and *Anthopleura* (Chagot et al. 2005a; Torres and Kuchel 2004). BDS and APETx1 have an additional mini-antiparallel β-sheet at the N terminus and a long extracellular loop that connects the first and second strands of the antiparallel β-sheet (Chagot et al. 2005a, Driscoll et al. 1989a). This structural family also includes other sea anemone peptides having the same structural scaffold, but with another functional target. This is the case for APETx2, which shares 64% sequence identity with APETx1 and blocks a Na^+ permeable Acid Sensing Ion Channel, ASIC3 (Chagot et al. 2005b; Diochot et al. 2004). A new toxin, BcIV, recently isolated from *Bunodosoma caissarum* is also structurally homologous to APETx1 and APETx2 (45% and 48% sequence identities and similar scaffold to the "disulfide rich all beta toxins") but its activity on potassium channels has not yet been demonstrated (Oliveira et al. 2006).

5   How to Analyze the Effects of SAK Toxins on K^+ Channels

5.1   Effect of SAK-I Toxins on Kv1 Channels

The demonstration of an effect of sea anemone toxins on K^+ channel was extensively investigated using binding experiments, by testing their ability to displace ^{125}I-dendrotoxin-I (probably the most potent and selective Kv inhibitor) (Harvey and Anderson 1985; Rehm et al. 1988; Rehm and Lazdunski 1988), from rat brain membrane preparations. BgK, ShK, HmK, AeK, AsKs, AsKC and AETX K were able to displace ^{125}I-DTX-I or ^{125}I-αDTX from their binding sites in brain synaptosomal membranes (Table 1 and Fig. 3) (Aneiros et al. 1993; Castaneda et al. 1995; Gendeh

Table 1   Affinities of different sea anemone toxins for potassium channels determined by electrophysiological studies

Toxin	Channel	IC_{50} (nM)	Ki (IC_{50}) nM[a]	References
BgK	Kv1.1	6 0.034[b]	0.7 (1.4) ^{125}I-DTX$_I$	Cotton et al. (1997), Racape et al. (2002), Gilquin et al. (2002), Aneiros et al. (1993)
	Kv1.2	15 0.066[b]		Cotton et al. (1997), Racape et al. (2002)
	Kv1.3	10–39 0.77[b]		Rauer et al. (1999), Cotton et al. (1997), Racape et al. (2002)
	Kv1.4	NA		Gilquin et al. (2005)
	Kv1.5	NA		Gilquin et al. (2005)
	Kv1.6	0.013[b]		Racape et al. (2002)
	IKCa1	172		Rauer et al. (1999)
ShK	Kv1.1	0.016	0.3 (3)[125]I-DTX$_I$	Kalman et al. (1998), Castaneda et al., (1995)
	Kv1.2	9	0.6 (0.7) ^{125}I-αDTX	Kalman et al. (1998), Castaneda et al., (1995)
	Kv1.3[c]	0.011–0.133		Pennington et al. (1995), Kalman et al. (1998)
	Kv1.4	0.31		Kalman et al. (1998)
	Kv1.5	NA		Kalman et al. (1998)
	Kv1.6	0.16		Kalman et al. (1998)
	Kv3.1	NA		Kalman et al. (1998)
	Kv3.2	0.3–6		Yan et al. (2005)
	Kv3.4	NA		Kalman et al. (1998)
	HKCa4	28		Kalman et al. (1998)
	IKCa1	30		Rauer et al. (1999)
AsKs	Kv1.2	140	10 (27) ^{125}I-DTX$_I$	Schweitz et al. (1995)

(continued)

Table 1 (continued)

Toxin	Channel	IC_{50} (nM)	Ki (IC_{50}) nM[a]	References
HmK	Kv1.2	<10 nM	1 (3.2) ^{125}I-αDTX	Gendeh et al. (1997a)
AsKC2	Kv1.2	1100	20 (60) ^{125}I-DTX$_I$	Schweitz et al. (1995)
BDS-I	Kv3.1	220		Yeung et al. (2005)
	Kv3.2	500		Yeung et al. (2005)
	Kv3.4	47–500[d]		Diochot et al. (1998), Yeung et al. (2005)
BDS-II	Kv3.1	750		Yeung et al. (2005)
	Kv3.2	500		Yeung et al. (2005)
	Kv3.4	56		Diochot et al. (1998)
APETx1	HERG1	34		Diochot et al. (2003)
	ERG2	NA		Restano-Cassulini et al. (2006)
	ERG3	NA		Restano-Cassulini et al. (2006)

[a]Ki and IC_{50} values for toxin against the binding of ^{125}I-αDTX or ^{125}I-DTXI on rat brain membranes
[b]Kd values determined by binding experiments on transfected cells
[c]Native current in T lymphocytes
[d]Values in different cell expression systems, calculated at +30 or +40 mV
NA: no activity

et al. 1997a; Hasegawa et al. 2006; Minagawa et al. 1998; Schweitz et al. 1995). Both kaliseptine (AsKs) and kalicludines (and most potently AsKC2) which have been purified from the extract of *A. sulcata* inhibit in a competitive way ^{125}I-DTX-I binding with values of concentration (IC_{50}) ranging from 27 nM to 500 nM (Schweitz et al. 1995).

BgK and ShK effects have also been tested on K^+ currents in a variety of cell preparations: [dorsal root ganglia (DRG)](Aneiros et al. 1993; Castaneda et al. 1995); T lymphocytes (Pennington et al. 1995); snail neurons (Garateix et al. 2000). These toxins were shown to inhibit a part of the outward current at nanomolar concentrations. More specific studies performed on cloned K^+ channels expressed in heterologous systems showed that the SAK-I toxins, preferentially block some members of the Kv1 subfamily (see Table 1). For example, among SAK peptides isolated from *A. sulcata*, AsKs seems to be the most potent blocker of Kv1.2 currents with IC_{50} around 100 nM (Fig. 4). Meanwhile, it is not excluded that other Kv1 channels could be targets for AsKs toxin, but this remains to be determined.

Other SAK-I peptides like ShK blocks Kv1.1, Kv1.3, Kv1.4 and Kv1.6 channels with picomolar affinities (Table 1) (Kalman et al. 1998), and BgK inhibits Kv1.1, Kv1.2 and Kv1.3 currents at nanomolar concentrations (Cotton et al. 1997). Binding studies using radioiodinated-peptides (^{125}I-BgK) also demonstrated a high affinity for Kv1.1, Kv1.2 and Kv1.6 channels (Table 1). BgK seems to be inactive on Kv1.4, Kv1.5 and Kv1.7 channels (Racape et al. 2002). In leukemia cells, ShK and BgK also block the intermediate conductance Ca^{2+} activated potassium current (IKCa1), a current regulating the membrane potential and modulating the calcium

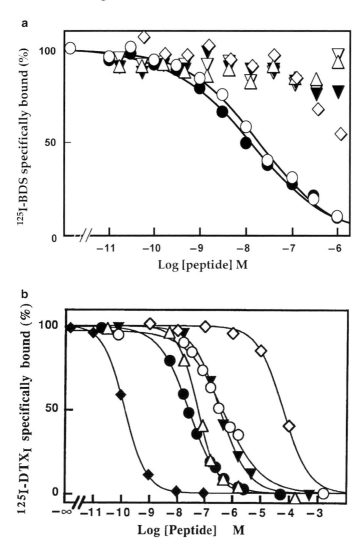

Fig. 3 Competitive inhibition by different venom peptides of specific ^{125}I-BDS-I or ^{125}I-DTXI binding to rat brain membranes. (**a**) Unlabeled BDS-I, BDS-II, and the other peptides were first incubated at different concentrations with rat brain synaptic membranes (40 μg/ml), and then, ^{125}I-BDS-I (9 pM) was added, and membranes were incubated for 3 h. Results are means of two experiments for BDS-I. ●, BDS-I; ○, BDS-II; ◇, AsII; ▽, DTX-I; ▼, MCD peptide; △, ChTx. (**b**) Unlabeled DTX$_I$ and the different peptides were first incubated at different concentrations with the membranes (20 μg/ml) and then ^{125}I-DTX$_I$ (3 pM) was added and membranes were incubated for 1 h at 25°C. Results are mean of two experiments. Nonspecific ^{125}I-DTX$_I$ binding was below 2% and was subtracted. ●, AsKS; ▼, AsKC1; △, AsKC2; ○, AsKC3; ◆, DTX$_I$; ◇, the Kunitz inhibitor BPTI

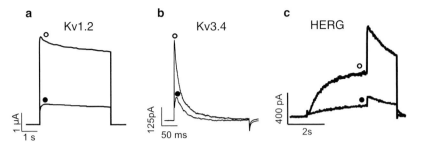

Fig. 4 Effects of different sea anemone toxins on K⁺ currents. (**a**) Inhibition of Kv1.2 currents by AsKS (600 nM), an SAK-I type peptide isolated from *Anemonia sulcata*. Channels are expressed in *Xenopus* oocytes. Currents are measured at +30 mV from a holding potential of −80 mV. (**b**) The transient Kv3.4 current is inhibited by BDS-I (75 nM) an SAK-III type peptide isolated from *A. sulcata*. Currents are measured at +50 mV from a holding potential of −80 mV on transfected COS-7 cells.(**c**) The cardiac HERG channel is blocked by APETx1 (1 μM), an SAK-III type peptide isolated from *Anthopleura elegantissima*. Currents are measured at 0 mV and repolarization at −40 mV from a holding potential of −80 mV on transfected COS-7 cells ○. Control current trace ● Current trace after toxin effect

signal in many different type of cells (for example, T lymphocytes, erythrocytes, colon) (Rauer et al. 1999).

A recent publication shows an extended interaction of ShK with other families of Kv channels than the Kv1 subfamily. For example, ShK blocks with high affinity (IC$_{50}$ 0.3–6 nM) the Kv3.2 current (Table 1), which is critical for high-frequency repetitive firing in cortical GABAergic fast-spiking interneurons (Yan et al. 2005).

5.2   Dual Activities of SAK-II Toxins

The three kalicludines described in the SAK-II group have two distinct biological activities. First, probably because of their extensive homologies with BPTI, they retain a potent protease inhibitory activity. Second, kalicludines inhibit competitively the binding of DTX-I to rat brain membranes, and inhibit the Kv1.2 current (Fig. 3) (Schweitz et al. 1995). Because of (i) their protease inhibitory property which protects the toxins from degradation and, (ii) their K⁺ channel inhibitory activity which contributes to paralysis of the preys, these toxins could have a dual function shared between hunting and the self-protection of sea anemones.

5.3   Effect of SAK-III Toxins on Different Subfamilies
of Kv Channels

The effects of SAK-III toxins were determined on a large variety of cloned K⁺ channels expressed in mammalian cells (Fig. 5). Slight inhibitions of Kv1 channels

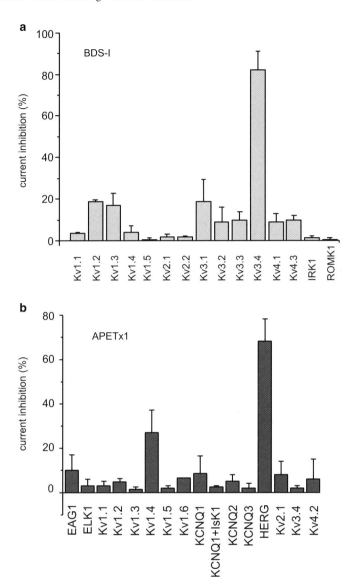

Fig. 5 Effect of BDSI (10 μM) and APETx1 (100 nM) on K⁺ channels. (**a**) Channels were expressed in *Xenopus* oocytes (for BDS-I) or (**b**) in COS-7 cells (for APETx1). Inhibitions were calculated at peak K⁺ current upon depolarizations to −10 mV to +30 mV from a holding potential of −80 mV before and during application of toxins. Inward rectifier IRK1 currents were evoked from 0 mV to −80 mV. N = 3 to 5 for each channel type

like Kv1.2 and Kv1.3 can be observed with high concentrations of BDS-I. This is not very surprising, as inhibition of different channel subtypes has already been described for high concentrations of some toxins, such as spider grammotoxins and hanatoxin, acting on both Ca²⁺ and K⁺ channels (Li-Smerin and Swartz 1998).

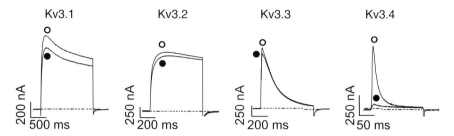

Fig. 6 Effect of BDS-I on the different Kv3 channel subtypes. Kv3 currents were measured at +30 mV from a holding potential of −80 mV. Channels were expressed in *Xenopus* oocytes. Kv3.4 current was almost completely inhibited (94%) by application of BDS-I (10 μM) and there was a slight effect on Kv3.1 (22 ± 10%, $N = 5$), and Kv3.2 (5 ± 4%, $N = 4$) currents

Until recently, the two BDS peptides were the only animal toxins known to block the Kv3 channels. BDS-I and BDS-II display a marked affinity for the transient Kv3.4 current but they also interact with the Kv3.1 and Kv3.2 channels (with a lower affinity) (Table 1, Figs. 4 and 6) (Diochot et al. 1998; Yeung et al. 2005). Both toxins induce a positive shift of the activation curve of Kv3.1 and Kv3.2 currents, indicating that BDS toxins act as gating modifiers similar to a number of spider toxins acting on Kv2 and Kv4 channel (Diochot et al. 1999; Swartz and MacKinnon 1997). Kv3 channels which generate rapidly activating and deactivating currents, play an important role in the firing frequency of central nervous system (CNS) neurons (Erisir et al. 1999; Rudy and McBain 2001). Kv3.4 channels may be of a particular importance in skeletal muscle periodic paralysis (Abbott et al. 2001), Alzheimer's disease where the Kv3.4 gene seems to be overexpressed in cerebral cortex (Angulo et al. 2004; Lien and Jonas 2003) and in Parkinson disease (Baranauskas et al. 2003).

Although sharing 54% sequence homology with BDS toxins, APETx1 has a different activity on K$^+$ channels. Despite slight effects on Kv1.4 currents at high concentrations, APETx1 is a specific blocker of the human *ether a go-go* related gene (human ERG, HERG) K$^+$ channel (IC$_{50}$ 34 nM) (Figs. 4 and 5). HERG is particularly expressed in mammalian heart (Diochot et al. 2003) where it contributes to the rapidly activating delayed rectifier potassium current (IKr) which controls the duration of the plateau phase of the action potential (Sanguinetti et al. 1995). Several mutations on the HERG gene are responsible for inherited disorders characterized by abnormal slow repolarization of action potentials associated with long QT intervals (Sanguinetti et al. 1996). APETx1 shifts the voltage-dependence of HERG activation towards depolarizing states (Diochot et al. 2003; Zhang et al. 2007). Only two other peptides are known to block ERG type channels: ErgTx1 and BeKm-1, two scorpion toxins isolated from *Centruroides noxius* and *Buthus eupeus* venoms, respectively. These scorpion toxins belong to the Csαβ fold family characterized by a α-helix linked by disulfide bridges to a triple-stranded β-sheet (Bontems et al. 1991; Mouhat et al. 2004). ErgTx1 and BeKm display different species-specific effects and affinities for the three neuronal members of ERG channel family (erg1, erg2 and erg3). APETx1, owing to its different and unique structure, has been reported to be selective for the neuronal and cardiac (human and rat) ERG1 channels and does not compete with the two other scorpion toxins (Chagot et al. 2005a; Restano-Cassulini et al. 2006; Wanke and Restano-Cassulini 2007). A recent publication indicates that inhibition of human ERG3 currents is accomplished by modifying the voltage dependence of the channel gating (Zhang et al. 2007).

Interestingly SAK-III peptides are not toxic "per se" after in vivo injection in animals (crabs or mice) contrary to sea anemone Nav toxins which are potent neurotoxic and cardiotoxic peptides. BDS toxins have been described for their antiviral and antihypertensive activities after intraveinous injections in mammal (Beress et al. 1985; Driscoll et al. 1989a) but neither cardiotoxic nor neurotoxic effects were observed (Diochot et al. 1998; Driscoll et al. 1989a).

6 Molecular Interaction of Sea Anemone Toxins with Kv Channels

6.1 Sea Anemone Toxins Interacting with Kv1 Channels

All SAK-I toxins contain two α-helices and no β-sheet. They block Kv1 channels by plugging the channel pore. Three residues (S20, K22, and Y23 in ShK; S23, K25 and Y26 in BgK) are strictly conserved in all SAK-I toxin sequences, and represent a common core of hot spot residues for pharmacological activity (Fig. 2a) (Alessandri-Haber et al. 1999; Castaneda et al. 1995; Gasparini et al. 2004; Gendeh et al. 1997b; Minagawa et al. 1998; Pennington et al. 1996a, b; Schweitz et al. 1995). Alanine mutation analyses have shown that other residues (I7, R11, H19, R24 and F27 in ShK), which are clustered around the three conserved residues, are also important for the binding activity of ShK to Kv1.2 channels in brain membranes, and to Kv1.3 channels in T lymphocytes (Pennington et al. 1996a, b; Rauer et al. 1999). For BgK, other residues such as F6 and N19 are also involved in the binding and selectivity to Kv1 channels (Racape et al. 2002). Molecular models of complexes between sea anemone toxins and Kv1 channels have been developed using the structure of KcsA (Doyle et al. 1998; Lanigan et al. 2002). They provide a molecular description for the interaction of toxin residues with the pore of the channel (Fig. 2c) (Doyle et al. 1998; Gilquin et al. 2002; Kalman et al. 1998; Lanigan et al. 2002; Norton et al. 2004). Since the different Kv1 subtypes are highly homologous (83% identity) in the P region, it is not really surprising that a single type of toxin can bind to several members of the Kv1 channel family (Alessandri-Haber et al. 1999). Double-mutant cycles experiments have shown that a diad, composed of an aromatic hydrophobic amino acid and a lysine, constitutes a conserved functional core and acts as a common anchor in different models of toxin-channel interaction (Dauplais et al. 1997; Gasparini et al. 2004; Gilquin et al. 2002). The toxin diad makes electrostatic interactions with carbonyl oxygen atoms of the conserved tyrosine residue (motif GYGD) in the channel selectivity filter. The functional diad K22-Y23 of ShK interacts with the Y in the GYGD motif of the Kv1.3 channel pore domain (Gasparini et al. 2004; Kalman et al. 1998; Lanigan et al. 2002; Norton et al. 2004). This key step is completed by hydrophobic interactions between residues F6 and Y26 in BgK, (Gilquin et al. 2002), Y23 in ShK, (Lanigan et al. 2002) (Fig. 2c) and the hydrophobic Y379 residue in the S5–S6 region of the Kv1.1 channel. Mutations of the Y379 residue in Kv1.1 modify the affinity and the selectivity of several scorpion and sea anemone toxins such as BgK and ShK (Gilquin et al. 2005).

All these toxin-channel interaction studies have assisted in the design of new types of toxins, such as ShK-Dap22 a mutant peptide where K22 has been replaced by a diaminopropionic acid. Binding and electrophysiological studies have shown that ShK-Dap22 is a highly potent and selective blocker of the Kv1.3 channel with a 100-fold decreased affinity for Kv1.1, Kv1.4 and Kv1.6 channels (Kalman et al. 1998). NMR studies have shown that the overall structures of ShK and ShK-Dap22 are quite similar, but there are differences in the side chains involved in Kv1.3 binding (Kalman et al. 1998; Norton et al. 2004). A high expression level of Kv1.3 is considered as a marker for activated effector memory T cells (T_{EM} cells), which are involved in the pathogenesis of autoimmune diseases. Therefore, the selective suppression of autoreactive T_{EM} cells with Kv1.3 blockers might constitute a novel approach for the treatment of multiple sclerosis (MS) and other autoimmune diseases such as type-1 diabetes mellitus or psoriasis. Both ShK and ShK-Dap22 were proven to prevent and treat rat autoimmune encephalomyelitis (Beeton et al. 2001; Norton et al. 2004). In addition, ShK-F6CA, a fluorescein-labeled analogue of ShK, has been reported to have potential applications in the diagnostic of autoimmune diseases (Beeton et al. 2003; Norton et al. 2004). This peptide, containing an additional negatively-charged moiety at the N-terminus, has a higher affinity and selectivity than ShK for Kv1.3 channels, allowing a specific detection of activated T_{EM} cells implicated in multiple sclerosis.

The kalicludines (AsKC1–3), which belong to the SAK-II family of toxins, also block Kv1.2 channels, although with less affinity (Schweitz et al. 1995); they do not contain the functional dyad "K-Y" conserved in SAK-I.

6.2   Molecular Interactions of SAK-III Toxins with Kv Channels

All of the sea anemone toxins within this group are gating modifier toxins that modify the voltage dependent properties of K^+ channels. The APETx1 binding site was identified in the S3C region of the HERG channel, referred to as the S3b region in some other publications (Yeung et al. 2005); it shares similar characteristics with gating modifier toxin binding sites on other Kv channels (Jiang et al. 2003b; Lee et al. 2003). A negative charge at position E518 in the S3C region of HERG channel is critical, likely by establishing an electrostatic interaction with a positively charged residue on the toxin (Zhang et al. 2007). Moreover, hydrophobic interactions between position G514 and APETx1 stabilize toxin binding. Positions G514 and E518 are on the same face of the S3C helix, and are exposed to the extracellular aqueous phase, allowing accessibility of APETx1 in the resting state of the channel and enabling gating paddle movement during channel activation (Zhang et al. 2007).

A comparison with other gating modifier toxins suggests that the interaction surface of APETx1 could be composed of K18, L34, F33, and Y32 (Zhang et al. 2007). A potential interaction surface composed of three aromatic residues (Y5, Y32, and F33), two basic residues (K8 and K18), and three aliphatic amino acids (G7, G31 and L34) has been described for APETx1 (Chagot et al. 2005a). In spite of a different fold, the APETx1 molecular surface shares some similarities (aromatic and basic residues) with the active surfaces of scorpion toxins (CnErg1 and BeKm1, Csαβ type of fold) which block erg channels. The main difference is the absence of a central lysine (K) in APETx1. APETx1, like spider and scorpion toxins which modify K$^+$ channel gating, probably binds to the same region (S3C) in the outer vestibule of the channel (Chagot et al. 2005a; Zhang et al. 2007).

BDS toxins bind to the external surface of Kv3 channels in both open and closed states (Diochot et al. 1998; Yeung et al. 2005). Electrophysiological studies have shown that BDS toxins are not pore blockers, but rather gating modifiers since they shift the Kv3 activation curve towards more depolarized membrane potentials. They alter the voltage-dependence of the K$^+$ currents, inducing more inhibition at voltages below +20 mV. Gating currents experiments also support this idea since the toxin shifts the voltage dependence of gating charge movements (Yeung et al. 2005). Mutagenesis studies of the Kv3.2 subunit indicates that residues (329–334) in the S3C region and, to a lesser extent, amino acids (347–353) near the S4 region, are good candidates for the interaction with BDS (Yeung et al. 2005). These residues are part of the voltage sensor of the channel, a crucial region that is also the target for spider gating modifier toxins interacting with the S3C helix of Kv2 and Kv4 channels (Diochot et al. 1999; Swartz and MacKinnon 1997; Wang et al. 2004). Studies performed on mutated K$^+$, Na$^+$, and Ca^{2+} channels with spider toxins suggest a conserved binding motif for gating modifier toxins within the S3C region of the voltage-sensing domain of these channels (Bourinet et al. 2001; Li-Smerin and Swartz 1998; Ruta and MacKinnon 2004; Winterfield and Swartz 2000). This could also be the case for sea anemone gating modifier toxins.

7 Concluding Comments

In the last 30 years, sea anemones have provided a battery of particularly interesting peptidic toxins acting on ionic channels. Some of them are specific for cardiac or neuronal Nav channels, others can distinguish Kv channel subfamilies. Very recently, a peptide which specifically blocks an Acid Sensing Ion Channel (ASIC3) has been discovered (Diochot et al. 2004). ASIC channels are proton-gated cationic channels mainly expressed in central and peripheric nervous system, and some functional channels like ASIC1a and ASIC3 which are expressed in nociceptive neurons, have been implicated in inflammation and in various pain processes (Mazzuca et al. 2007). Voltage dependent K$^+$ channels are key elements in the repolarization phase and duration of action potentials, consequently they are very important in a number of physiological processes. Dysfunctions of K$^+$ channels often correspond to diseases such as episodic ataxia (Kv1.1), cardiovascular disorders (KCNQ1, HERG), neonatal convulsions (KCNQ2, KCNQ3) and deafness (KCNQ1, KCNQ4). It is very important to find molecules that are able to selectively interact with these channels. The most important sources of such molecules are animal venoms. SAK toxins which block some Kv channels, by their original mode of action and their particular structural properties, are important additions to the currently known panel of natural Kv inhibitors isolated from other animal venoms. SAK toxins target Kv1, HERG and Kv3 channels subfamilies, they are very useful molecular tools to study their implication in neuronal and cardiovascular diseases. ShK and its more selective analogs might turn out to be useful drugs for the diagnosis and treatment of certain autoimmune diseases such as multiple sclerosis.

Sea anemone venoms will hopefully continue to provide a multitude of toxins acting on ion channels, particularly on "orphan" K$^+$ channels for which specific toxins have yet to be discovered, as for example, the 2P domain K$^+$ channels. Thus, it is essential to continue to search, with the now very elaborate techniques of ion channel assays, chemical analyses and syntheses, for the presence of even minor toxic components that may become useful tools in future investigations concerning the physiopathological importance of particular K$^+$ channels.

Acknowledgments We are grateful to Dr A. Baron and M. Borsotto for a critical reading of the manuscript. A considerable part of the work done in the laboratory was achieved thanks to collaboration with Professor Laszlo Béress of Kiel, Germany. We thank the Association Française contre les Myopathies (AFM), the Fondation pour la Recherche Médicale (FRM), the Institut Paul Hamel, and the Agence Nationale de la Recherche (ANR) for financial support.

References

Abbott GW, Butler MH, Bendahhou S, Dalakas MC, Ptacek LJ, Goldstein SA (2001) MiRP2 forms potassium channels in skeletal muscle with Kv3.4 and is associated with periodic paralysis. Cell 104:217–231.

Abita JP, Chicheportiche R, Schweitz H, Lazdunski, M (1977) Effects of neurotoxins (veratridine, sea anemone toxin, tetrodotoxin) on transmitter accumulation and release by nerve terminals in vitro. Biochemistry 16:1838–1844.

Aiyar J, Withka JM, Rizzi JP, Singleton DH, Andrews GC, Lin W, Boyd J, Hanson DC, Simon M, Dethlefs B, Lee C-L, Hall JE, Gutman GA, Chandy KG (1995) Topology of the pore-region of a K$^+$ channel revealed by the NMR-derived structures of scorpion toxins. Neuron 15: 1169–1181.

Alessandri-Haber N, Lecoq A, Gasparini S, Grangier-Macmath G, Jacquet G, Harvey AL, de Medeiros C, Rowan EG, Gola M, Menez A, Crest M (1999) Mapping the functional anatomy of BgK on Kv1.1, Kv1.2, and Kv1.3. Clues to design analogs with enhanced selectivity. J Biol Chem 274:35653–35661.

Alsen C, Beress L, Tesseraux I (1978) Toxicities of sea anemone (*Anemonia sulcata*) polypeptides in mammals. Toxicon 16:561–566.

Aneiros A, Garcia I, Martinez JR, Harvey AL, Anderson AJ, Marshall DL, Engstrom A, Hellman U, Karlsson E (1993) A potassium channel toxin from the secretion of the sea anemone *Bunodosoma granulifera*. Isolation, amino acid sequence and biological activity. Biochim Biophys Acta 1157:86–92.

Angulo E, Noe V, Casado V, Mallol J, Gomez-Isla T, Lluis C, Ferrer I, Ciudad CJ, Franco R (2004) Up-regulation of the Kv3.4 potassium channel subunit in early stages of Alzheimer's disease. J Neurochem 91:547–557.

Baranauskas G, Tkatch T, Nagata K, Yeh JZ, Surmeier DJ (2003) Kv3.4 subunits enhance the repolarizing efficiency of Kv3.1 channels in fast-spiking neurons. Nat Neurosci 6:258–266.

Barhanin J, Hugues M, Schweitz H, Vincent JP, Lazdunski M (1981) Structure-function relationships of sea anemone toxin II from *Anemonia sulcata*. J Biol Chem 256:5764–5769.

Beeton C, Wulff H, Barbaria J, Clot-Faybesse O, Pennington M, Bernard D, Cahalan MD, Chandy KG, Beraud E (2001) Selective blockade of T lymphocyte K+ channels ameliorates experimental autoimmune encephalomyelitis, a model for multiple sclerosis. Proc Natl Acad Sci USA 98:13942–13947.

Beeton C, Wulff H, Singh S, Botsko S, Crossley G, Gutman GA, Cahalan MD, Pennington M, Chandy KG (2003) A novel fluorescent toxin to detect and investigate Kv1.3 channel up-regulation in chronically activated T lymphocytes. J Biol Chem 278:9928–9937.

Benzinger GR, Kyle JW, Blumenthal KM, Hanck DA (1998) A specific interaction between the cardiac sodium channel and site-3 toxin anthopleurin B. J Biol Chem 273:80–84.

Beress L, Beress R, Wunderer G (1975) Isolation and characterisation of three polypeptides with neurotoxic activity from *Anemonia sulcata*. FEBS Lett 50:311–314.

Beress L, Doppelfeld I-S, Etschenberg E, Graf E, Henschen A, Zwick J (1985) Federal Republic of Germany Patent DE 3324689 A1.

Bergman C, Dubois JM, Rojas E, Rathmayer W (1976) Decreased rate of sodium conductance inactivation in the node of Ranvier induced by a polypeptide toxin from sea anemone. Biochim Biophys Acta 455:173–184.

Bezanilla F (2000) The voltage sensor in voltage-dependent ion channels. Physiol Rev 80:555–592.

Bidard JN, Mourre C, Lazdunski M (1987) Two potent central convulsant peptides, a bee venom toxin, the MCD peptide, and a snake venom toxin, dendrotoxin I, known to block K+ channels, have interacting receptor sites. Biochem Biophys Res Commun 143:383–389.

Bontems F, Roumestand C, Gilquin B, Menez A, Toma, F (1991) Refined structure of charybdotoxin: common motifs in scorpion toxins and insect defensins. Science 254:1521–1523.

Bourinet E, Stotz SC, Spaetgens RL, Dayanithi G, Lemos J, Nargeot J, Zamponi GW (2001) Interaction of SNX482 with domains III and IV inhibits activation gating of $a_{1E}(Ca_V2.3)$ calcium channels. Biophys J 81:79–88.

Braud S, Belin P, Dassa J, Pardo, L, Mourier G, Caruana A, Priest BT, Dulski P, Garcia ML, Menez A, Boulain JC, Gasparini S (2004) BgK, a disulfide-containing sea anemone toxin blocking K+ channels, can be produced in *Escherichia coli* cytoplasm as a functional tagged protein. Protein Expr Purif 38:69–78.

Bruhn T, Schaller C, Schulze C, Sanchez-Rodriguez J, Dannmeier C, Ravens U, Heubach JF, Eckhardt K, Schmidtmayer J, Schmidt H, Aneiros A, Wachter E, Beress L (2001) Isolation and characterisation of five neurotoxic and cardiotoxic polypeptides from the sea anemone *Anthopleura elegantissima*. Toxicon 39:693–702.

Castaneda O, Sotolongo V, Amor AM, Stocklin R, Anderson AJ, Harvey AL, Engstrom A, Wernstedt C, Karlsson E (1995) Characterization of a potassium channel toxin from the Caribbean sea anemone *Stichodactyla helianthus*. Toxicon 33:603–613.

Cestele S, Catterall WA (2000) Molecular mechanisms of neurotoxin action on voltage-gated sodium channels. Biochimie 82:883–892.

Chagot B, Diochot S, Pimentel C, Lazdunski M, Darbon H (2005a) Solution structure of APETx1 from the sea anemone *Anthopleura elegantissima*: a new fold for an HERG toxin. Proteins 59:380–386.

Chagot B, Escoubas P, Diochot S, Bernard C, Lazdunski M, Darbon H (2005b) Solution structure of APETx2, a specific peptide inhibitor of ASIC3 proton-gated channels. Protein Sci 14: 2003–2010.

Coetzee WA, Amarillo Y, Chiu J, Chow A, Lau D, McCormack T, Moreno H, Nadal MS, Ozaita A, Pountney D, Saganich M, Vega-Saenz de Miera E, Rudy B (1999) Molecular diversity of K⁺ channels. Ann NY Acad Sci 868:233–285.

Cotton J, Crest M, Bouet F, Alessandri N, Gola M, Forest E, Karlsson E, Castaneda O, Harvey AL, Vita C, Menez A (1997) A potassium-channel toxin from the sea anemone *Bunodosoma granulifera*, an inhibitor for Kv1 channels. Revision of the amino acid sequence, disulfide-bridge assignment, chemical synthesis, and biological activity. Eur J Biochem 244:192–202.

Dauplais M, Lecoq A, Song J, Cotton J, Jamin N, Gilquin B, Roumestand C, Vita C, de Medeiros CL, Rowan EG, Harvey AL, Menez A (1997) On the convergent evolution of animal toxins. Conservation of a diad of functional residues in potassium channel-blocking toxins with unrelated structures. J Biol Chem 272:4302–4309.

Diochot S, Schweitz H, Beress L, Lazdunski M (1998) Sea anemone peptides with a specific blocking activity against the fast inactivating potassium channel Kv3.4. J Biol Chem 273: 6744–6749.

Diochot S, Drici MD, Moinier D, Fink M, Lazdunski M, Schweitz H, Beress L (1999) Effects of phrixotoxins on the Kv4 family of potassium channels and implications for the role of Ito1 in cardiac electrogenesis. Brit J Pharmacol 126:251–263.

Diochot S, Loret E, Bruhn T, Beress L, Lazdunski M (2003) APETx1, a new toxin from the sea anemone *Anthopleura elegantissima*, blocks voltage-gated human ether-a-go-go-related gene potassium channels. Mol Pharmacol 64:59–69.

Diochot S, Baron A, Rash LD, Deval E, Escoubas P, Scarzello S, Salinas M, Lazdunski M (2004) A new sea anemone peptide, APETx2, inhibits ASIC3, a major acid-sensitive channel in sensory neurons. EMBO J 23:1516–1525.

Doyle DA, Morais Cabral J, Pfuetzner RA, Kuo A, Gulbis JM, Cohen SL, Chait BT, MacKinnon R (1998) The structure of the potassium channel: molecular basis of K⁺ conduction and selectivity. Science 280:69–77.

Driscoll PC, Clore GM, Beress L, Gronenborn AM (1989a) A proton nuclear magnetic resonance study of the antihypertensive and antiviral protein BDS-I from the sea anemone *Anemonia sulcata*: sequential and stereospecific resonance assignment and secondary structure. Biochemistry 28:2178–2187.

Driscoll PC, Gronenborn AM, Beress L, Clore GM (1989b) Determination of the three-dimensional solution structure of the antihypertensive and antiviral protein BDS-I from the sea anemone Anemonia sulcata: a study using nuclear magnetic resonance and hybrid distance geometry-dynamical simulated annealing. Biochemistry 28:2188–2198.

Erisir A, Lau D, Rudy B, Leonard CS (1999) Function of specific K(+) channels in sustained high-frequency firing of fast-spiking neocortical interneurons. J Neurophysiol 82:2476–2489.

Gallagher MJ, Blumenthal KM (1994) Importance of the unique cationic residues arginine 12 and lysine 49 in the activity of the cardiotonic polypeptide anthopleurin B. J Biol Chem 269: 254–259.

Garateix A, Vega R, Salceda E, Cebada J, Aneiros A, Soto E (2000) BgK anemone toxin inhibits outward K⁺ currents in snail neurons. Brain Res 864:312–314.

Gasparini S, Gilquin B, Menez A (2004) Comparison of sea anemone and scorpion toxins binding to Kv1 channels: an example of convergent evolution. Toxicon 43:901–908.

Gendeh GS, Chung MC, Jeyaseelan K (1997a) Genomic structure of a potassium channel toxin from *Heteractis magnifica*. FEBS Lett 418:183–188.

Gendeh GS, Young LC, de Medeiros CL, Jeyaseelan K, Harvey AL, Chung MC (1997b) A new potassium channel toxin from the sea anemone *Heteractis magnifica*: isolation, cDNA cloning, and functional expression. Biochemistry 36:11461–11471.

Gilquin B, Racape J, Wrisch A, Visan V, Lecoq A, Grissmer S, Menez A, Gasparini S (2002) Structure of the BgK-Kv1.1 complex based on distance restraints identified by double mutant cycles. Molecular basis for convergent evolution of Kv1 channel blockers. J Biol Chem 277:37406–37413.

Gilquin B, Braud S, Eriksson MA, Roux B, Bailey TD, Priest BT, Garcia ML, Menez A, Gasparini S (2005) A variable residue in the pore of Kv1 channels is critical for the high affinity of blockers from sea anemones and scorpions. J Biol Chem 280:27093–27102.

Grissmer S, Nguyen AN, Aiyar J, Hanson DC, Mather RJ, Gutman GA, Karmilowicz MJ, Auperin DD, Chandy KG (1994) Pharmacological characterization of five cloned voltage-gated K+ channels, types Kv1.1, 1.2, 1.3, 1.5, and 3.1, stably expressed in mammalian cell lines. Mol. Pharmacol 45:1227–1134.

Gross A, Abramson T, MacKinnon R (1994) Transfer of the scorpion toxin receptor to an insensitive potassium channel. Neuron 13:961–966.

Halliwell JV, Othman IB, Pelchen-Matthews A, Dolly JO (1986) Central action of dendrotoxin: selective reduction of a transient K conductance in hippocampus and binding to localized acceptors. Proc Natl Acad Sci USA 83:493–497.

Hanck DA, Sheets MF (1995) Modification of inactivation in cardiac sodium channels: ionic current studies with anthopleurin-A toxin. J Gen Physiol 106:601–616.

Harvey AL, Anderson AJ (1985) Dendrotoxins: snake toxins that block potassium channels and facilitate neurotransmitter release. Pharmacol Ther 31:33–55.

Harvey AL, Rowan EG, Vatanpour H, Fatehi M, Castaneda O, Karlsson E (1994) Potassium channel toxins and transmitter release. Ann NY Acad Sci 710:1–10.

Hasegawa Y, Honma T, Nagai H, Ishida M, Nagashima Y, Shiomi K (2006) Isolation and cDNA cloning of a potassium channel peptide toxin from the sea anemone *Anemonia erythraea*. Toxicon 48:536–542.

Jan LY, Jan YN (1997) Voltage-gated and inwardly rectifying potassium channels. J Physiol 505 (Pt. 2):267–282.

Jiang Y, Lee A, Chen J, Cadene M, Chait BT, MacKinnon, R (2002) The open pore conformation of potassium channels. Nature 417:523–526.

Jiang Y, Lee A, Chen J, Ruta V, Cadene M, Chait BT, MacKinnon R (2003a) X-ray structure of a voltage-dependent K+ channel. Nature 423:33–41.

Jiang Y, Ruta V, Chen J, Lee A, MacKinnon R (2003b) The principle of gating charge movement in a voltage-dependent K+ channel. Nature 423:42–48.

Jin W, Lu Z (1998) A novel high-affinity inhibitor for inward-rectifier K+ channels. Biochemistry 37:13291–13299.

Kalman K, Pennington MW, Lanigan MD, Nguyen A, Rauer H, Mahnir V, Paschetto K, Kem WR, Grissmer S, Gutman GA, Christian EP, Cahalan MD, Norton RS, Chandy KG (1998) ShK-Dap22, a potent Kv1.3-specific immunosuppressive polypeptide. J Biol Chem 273:32697–32707.

Kauferstein S, Huys I, Lamthanh H, Stocklin R, Sotto F, Menez A, Tytgat J, Mebs D (2003) A novel conotoxin inhibiting vertebrate voltage-sensitive potassium channels. Toxicon 42:43–52.

Kelso GJ, Blumenthal KM (1998) Identification and characterization of novel sodium channel toxins from the sea anemone *Anthopleura xanthogrammica*. Toxicon 36:41–51.

Kelso GJ, Drum CL, Hanck DA, Blumenthal KM (1996) Role for Pro-13 in directing high-affinity binding of anthopleurin B to the voltage-sensitive sodium channel Biochemistry 35:14157–14164.

Khera PK, Benzinger GR, Lipkind G, Drum CL, Hanck DA, Blumenthal KM (1995) Multiple cationic residues of anthopleurin B that determine high affinity and channel isoform discrimination. Biochemistry 34:8533–8541.

Lanigan MD, Kalman K, Lefievre Y, Pennington MW, Chandy KG, Norton RS (2002) Mutating a critical lysine in ShK toxin alters its binding configuration in the pore-vestibule region of the voltage-gated potassium channel, Kv1.3. Biochemistry 41:11963–11971.

Laraba-Djebari F, Legros C, Crest M, Ceard B, Romi R, Mansuelle P, Jacquet G, van Rietschoten J, Gola M, Rochat H, Bougis PE, Martin-Eauclaire MF (1994) The kaliotoxin family enlarged. Purification, characterization, and precursor nucleotide sequence of KTX2 from *Androctonus australis* venom. J Biol Chem 269:32835–32843.

Lee HC, Wang JM, Swartz KJ (2003) Interaction between extracellular hanatoxin and the resting conformation of the voltage-sensor paddle in Kv channels. Neuron 40:527–536.

Legros C, Pollmann V, Knaus HG, Farrell AM, Darbon H, Bougis PE, Martin-Eauclaire MF, Pongs O (2000) Generating a high affinity scorpion toxin receptor in KcsA-Kv1.3 chimeric potassium channels. J Biol Chem 275:16918–16924.

Lesage F, Lazdunski M (2000) Molecular and functional properties of two-pore-domain potassium channels. Am J Physiol Renal Physiol 279:793–801.

Li-Smerin Y, Swartz KJ (1998) Gating modifier toxins reveal a conserved structural motif in voltage-gated Ca^{2+} and K^+ channels. Proc Natl Acad Sci USA 95:8585–8589.

Lien CC, Jonas P (2003) Kv3 potassium conductance is necessary and kinetically optimized for high-frequency action potential generation in hippocampal interneurons. J Neurosci 23:2058–2068.

Loret EP, del Valle RM, Mansuelle P, Sampieri F, Rochat H (1994) Positively charged amino acid residues located similarly in sea anemone and scorpion toxins. J Biol Chem 269:16785–16788.

MacKinnon R (1991) Determination of the subunit stoichiometry of a voltage-activated potassium channel. Nature 350:232–235.

Mazzuca M, Heurteaux C, Alloui A, Diochot S, Baron A, Voilley N, Blondeau N, Escoubas P, Gelot A, Cupo A, Zimmer A, Zimmer AM, Eschalier A, Lazdunski M (2007) A tarantula peptide against pain via ASIC1a channels and opioid mechanisms. Nat Neurosci 10:943–945.

Miller C (1995) The charybdotoxin family of K^+ channel-blocking peptides. Neuron 15:5–10.

Minagawa S, Ishida M, Nagashima Y, Shiomi, K (1998) Primary structure of a potassium channel toxin from the sea anemone Actinia equina. FEBS Lett 427:149–151.

Mouhat S, Jouirou B, Mosbah A, De Waard M, Sabatier JM (2004) Diversity of folds in animal toxins acting on ion channels. Biochem J 378:717–726.

Mourre C, Hugues M, Lazdunski M (1986) Quantitative autoradiographic mapping in rat brain of the receptor of apamin, a polypeptide toxin specific for one class of Ca^{2+}-dependent K^+ channels. Brain Res 382:239–249.

Mourre C, Lazdunski M, Jarrard LE (1997) Behaviors and neurodegeneration induced by two blockers of K^+ channels, the mast cell degranulating peptide and dendrotoxin I. Brain Res 762:223–227.

Norton RS (1991) Structure and structure-function relationships of sea anemone proteins that interact with the sodium channel. Toxicon 29:1051–1084.

Norton RS, Pennington MW, Wulff H (2004) Potassium channel blockade by the sea anemone toxin ShK for the treatment of multiple sclerosis and other autoimmune diseases. Curr Med Chem 11:3041–3052.

Oliveira JS, Zaharenko AJ, Ferreira WA Jr, Konno K, Shida CS, Richardson M, Lucio AD, Beirao PS, de Freitas JC (2006) BcIV, a new paralyzing peptide obtained from the venom of the sea anemone Bunodosoma caissarum. A comparison with the Na^+ channel toxin BcIII. Biochim Biophys Acta 1764:1592–1600.

Papazian DM, Timpe LC, Jan YN, Jan LY (1991) Alteration of voltage-dependence of Shaker potassium channel by mutations in the S4 sequence. Nature 349:305–310.

Patel AJ, Lazdunski M, Honore E (2001) Lipid and mechano-gated 2P domain K^+ channels. Curr Opin Cell Biol 13:422–428.

Pennington MW, Byrnes ME, Zaydenberg I, Khaytin I, de Chastonay J, Krafte DS, Hill R, Mahnir VM, Volberg WA, Gorczyca W, Kem WR (1995) Chemical synthesis and characterization of ShK toxin: a potent potassium channel inhibitor from a sea anemone. Int J Pept Protein Res 46:354–358.

Pennington MW, Mahnir VM, Khaytin I, Zaydenberg I, Byrnes ME, Kem WR (1996a) An essential binding surface for ShK toxin interaction with rat brain potassium channels. Biochemistry 35:16407–16411.

Pennington MW, Mahnir VM, Krafte DS, Zaydenberg I, Byrnes ME, Khaytin I, Crowley K, Kem WR (1996b) Identification of three separate binding sites on SHK toxin, a potent inhibitor of voltage-dependent potassium channels in human T-lymphocytes and rat brain. Biochem Biophys Res Commun 219:696–701.

Pongs O (1992) Molecular biology of voltage-dependent potassium channels. Physiol Rev 72:S69–S88.

Racape J, Lecoq A, Romi-Lebrun R, Liu J, Kohler M, Garcia ML, Menez A, Gasparini S (2002) Characterization of a novel radiolabeled peptide selective for a subpopulation of voltage-gated potassium channels in mammalian brain. J Biol Chem 277:3886–3893.

Rauer H, Pennington M, Cahalan M, Chandy KG (1999) Structural conservation of the pores of calcium-activated and voltage-gated potassium channels determined by a sea anemone toxin. J Biol Chem 274:21885–21892.

Rehm H, Lazdunski M (1988) Purification and subunit structure of a putative K$^+$-channel protein identified by its binding properties for dendrotoxin I. Proc Natl Acad Sci USA 85:4919–4923.

Rehm H, Bidard JN, Schweitz H, Lazdunski M (1988) The receptor site for the bee venom mast cell degranulating peptide. Affinity labeling and evidence for a common molecular target for mast cell degranulating peptide and dendrotoxin I, a snake toxin active on K$^+$ channels. Biochemistry 27:1827–1832.

Reimer NS, Yasunobu CL, Yasunobu KT, Norton TR (1985) Amino acid sequence of the *Anthopleura xanthogrammica* heart stimulant, anthopleurin-B. J Biol Chem 260:8690–8693.

Renaud JF, Fosset M, Schweitz H, Lazdunski M (1986) The interaction of polypeptide neurotoxins with tetrodotoxin-resistant Na$^+$ channels in mammalian cardiac cells. Correlation with inotropic and arrhythmic effects. Eur J Pharmacol 120:161–170.

Restano-Cassulini R, Korolkova YV, Diochot S, Gurrola G, Guasti L, Possani LD, Lazdunski M, Grishin EV, Arcangeli A, Wanke E (2006) Species diversity and peptide toxins blocking selectivity of ether-a-go-go-related gene subfamily K$^+$ channels in the central nervous system. Mol Pharmacol 69:1673–1683.

Rogers JC, Qu Y, Tanada TN, Scheuer T, Catterall WA (1996) Molecular determinants of high affinity binding of a-scorpion toxin and sea anemone toxin in the S3–S4 extracellular loop in domain IV of the Na$^+$ channel a subunit. J Biol Chem 271:15950–15962.

Romey G, Abita JP, Schweitz H, Wunderer G, Lazdunski M (1976) Sea anemone toxin: a tool to study molecular mechanisms of nerve conduction and excitation-secretion coupling. Proc Natl Acad Sci USA 73:4055–4059.

Rudy B, McBain CJ (2001) Kv3 channels: voltage-gated K$^+$ channels designed for high-frequency repetitive firing. Trends Neurosci 24:517–526.

Ruta V, MacKinnon R (2004) Localization of the voltage-sensor toxin receptor on KvAP. Biochemistry 43:10071–10079.

Ruta V, Jiang Y, Lee A, Chen J, MacKinnon R (2003) Functional analysis of an archaebacterial voltage-dependent K$^+$ channel. Nature 422:180–185.

Sanguinetti MC, Jiang C, Curran ME, Keating MT (1995) A mechanistic link between an inherited and an acquired cardiac arrhythmia: HERG encodes the IKr potassium channel. Cell 81:299–307.

Sanguinetti MC, Curran ME, Spector PS, Keating MT (1996) Spectrum of HERG K$^+$-channel dysfunction in an inherited cardiac arrhythmia. Proc Natl Acad Sci USA 93:2208–2212.

Schweitz H (1984) Lethal potency in mice of toxins from scorpion, sea anemone, snake and bee venoms following intraperitoneal and intracisternal injection. Toxicon 22:308–311.

Schweitz H, Vincent JP, Barhanin J, Frelin C, Linden G, Hugues M, Lazdunski M (1981) Purification and pharmacological properties of eight sea anemone toxins from *Anemonia sulcata*, *Anthopleura xanthogrammica*, *Stoichactis giganteus*, and *Actinodendron plumosum*. Biochemistry 20:5245–5252.

Schweitz H, Bruhn T, Guillemare E, Moinier D, Lancelin JM, Beress L, Lazdunski M (1995) Kalicludines and kaliseptine. Two different classes of sea anemone toxins for voltage sensitive K$^+$ channels. J Biol Chem 270:25121–25126.

Shibata S, Norton TR, Izumi T, Matsuo T, Katsuki S (1976) A polypeptide (AP-A) from sea anemone (*Anthopleura xanthogrammica*) with potent positive inotropic action. J Pharmacol Exp Ther 199:298–309.

Shieh CC, Coghlan M, Sullivan JP, Gopalakrishnan M (2000) Potassium channels: molecular defects, diseases, and therapeutic opportunities. Pharmacol Rev 52:557–594.

Shon KJ, Stocker M, Terlau H, Stuhmer W, Jacobsen R, Walker C, Grilley M, Watkins M, Hillyard DR, Gray WR, Olivera BM (1998) k-Conotoxin PVIIA is a peptide inhibiting the shaker K$^+$ channel. J Biol Chem 273:33–38.

Swartz KJ (2004) Towards a structural view of gating in potassium channels. Nat Rev Neurosci 5:905–916.

Swartz KJ, MacKinnon R (1995) An inhibitor of the Kv2.1 potassium channel isolated from the venom of a Chilean tarantula. Neuron 15:941–949.

Swartz KJ, MacKinnon R (1997) Hanatoxin modifies the gating of a voltage-dependent K⁺ channel through multiple binding sites. Neuron 18:665–673.

Torres AM, Kuchel PW (2004) The b-defensin-fold family of polypeptides. Toxicon 44:581–588.

Tudor JE, Pallaghy PK, Pennington MW, Norton RS (1996) Solution structure of ShK toxin, a novel potassium channel inhibitor from a sea anemone. Nat Struct Biol 3:317–320.

Tytgat J, Debont T, Carmeliet E, Daenens P (1995) The a-dendrotoxin footprint on a mammalian potassium channel. J Biol Chem 270:24776–24781.

Vincent JP, Balerna M, Barhanin J, Fosset M, Lazdunski M (1980) Binding of sea anemone toxin to receptor sites associated with gating system of sodium channel in synaptic nerve endings in vitro. Proc Natl Acad Sci USA 77:1646–1650.

Wang JM, Roh SH, Kim S, Lee CW, Kim JI, Swartz KJ (2004) Molecular surface of tarantula toxins interacting with voltage sensors in K(v) channels. J Gen Physiol 123:455–467.

Wanke E, Restano-Cassulini R (2007) Toxins interacting with ether-a-go-go-related gene voltage-dependent potassium channels. Toxicon 49:239–248.

Winterfield JR, Swartz KJ (2000) A hot spot for the interaction of gating modifier toxins with voltage-dependent ion channels. J Gen Physiol 116:637–644.

Yan L, Herrington J, Goldberg E, Dulski PM, Bugianesi RM, Slaughter RS, Banerjee P, Brochu RM, Priest BT, Kaczorowski GJ, Rudy B, Garcia ML (2005) *Stichodactyla helianthus* peptide, a pharmacological tool for studying Kv3.2 channels. Mol Pharmacol 67:1513–1521.

Yeung SY, Thompson D, Wang Z, Fedida D, Robertson B (2005) Modulation of Kv3 subfamily potassium currents by the sea anemone toxin BDS: significance for CNS and biophysical studies. J Neurosci 25:8735–8745.

Zhang M, Liu XS, Diochot, Lazdunski M, Tseng GN (2007) APETx1 from sea anemone *Anthopleura elegantissima* is a gating modifier peptide toxin of the human ether-a-go-go-related potassium channel. Mol Pharmacol 72:259–268.

Ligands for Ionotropic Glutamate Receptors

Geoffrey T. Swanson and Ryuichi Sakai

Abstract Marine-derived small molecules and peptides have played a central role in elaborating pharmacological specificities and neuronal functions of mammalian ionotropic glutamate receptors (iGluRs), the primary mediators of excitatory synaptic transmission in the central nervous system (CNS). As well, the pathological sequelae elicited by one class of compounds (the kainoids) constitute a widely-used

G.T. Swanson (✉)
Department of Molecular Pharmacology and Biological Chemistry, Northwestern University, Feinberg School of Medicine, 303 E. Chicago Ave., Chicago, IL, 60611
email: gtswanson@northwestern.edu

R. Sakai
Faculty of Fisheries Sciences, Hokkaido University, Hakodate 041-8611, Japan

N. Fusetani and W. Kem (eds.), *Marine Toxins as Research Tools*,
Progress in Molecular and Subcellular Biology, Marine Molecular Biotechnology 46,
DOI: 10.1007/978-3-540-87895-7, © Springer-Verlag Berlin Heidelberg 2009

animal model for human mesial temporal lobe epilepsy (mTLE). New and existing molecules could prove useful as lead compounds for the development of therapeutics for neuropathologies that have aberrant glutamatergic signaling as a central component. In this chapter we discuss natural source origins and pharmacological activities of those marine compounds that target ionotropic glutamate receptors.

1 Introduction

Marine organisms have provided some of the most well-known and widely used ligands for mammalian iGluRs (Fig. 1). Indeed, one family of iGluRs, the kainate receptors, was named for a molecule derived from a common variety of seaweed that was used extensively in Japanese native medicine as an anthelmintic to treat ascariasis. Despite the prominence in neuroscience research of a limited set of marine natural products that target iGluRs, in sheer numbers far more iGluR-active secondary metabolites, particularly amino acid derivatives, have been derived from terrestrial sources (Moloney 1998, 1999, 2002). Notable examples include quisqualic acid, from

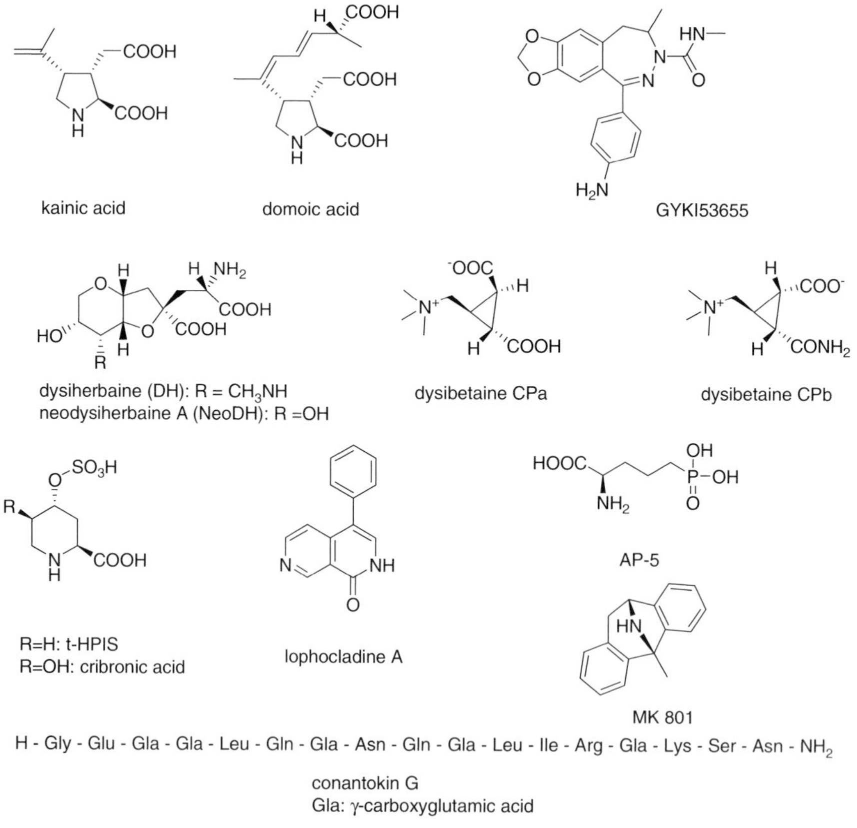

Fig. 1 Chemical structures of representative iGluR ligands discussed in this chapter

the fruit of the Rangoon creeper *Quisqualis indica* (Takemoto et al. 1975; Takemoto 1978), and willardiine, from the pea seedling of *Acacia willardinia* (Gmelin 1959; Ashworth et al. 1972), both of which have played important roles in the pharmacological characterization of iGluRs. Wasp toxins also are sources of metabolites with affinity for iGluRs. Philanthotoxins, which are polyamine-containing toxins from the digger wasp *Philanthus triangulum*, and Joro spider toxins (JSTX) from the Joro spider *Nephila clavata* are open-channel blockers for a subset of glutamate receptors (Clark et al. 1982; Bruce et al. 1990; Blagbrough et al. 1994; Usherwood 2000; Estrada et al. 2007). Numerous other examples of molecules derived from terrestrial organisms exist whose activity on iGluRs have been characterized to varying degrees (e.g., Takemoto et al. 1964; Evans and Usherwood 1985; Konno et al. 1988; Shin-ya et al. 1997a, b; McCormick et al. 1999; Watanabe and Kitahara 2007). This rich abundance of iGluR ligands from terrestrial sources perhaps explains why comparatively few new molecules have been isolated from marine organisms, which are often more difficult to collect and exhibit more limited diversity at the species level. Nevertheless, molecules isolated from marine sources are structurally novel in many cases and for that reason could serve both as important tools in neurobiological research and as templates for development of clinically relevant drugs.

In this chapter we will discuss several broad families of marine-derived iGluR ligands, including peptides isolated from Conus snails that target *N*-methyl-D-aspartate (NMDA) receptors and rigid analogs of the excitatory amino acid L-glutamate that predominantly activate non-NMDA receptors, which consist of (S)-α-amino-3-hydroxy-5-methyl-4-isoxazolepropionic acid (AMPA) and kainate receptor families. Molecules that target non-NMDA receptors include kainic acid (KA) and domoic acid (DOM), which are algal products collectively referred to as kainoids because they share a 2,3,4-trisubstituted pyrrolidine core structure (Laycock et al. 1989). The terrestrial excitotoxin acromelic acid, from the Japanese mushroom *Clitocybe acromelalga*, also falls into this structural class. More recent studies discovered two new natural iGluR ligands, dysiherbaine and neodysiherbaine A (Sakai et al. 1997, 2001a), underscoring the potential utility of screening marine benthic organisms for neuroactive molecules. These molecules are structurally distinct from the kainoids and thus constitute a third family of marine-derived ligands for ionotropic glutamate receptors. As a necessary introduction to the molecules themselves, we first briefly review the genetics, pharmacology, and neurophysiology of the three primary members of the iGluR family of subunits, the AMPA, kainate and NMDA receptors.

2   Ionotropic Glutamate Receptors

Ionotropic glutamate receptors are essential to the appropriate function of the mammalian CNS. They mediate chemical synaptic transmission at the vast majority of excitatory synapses, underlie well-characterized cellular models of learning and memory, modulate excitability of neuronal networks, and are required for maturation of synaptic connections during early development (reviewed in Mayer et al. 1992; Hollmann and Heinemann 1994; Aamodt and Constantine-Paton 1999; Dingledine et al. 1999; Huettner 2003; Lerma 2006; Paoletti and Neyton 2007). Modulation of the strength of excitatory synaptic transmission by enhancing or inhibiting glutamate receptor function is under active investigation for therapeutic benefits in a number of neuropathologies, including mild to moderate cognitive impairment and chronic pain (Bleakman et al. 2006; Lynch and Gall 2006; Planells-Cases et al. 2006).

Ionotropic glutamate receptors have been highly conserved during evolution of marine and terrestrial organisms. They subserve similar functional roles in neurotransmission in higher and lower vertebrates, including fish (Nawy and Copenhagen 1987; Kung et al. 1996), and structurally related homologues have been identified from invertebrates such as *Drosophila melanogaster* (Schuster et al. 1991) and *Caenorhabditis elegans* (Hart et al. 1995; Maricq et al. 1995). Their central role in excitatory neurotransmission in a wide variety of organisms in part accounts for the occurrence of natural iGluR ligands used aggressively for prey immobilization (e.g., conantokins). In the following section we confine our brief discussion to the structure, physiology and pharmacology of mammalian iGluRs, because the vast majority of such research has been focused on these molecules. The general principles of receptor function, however, are likely applicable to the true targets of marine excitotoxins in fish and other predatory marine organisms. Reviews with significantly greater detail on the structure, function and significance of each type of mammalian iGluR are available in the recently published book *The Glutamate Receptors* (Gereau and Swanson 2008).

2.1   iGluR Gene Families and Structure

Ionotropic glutamate receptors are ligand-gated cation channels formed from subunit proteins of the AMPA, kainate, NMDA or delta receptor gene families. In mammalians, these gene families are known by the acronyms *GRIA*, *GRIK*, *GRIN*, and *GRID*, for *G*lutamate *R*eceptor, *I*onotropic, *A*MPA (*K*ainate, *N*MDA, or *D*elta) (Fig. 2a). While these subunits have shared structural features and similar primary amino acid sequences, assembly of functional tetrameric receptors is strictly controlled so that each subunit only oligomerizes with partners within its gene family. This restriction ensures, for example, that functional AMPA receptors in the mammalian CNS only contain AMPA receptor subunits; similar stoichiometric restrictions exist for other iGluRs.

A functional iGluR is composed of four subunit proteins, which can be identical (homomeric receptors) or heterogeneous (heteromeric receptors). The secondary structure of a single iGluR subunit has a generally modular design, with different domains in the proteins playing distinct roles in receptor biogenesis, trafficking, or function (Madden 2002; Greger and Esteban 2007) (Fig. 2b). All iGluR subunits have three trans-membrane domains (M1, M3 and M4) and one re-entrant P-loop (M2) similar to that found in voltage-gated channels. The M2 domains form the pore of the channel, whereas other membrane domains are intimately involved in channel gating processes. The large extracellular amino-terminal domain (NTD) is critical for appropriately restricted oligomerization during early steps in receptor assembly and contains allosteric modulatory sites in NMDA receptor subunits

Fig. 2 **a** The mammalian ionotropic glutamate receptor gene families represented in a dendrogram, with the distance of the connecting lines proportional to primary sequence identity. **b** The structure of a representative iGluR subunit. The N-terminal domain (NTD) is involved in assembly of tetrameric receptors. The ligand-binding domain (LBD) is extracellular and a bi-lobate structure composed of two distinct domains (D1 and D2). Four membrane domains are noted (M1–M4), with the M2 domain constituting a P-loop. The c-terminal domain contains trafficking, targeting and modulatory determinants.

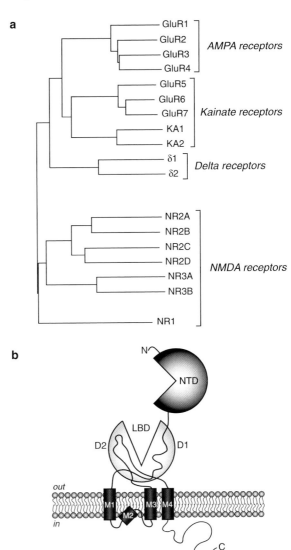

(Paoletti and Neyton 2007). The extracellular ligand-binding domain (LBD) is formed from two discontiguous segments of the protein, with the first (S1) located immediately before M1 and the second (S2) between M3 and M4 (Stern-Bach et al. 1994). Each subunit protein contains one binding site for its primary ligand, which in most cases is the excitatory neurotransmitter L-glutamate. While several subunits bind ligands other than glutamate, the tertiary structures of the ligand binding domains of each iGluR subunits are remarkably conserved (e.g., Armstrong et al. 1998; Furukawa et al. 2005; Mayer 2005; Naur et al. 2007), suggesting the fundamental mechanisms for ligand binding and channel gating are common to the different receptor families. Finally, the intracellular carboxy-terminal domains of receptor subunits interact with signaling systems and other protein complexes to modulate function and control trafficking and targeting to synaptic and non-synaptic sites in neurons (Perez-Otaño and Ehlers 2004; Jaskolski et al. 2005; Greger and Esteban 2007).

2.2 AMPA Receptors

AMPA receptors are the "workhorses" of mammalian excitatory synapses. Glutamate release from presynaptic vesicles binds to closely apposed receptors, resulting in channel gating and a brief, localized depolarization of the postsynaptic neuron. Typically, AMPA receptor-mediated excitatory postsynaptic potentials (EPSPs) have a half-time of less than 10 ms, ensuring that receptors are available for re-activation during periods of relatively high frequency input. Summation and propagation of these depolarizing signals to the cell soma can result in initiation of action potentials. Generalized inhibition of AMPA receptors results in cessation of excitatory neurotransmission and, essentially, brain activity.

AMPA receptors are formed from a combination of four individual gene products known as GluR1–4 (with the corresponding genes named *GRIA1–4*) (Hollmann and Heinemann 1994). All AMPA receptor subunits contain a binding site for glutamate and form channels that are permeable primarily to monovalent cations. Notably, receptors lacking the GluR2 subunit are also weakly permeable to calcium (Hollmann et al. 1991). Incorporation of a GluR2 subunit restricts divalent cation permeability as a result of a single amino acid difference from GluR1, GluR3, and GluR4 subunits in the critical pore-forming M2 domain (Dingledine et al. 1992). This amino acid difference (an arginine instead of a glutamine) arises not from a difference in the gene sequence, but rather from a post-transcriptional enzymatic alteration (RNA editing) of an adenosine within the glutamine codon in the mRNA, resulting in its translation as an arginine (Sommer et al. 1991). Functional diversity between receptor subunits is further introduced by an additional RNA editing event, as well as alternative splicing in both extracellular and intracellular domains (Sommer et al. 1990; Lomeli et al. 1994).

Neurons in the mammalian brain appear to use AMPA receptors with distinct stoichiometric combinations of subunits. For example, AMPA receptor excitatory postsynaptic currents (EPSCs) at hippocampal Schaffer collateral–CA1 pyramidal neuron synapses are thought to arise primarily from GluR2/GluR3 receptors. This is altered during periods of strong synaptic stimulation, which induce insertion of GluR1/GluR2 or GluR2/GluR4 AMPA receptors (Hayashi et al. 2000; Zhu et al. 2000; Shi et al. 2001). In contrast to pyramidal neurons, hippocampal interneurons tend to express AMPA receptors that lack the GluR2 subunit and thus have calcium-permeable populations of receptors (Geiger et al. 1995), which make them sensitive to open-channel blockage by natural polyamine toxins such as philanthotoxin and Joro spider toxin (Iino et al. 1996; Washburn and Dingledine 1996). In principle, this diversity suggests that specific populations of receptors could be targeted pharmacologically, but in actuality there are few identified ligands that exhibit high degrees of selectivity between AMPA receptor subunits.

As will be discussed in subsequent sections, a number of marine-derived agonists that activate kainate receptors with high affinity also act upon AMPA receptors with lower affinity. Kainic acid itself is the best-known example of an agonist with overlapping but divergent affinities for AMPA and kainate receptors. It is clear now, however, that the potent excitant activity elicited by kainoids and dysiherbaines arise largely (though perhaps not exclusively) from their affinity for and activation of kainate rather than AMPA receptors (Mulle et al. 1998; Sakai et al. 2001b).

2.3   Kainate Receptors

Kainate receptors play a variety of roles in the mammalian CNS (Lerma 2006). They modulate excitatory and inhibitory synaptic transmission (Rodriguez-Moreno et al. 1997; Contractor et al. 2000; Kamiya and Ozawa 2000; Schmitz et al. 2000; Frerking et al. 2001; Jiang et al. 2001), modulate some forms of synaptic plasticity (Bortolotto et al. 1999; Contractor et al. 2001; Lauri et al. 2001; Schmitz et al. 2003), control neuronal excitability through inhibitory actions on intrinsic conductances (Melyan et al. 2002, 2004; Fisahn et al. 2005), and can contribute to temporal summation of postsynaptic depolarization in response to bursts of action potentials (Castillo et al. 1997; Vignes and Collingridge 1997; Frerking and Ohliger-Frerking 2002; Jin et al. 2006). Despite these widespread actions in neuronal function, inhibition of kainate receptors (or genetic ablation of one or more subunits) does not have the profound impact on brain activity observed upon inhibition of AMPA receptors (Mulle et al. 1998, 2000; Simmons et al. 1998; Smolders et al. 2002; Contractor et al. 2003; Alt et al. 2007; Pinheiro et al. 2007). This has led to the hypothesis that the roles subserved by kainate receptors are largely modulatory, fine-tuning the balance between excitation and inhibition in the CNS, rather than being obligatory constituents of central synaptic transmission.

Kainate receptors are formed from a combination of five individual gene products, which are further subdivided into two groups based on primary sequence identity and pharmacological specificity. GluR5, GluR6, and GluR7 comprise the first sub-family to be isolated (with the corresponding genes *GRIK1–3*) (Hollmann and Heinemann 1994). These three subunits are collectively referred to as "low-affinity" kainate receptor subunits, because they have a lower affinity for the eponymous ligand, kainic acid, than do the two members of the second sub-family, KA1 and KA2 (*GRIK4 and GRIK5*), which consequently have been referred to as the "high-affinity" kainate receptor subunits. There are important differences in the physiological function of these receptor subunits as well: low-affinity subunits can assemble into functional homo-oligomeric receptors, whereas high-affinity subunits KA1 and KA2 must combine with GluR5, GluR6 or GluR7 to form functional hetero-oligomeric receptors (Egebjerg et al. 1991; Herb et al. 1992; Sommer et al. 1992;

Schiffer et al. 1997; Ren et al. 2003). For this reason, KA1 and KA2 are also denoted "auxiliary" subunits to the "principal" GluR5, GluR6 and GluR7 subunits. This concept is somewhat misleading, however, because the majority of neuronal kainate receptors are likely composed of one or more members of both sub-families of subunits as heteromeric receptors (Wisden and Seeburg 1993; Bahn et al. 1994), and thus both "principal" and "auxiliary" subunits are obligatory to the appropriate functioning of these receptors in the brain.

While inhibition of kainate receptors appears to be well-tolerated by mammalian nervous systems, activation of neuronal kainate receptors with potent and high-affinity agonists, such as kainic or domoic acid, elicits characteristic stereotyped behaviors, tonic-clonic seizures, or even death at high concentrations (Nadler 1979; Ben-Ari 1985). Long-term pathological consequences of sub-lethal exposure to kainate receptor agonists include deterioration of hippocampal pyramidal neurons and lesions similar to that observed in human patients with mTLE (Nadler 1981; Ben-Ari 1985). A similar neuropathology can arise from ingestion of marine organisms (fish or shellfish) containing highly concentrated domoic acid, as is discussed in more detail in the subsequent section.

2.4 NMDA Receptors

NMDA receptors are essential mediators of many forms of learning and memory, and NMDA receptor activation is a requisite early step in most models of long-term synaptic plasticity of excitatory neurotransmission in the mammalian CNS (Dingledine et al. 1999). These receptors have a number of unusual functional features central to their important roles at excitatory synapses. For example, they are the only type of glutamate receptor that requires binding of two distinct agonists, glutamate and glycine, for channel gating (Johnson and Ascher 1987).

As well, NMDA receptors are occluded at physiological membrane potentials by the presence of a Mg^{2+} ion bound to a high-affinity site within the channel pore (Nowak et al. 1984). Voltage-dependent channel block by Mg^{2+} is relieved upon strong depolarization of the postsynaptic membrane, which can occur as a result of robust AMPA or kainate receptor activation or from back-propagating action potentials in the dendritic arbor (Bliss and Collingridge 1993; Spruston et al. 1995; Magee and Johnston 1997). Thus, NMDA receptors function as "coincidence detectors": in order to gate current, they must receive both a presynaptic signal (glutamate released from the synaptic vesicle) and a postsynaptic signal (depolarization). If these conditions are met, NMDA receptors will open and allow permeation of both monovalent and divalent cations. Relative to AMPA and kainate receptors, NMDA receptors are highly Ca^{2+} permeable (MacDermott et al. 1986; Mayer and Westbrook 1987; Burnashev et al. 1995), and this Ca^{2+} entry through the channel plays a unique role in mediating downstream signals that lead to alterations in synaptic strength. Calcium- and calmodulin-dependent kinase II (CaMKII) and the Ca^{2+} dependent phosphatase calcineurin are perhaps the best-characterized signaling proteins that play central roles in synaptic plasticity downstream of Ca^{2+} entry through NMDA receptors. In addition, Ca^{2+} entry stimulates gene expression to effect protein synthesis-dependent stabilization of alterations in synaptic strength (Nicoll and Malenka 1999; Xia and Storm 2005).

NMDA receptors are formed from heteromeric combinations of three sub-families of gene products: the NR1, NR2, and NR3 subunits. A single NR1 gene exists in mammals (named *GRIN1*). NR2 subunits are encoded by four distinct genes (*GRIN2A–GRIN2D*). NR3 subunits, the most recent sub-family to be cloned and characterized, are produced by two distinct genes (*GRIN3A and -3B*). Further diversity is introduced into the NR1 family of subunits by alternate splicing events in both the amino-and carboxy-terminal domains, such that a total of eight unique transcripts are utilized in mammalian brains. Functional NMDA receptors are composed of NR1 in combination with either NR2 or NR3 subunits (or, potentially, both types of subunits) (Dingledine et al. 1999; Chatterton et al. 2002; Awobuluyi et al. 2007; Smothers and Woodward 2007). As with all ionotropic glutamate receptor subunits, each NMDA receptor subunit contains a single binding site for a neurotransmitter, but the identity of the endogenous agonist molecule differs between NMDA receptor subunits. Glycine is the native ligand for NR1 and NR3 subunits (Kuryatov et al. 1994; Chatterton et al. 2002), whereas glutamate binds selectively to NR2 subunits (Laube et al. 1997). Thus, NR1/NR2 NMDA receptors (likely the predominant form found in the brain) have glutamate and glycine as obligate coagonists and are the primary contributors to plasticity in the CNS. NR1/NR3 "NMDA" receptors, in contrast, can be gated by glycine alone (Chatterton et al. 2002); the importance of these unusual receptors in excitatory neurotransmission is not well-characterized.

2.5 Delta Receptors

Delta receptors are a fourth family of receptors that are classified as iGluRs based on structural, rather than functional, similarity (Araki et al. 1993; Lomeli et al. 1993). Glutamate does not bind to or activate either delta-1 (δ1) or -2 (δ2) receptors to produce a current. Indeed, the endogenous ligand remains unknown for these receptors, which were classified for many years after their cloning as "orphans." Recent data suggests that glycine or D-serine could represent physiological ligands (Naur et al. 2007), based on binding studies and resolved crystal structures, but neither elicit a detectable current from the receptors when applied in voltage-clamp experiments. Thus, their mechanism of action remains a mystery. No marine-derived molecules have been identified that target delta receptors, and for that reason we limit our discussion here of these interesting molecules. Several in-depth reviews discuss their potential function and role in development and neuropathology (Yuzaki 2004; Hirano 2006).

3   AMPA/Kainate Receptor Ligands: Kainoids

3.1   *Natural Sources and Synthetic Analogs of Kainoids*

Kainic acid (KA), the original member of the kainoid family of molecules, was first found from marine red alga *Digenea simplex* (Ceramiales, Rhodomelaceae) (Fig. 3a, b). Aqueous extracts of *D. simplex* were used as anthelmintics in traditional

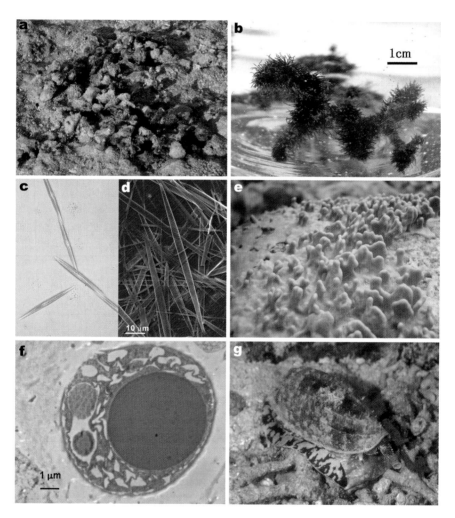

Fig. 3 Marine organisms that contain iGluR compounds. **a** *Digenea simplex* (Iriomote, Okinawa). The algae are often covered by sand. **b** *Digenea simplex* in culture. **c** Light micrograph of *Pseudonitzschia multiseries*. **d** Scanning electron micrograph of *P. multiseries*. **e** *Lendenfeldia chondrodes* (Yap, Micronesia). **f** Transparent electron micrograph of a symbiotic *Synecocystis sp.* in mesohyl of *L. chondrodes*. **g** *Conus geographus* in Palau. Images acquired by G. T. Swanson (A), R. Sakai (B, E, F) Y. Kotaki (C), K. Koike (D), and K. Nomura (G)

Chinese and Japanese medicine for many centuries (Nitta et al. 1958; Pei-Gen and Shan-Lin 1986), but the isolation and structural determination of KA as the vermicidal principle was first achieved by Takemoto in 1953 (Murakami et al. 1953; Takemoto and Daigo 1958). KA has also been isolated from other species of Ceramiales red algae such as *Alsidium helmithochorton* (Calaf et al. 1989), *Caloglossa leprieurii* (Pei-Gen and Shan-Lin 1986), *Centroceras clavulatum* (Impellizzeri et al. 1975), and certain variety of non-Ceramiales red algae *Palmaria palmata* (Laycock et al. 1989). Tank cultures of a naturally occurring "dwarf" mutant of *P. palmata* provided a source for isolation of KA and other excitatory amino acids such as D-homocysteic acid and glutamate (Laycock et al. 1989). A survey of 46 marine red and green algae found KA or the structural analog domoic acid (DOM) in four and five Rhodomelaceae species, respectively. Interestingly, *D. simplex* itself contained a small amount of DOM (100-fold less than KA), which had not been noted previously (Sato et al. 1996).

In addition to use as a veterinary anthelmintic, KA is well-established as a standard probe in neurological research to elicit currents from neuronal or recombinant kainate receptors and to induce seizure-related behaviors and pathology in animal models of epilepsy. Production of KA for these purposes has relied on both total synthesis and isolation of natural product from algae. The latter process provided a stable and economical resource until the last decade, when isolation from *D. simplex* for commercial purposes ceased, leading to a widespread shortage of KA (Tremblay 2000). Recently, cultivation of *P. palmata* by Ocean Produce International Inc. and synthetic production by other commercial agents re-established a stable production of KA.

Two natural congeners of KA have been isolated from marine organisms. Allokainic acid is the C-4 epimer of KA and was isolated from *D. simplex*. 1'-hydroxykainic acid was found from a KA-producing mutant of *P. palmata* (Ramsey et al. 1994). In addition, a "kainic-peptide" was isolated from a red alga *A. helminthocorton*; this molecule appears to be a naturally occurring peptide that contains kainic acid as two of its 37 amino acids residues (Calaf et al. 1989), but no further structural information, beyond amino acid analysis, or biological activities have been described.

Recently Sakai and co-workers determined the cellular and subcellular localization of KA in *D. simplex* using immunohistochemical and immunocytochemical techniques (Sakai et al. 2005). A KA-specific antibody localized immunoreactivity within the outmost layer cells in the algal thallus. In subcellular observations using transmission electron microscopy, KA immunoreactivity was found in electron dense cytosolic granule bodies, nuclei, and pit plugs of the cells. No immunoreactivity was observed in epibionts, including bacteria attached to the outer surface of the thallus. Localization of KA in nuclei is of particular interest because the accumulation of secondary metabolites in this structure had not previously been reported. The function of KA in the nucleus, however, remains unknown. The pit plugs are cell-to-cell connective apparatus unique in Rhodophyceae; their physiological role(s) have been elusive but could involve transport of nutritive materials based on morphological observations. Localization of KA immunoreactivity in the pit plug constitutes the first evidence for translocation of cellular material through this structure.

The distribution and occurrence of another marine-derived kainoid, domoic acid (DOM), bears significant importance in the realms of public health and food hygiene. This kainoid is produced by algae that enter the food chain of marine mammals and seabirds, and potentially humans, through accumulation in marine primary producers and higher filter feeders, such as anchovies and certain shellfish (Olney 1994; Watters 1995; Clark et al. 1999; Lefebvre et al. 1999; Mos 2001). DOM was first isolated from red alga *Chondria armata* (Takemoto 1978), and several related compounds – isodomoic acids A to D, isodomoic acid G and H, and domoilactones A and B – were later identified from the same source (Takemoto 1978; Maeda et al. 1986, 1987; Zaman et al. 1997). To date, ten stereo- and regio-isomers or congeners of DOM have been detected (Clayden et al. 2005).

The ecological and toxicological threat posed by DOM has become evident over the last two decades. DOM-containing algae were also used as vermifugal agents in Japanese folk medicine, similar to *D. simplex,* because DOM is a potent neuronal excitant in both vertebrates and invertebrates arising from its high affinity for kainate receptors (indeed, significantly greater affinity than that of KA itself) (Debonnel et al. 1989; Lomeli et al. 1992). The use of DOM-containing algae for medicinal purposes in humans did not result in any reported incidences of severe toxicity. However, an outbreak of food poisoning resulting from ingestion of DOM-containing blue mussels on Prince Edward Island Canada in 1987 demonstrated its potential for toxicity in humans and was the proximal cause of three deaths (Bates et al. 1989; Wright et al. 1989; Perl et al. 1990b). The clinical manifestation of DOM toxicity included moderate to severe gastrointestinal disorders and neurological symptoms that included disorientation, seizure and memory deficits in a subset of individuals (Perl et al. 1990b; Teitelbaum et al. 1990). One elderly individual who survived the initial intoxication later developed complex partial seizures and exhibited hippocampal neuronal loss and sclerosis similar to that observed in animals following KA-induced toxicity (Cendes et al. 1995). As a result of the striking degree of acute and, in some cases, long-term anterograde amnesia, the term "Amnesiac Shellfish Poisoning" (ASP) was used to describe the clinical consequences of DOM intoxication resulting from consumption of contaminated shellfish (Perl et al. 1990a; Jeffery et al. 2004).

Pennate diatoms belonging to *Pseudo-nitzschia multiseries* species were identified initially as the source organisms that led to ASP after consumption of blue mussels in Canada (Bates et al. 1989). The toxic events localized to the western coast of North America are attributed to *Pseudo-nitzschia australis* (Fritz et al. 1992; Lefebvre et al. 1999; Scholin et al. 2000). Additionally, DOM has been detected in diatoms in Japan (*P. multiseries*) (Fig. 3c, d) (Kotaki et al. 1999), the United Kingdom (*P. australis* Frenguelli, *P. seriata f. seriata*) (Cusack et al. 2002; Fehling et al. 2004), and Vietnam (*Nitzschia navis-varingica*) (Kotaki et al. 2000; Lundholm and Moestrup 2000; Kotaki et al. 2004), demonstrating that DOM-producing diatoms can occur throughout many marine ecologies (Bates 2000). Accumulation and depuration of DOM following ingestion of *Pseudo-nitzschia* algae occur at variable rates and in distinct tissues in marine organisms, with the highest concentrations found in anchovies, razor clams and blue mussels following algal blooms on the Pacific coast of North America (Wekell et al. 1994; Lefebvre et al. 2002a, b, 2007).

Mass mortalities of sea mammals and coastal birds of California and Baja California, including the sea lions *Zalophus californianus* (Lefebvre et al. 1999; Scholin et al. 2000), brown pelicans *Pelecanus occidentalis* (Fritz et al. 1992; Work et al. 1993; Sierra Beltran et al. 1997), and Brant's cormorants *Phalacrocorax penicillatus* (Fritz et al. 1992; Work et al. 1993), were attributed to consumption of DOM-containing anchovies *Engraulis mordax*. A decade long monitoring study suggested that increasing numbers of California sea lions with neurological dysfunction and neuroanatomical damage (hippocampal atrophy and sclerosis) was attributable to chronic sub-lethal exposure to DOM (Goldstein et al. 2008). As well, krill (*Euphasia pacifica*) have been identified as a potential source of DOM toxicity that pose a risk to planktivorous organisms (Bargu et al. 2002; Lefebvre et al. 2002a). Governmental and fisheries organizations now routinely screen marine food sources for DOM levels. The US Food and Drug Safety sets a critical limit of 20 ppm DOM in the "edible portion of raw shellfish" (Guide for the Control of Molluscan Shellfish, 2005, available at http://www.cfsan.fda.gov), which is well below the levels toxic to humans (Iverson and Truelove 1994). Detection of suprathreshold DOM accumulation in marine organisms has resulted in several instances of temporary bans on fishing of particular species or within affected geographical areas (Trainer et al. 1998; Lefebvre et al. 2002a; Bill et al. 2006).

The physiological function of unusual secondary metabolites such as the kainoids within their marine ecosystem remains a matter of speculation. The most obvious possibility is that their potent and excitotoxic activity on vertebrate and invertebrate iGluRs serves in a defensive capacity to discourage attack.

Intraperitoneal or intracoelomic injection reproduces neurotoxicity observed in vertebrates, but oral gavage or ingestion of DOM is not neurotoxic in fish; (Hardy et al. 1995; Lefebvre et al. 2001, 2007). Furthermore, DOM accumulates without apparent lethality in a variety of benthic organisms (Lefebvre et al. 2002a, b, 2007), although more subtle neurological effects have been observed following exposure during development (Tiedeken et al. 2005). DOM also does not appear to subserve an allelopathic role to discourage competition between algal species (Lundholm et al. 2005). More recently, it was suggested that the excitotoxin might serve as a physiological defense mechanism against krill, which consume the diatoms (Bargu et al. 2006). In this intriguing study, the authors found that DOM effectively reduced krill grazing behaviors and thereby could serve to perpetuate algal blooms.

3.2   Biological Activities of Kainoids

All kainoids elicit currents from both kainate and AMPA receptors. In general, they exhibit higher affinity and potency for kainate receptors, depending on structure and stereochemistry of the side chain functional groups. KA itself exhibits >10-fold higher binding affinity for kainate receptors as compared to AMPA receptors (reviewed in Hollmann and Heinemann 1994). It is a full (or nearly so) agonist that elicits desensitizing currents from most combinations of kainate receptors (see examples in Fig. 4), but only weakly activates AMPA receptors to produce steady-state currents with a very small desensitizing component. Within the kainate receptor subunit family, KA1 and KA2 kainate receptor subunits have a higher binding affinity for kainate (K_D values of 5–15 nM) than do GluR5, GluR6, and GluR7 (K_D values of 13–95 nM) (Bettler et al. 1990, 1992; Werner et al. 1991; Herb et al. 1992; Lomeli et al. 1992; Swanson et al. 1997). As mentioned previously, KA1 and KA2 have been referred to as "high-affinity" kainate receptor subunits and GluR5–7 as "low-affinity" subunits; it is important to keep in mind that this nomenclature, while useful for categorizing the two sub-families of subunits, is not relevant to their sensitivities to the endogenous neurotransmitter, glutamate.

Application of KA to almost all types of neurons in the mammalian brain causes marked depolarization through the activation of kainate and AMPA receptors, in a concentration- and receptor composition-dependent manner. Thus, KA is only a moderately selective agonist and, in general, currents evoked by high concentrations of kainate (>100 µM) will arise largely from AMPA receptors, because these receptors tend to be present at much higher density in neuronal membranes compared to kainate receptors. The respective contributions of AMPA and kainate receptors to KA-evoked currents can be differentiated more effectively using a selective AMPA receptor antagonist, such as GYKI 53655, to isolate kainate receptor currents (Paternain et al. 1995, 1996; Wilding and Huettner 1997). Currents evoked by low concentrations of KA (<5 µM) are largely carried by kainate receptors because of their significantly higher affinity for the marine toxin. This approach has been used, for example, to characterize the modulatory action of presynaptic kainate receptors on AMPA receptor-mediated synaptic currents in the hippocampus (e.g., Kamiya and Ozawa 1998; Contractor et al. 2000). KA played an important role in the early pharmacological and structural differentiation of iGluRs. It was central to definitively establishing its cognate receptor family as a distinct pharmacological entity from AMPA receptors (or Quisqualate receptors, as they initially were denoted) in seminal studies from dorsal root ganglion sensory neurons (Agrawal and Evans 1986; Huettner 1990), which constitute a relatively unique population of neurons that predominantly express kainate receptors as their sole type of iGluR. KA was a critical tool in the iGluR pharmacologist's armamentarium for nearly a decade because of its unparalleled selectivity, commercial availability and low cost.

A variety of synthetic analogs of KA have been generated, but in large part these molecules have not been characterized as extensively on defined combinations of recombinant kainate or AMPA receptors or on neuronal iGluRs. Structure-activity relationship studies with the KA template indicated that the configurations at C2 and C4 and the composition of the C4 side-chain are particularly critical determinants of receptor selectivity. For example, dihydrokainate, which has a fully saturated C4 isopropenyl group, is a potent competitive substrate for electrogenic glutamate transporters rather than a high-affinity kainate receptor agonist

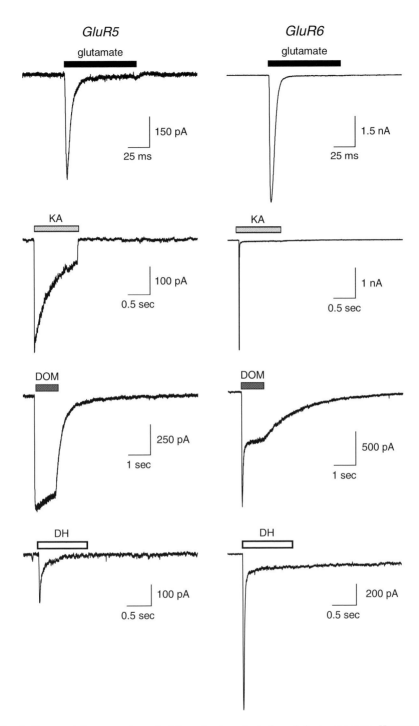

Fig. 4 Representative currents evoked by natural products from kainate receptors. Homomeric GluR5–2a (left column) or GluR6a (right column) receptors were expressed in HEK 293 cells. L-glutamate (10 mM), kainate (KA, 1 mM), domoate (DOM, 30 μM), or dysiherbaine (DH, 100 μM) were rapidly applied to cells in whole-cell voltage clamp recordings

(Johnston et al. 1979; Shinozaki 1988), whereas *trans*-2-carboxy-3-pyrrolidineacetic acid, which lacks the C4 group entirely, exhibits agonist activity on NMDA receptors (Tsai et al. 1988). Also, high affinity binding to and agonist activity on human recombinant GluR6 receptors was maintained in a variety of analogs with aryl substitutions of the C4 side-chain, but stereochemical reversal of the C4 position greatly reduced affinity for the receptor subunit (Cantrell et al. 1996). The natural terrestrial toxin acromelic acid is a C4 aryl-substituted KA analog (Konno et al. 1988) and thus may show a similar high affinity for a subset of kainate receptor subunits, although this has not been demonstrated formally.

The pharmacological actions of DOM are similar to that of KA; it is a high-affinity kainate receptor agonist and somewhat lower affinity AMPA receptor agonist. It binds, activates and desensitizes all homomeric and heteromeric kainate receptors, albeit to differing degrees and with distinct potencies. GluR5–2a receptors, which exhibit a particularly high affinity for domoate, gate a slowly desensitizing current upon activation by the marine toxin, whereas GluR6a receptors rapidly desensitize to a stable equilibrium current (Fig. 4) (Herb et al. 1992; Swanson et al. 1997). DOM is more potent and has a higher binding affinity for kainate receptors than KA (K_i values of ~2–60 nM) (reviewed in Hollmann and Heinemann 1994). Like KA, it is a partial agonist for AMPA receptors that elicits steady-state currents with a minimal desensitizing component. The actions of KA and DOM on non-NMDA receptors have been compared in a detailed review (Hampson and Manalo 1998).

Those natural analogs of DOM (the isodomoic acids) that have been examined generally display a lower potency and affinity for kainate and AMPA receptors. The radioligand binding affinities of isodomoic acid A and C for the GluR6 subunit are ~40- and ~240-fold lower than that that of DOM, respectively (Holland et al. 2005; Sawant et al. 2007); consistent with this observation, isodomoic acids A and C were less effective than DOM in reducing hippocampal population spike amplitudes (an effect known to be mediated predominantly by neuronal kainate receptors) (Sawant et al. 2007). Isodomoic acids D, E, and F have 5–280-fold lower affinities for high-affinity KA binding sites in the rat brain, which (at low radioligand concentrations) primarily arise from kainate receptors (Hampson et al. 1992).

In addition to its central importance to iGluR pharmacology research, KA has been widely used as an excitotoxic agent in behavioral and neuropathological studies (Nadler 1979, 1981; Ben-Ari 1985; Ben-Ari and Cossart 2000). Acute administration of kainoids induces characteristic acute behavioral changes in rodents, including stereotyped movement such as scratching behavior, head bobbing, and frequent grooming after ~10 min (Sperk 1994). The symptoms progress into more frequent and violent behaviors and, dependent upon the concentration of the toxins, animals display clonic whole body convulsions. At high doses, animals die after severe seizure episodes similar to those observed in humans. Repeated administration of KA will induce a permanent hyperexcitable state in animals marked by recurring convulsions, thought to mimic status epilepticus in humans (Ben-Ari and Cossart 2000). The seizurogenic action of DOM is more potent than that of KA, and it can induce long-lasting status epilepticus persisting for hours in mice (Chiamulera et al. 1992; Sakai et al. 2001b).

The neuroanatomical alterations and damage to the limbic regions produced by kainoid injection into rodents or ingestion of environmental DOM by some marine organisms partially reproduces that observed in patients with mTLE (Ben-Ari 1985). The rodent kainate-induced neuropathology model continues to be used as one diagnostic assay in screening potential anti-epileptic drugs (Loscher 2002). A discussion of the extensive literature on this model is beyond the scope of this chapter, but it has been the subject of a number of excellent reviews and book chapters (Sperk 1994; Dudek et al. 2006; Ratte and Lacaille 2006).

4 AMPA/Kainate Receptor Ligands: Dysiherbaines

4.1 Natural Sources and Synthetic Analogs of Dysiherbaines

Dysiherbaine (DH) is the first member of a new structural class of marine-derived iGluR agonists with a high degree of specificity for kainate receptors. In the course of screening for new excitatory amino acids from marine benthic organisms, Sakai and co-workers found that an aqueous extract of a sponge initially identified as *Dysidea herbacea*, which later analysis revealed was instead *Lendenfeldia chodrodes* (Fig. 3e), exhibited potent convulsant activity when injected into mice. The active principal isolated from the sponge extract was unprecedented and was comprised of a functionalized perhydro furanopyrane skeleton furnished with a 2-aminopropanoic acid side-chain (Sakai et al. 1997, 2006). Similar to kainoids, the structure of L-glutamate was embedded in DH and thus the toxin can be considered a C di-substituted analogue of L-glutamate. Subsequent searches for related compounds from the same sponge species resulted in an isolation of neodysiherbaine A (NDH A), an analog of DH with similar convulsant activity in mice (Sakai et al. 2001a). In addition to DH and NDH A, several structurally novel betaines, denoted dysibetaine PP, CPa and CPb, were isolated from the same sponge (Sakai et al. 2004). Weak affinity for NMDA and kainate receptors was observed in ligand-binding assays, but the pharmacological activity of the dysibetaines is not well-characterized beyond these preliminary results. The production of this array of structurally unusual molecules in the marine sponge underscores its diverse biosynthetic machinery and suggests that additional bioactive molecules await discovery (Sakai et al. 2006).

Recently Sakai and co-workers examined the localization of DH within the sponge tissue using immunohisto- and immunocytochemical techniques. Molecular analysis of the ribosomal DNA sequence resulted revealed that the taxonomy of the sponge was in fact *Lendenfeldia chodrodes* rather than *D. herbacea*. Moreover, localization of DH using a selective antibody found the toxin exclusively in the cells of endosymbiotic cyanobacteria, of *Synechosystis* sp. (Fig. 3f), suggesting that DH is in fact a metabolite of the cyanobacteria rather than the sponge itself (Sakai et al. 2008).

Because of its intriguing structural and biological features, intense efforts were undertaken towards the synthesis of DH. The first total synthesis by Hatakeyama and co-workers confirmed the proposed structure and absolute stereochemistry of DH (Masaki et al. 2000). To date, four total and one formal syntheses of DH, four total syntheses of NDH A, and structure-activity relationship (SAR) studies of NDH A have been reported (Masaki et al. 2000; Sasaki et al. 2000, 2007; Snider and Hawryluk 2000; Phillips and Chamberlin 2002; Lygo et al. 2005; Takahashi et al. 2006). A variety of DH analogues have been described, although only DH and NDH A are natural products derived from the sponge (Sasaki et al. 1999, 2006; Cohen et al. 2006; Shoji et al. 2006). Extensive molecular pharmacological and electro-physiological characterizations, as well as in vivo pharmacology, demonstrate that DH and its structural analogues are a new generation of excitatory amino acids with distinct receptor selectivity and agonist actions as described in the next section.

4.2 Biological Activities of Dysiherbaines

DH and neoDH are extraordinarily potent convulsants with high-affinity agonist activity on mammalian kainate receptors, a lesser potency for AMPA receptors, and (in the case of DH) an extremely weak activity on mGlu5 metabotropic glutamate receptors (Sakai et al. 2001b). Their pharmacological specificity for different KAR subunits diverges significantly from kainoids, which likely underlies their marked convulsant activity. Whereas KA exhibits highest affinity for the KA1 and KA2 subunits, DH and neoDH instead bind with very high affinity to GluR5, GluR6 and GluR7 KAR subunits (K_i values of ~0.5–1.5 nM) (Sakai et al. 2001b) but only have very weak interaction with KA2 subunits (K_i value of 4.3 μM, comparable to their affinity for AMPA receptor subunits) (Swanson et al. 2002). The action of DH on homomeric GluR5 and GluR6 receptors is unusually long-lived because the marine toxin effectively promotes a stable, desensitized conformation of the receptor, which can prevent unbinding of the agonist and subsequent re-activation by agonists (Swanson et al. 2002). While the nanomolar binding affinities exhibited by these molecules for the "primary" KAR subunits are indeed quite high, they are not so high as to suggest that the ligand-receptor interaction would be effectively irrevers-ible (as is the case of DH and homomeric GluR5 receptors, for example). Receptors composed of both high- and low-affinity subunits, and in particular GluR5/KA2 receptors, exhibit a further twist in their biophysical response to DH. Application of DH to GluR5/KA2 receptors elicits a slowly desensitizing current, as is typical for many agonists, but upon removal of the agonist from the bathing solution a slowly developing, steady-state current emerges that arises from the stable association of DH with GluR5 subunits, resulting in a tonic partial activation of the heteromeric GluR5/KA2 receptors (Swanson et al. 2002). Thus, the pharmacological activity of the DH molecules on KARs is critically determined by the subunit composition of the receptors, which can be complex and which is not well understood at the molecular level. A more detailed review of this topic can be found in Sakai et al.

(2006). Kainate receptors have diverse compositions in the mammalian brain, and therefore DH will impact neuronal function dependent upon a variety of factors, including toxin concentration and neuronal site of action (Sakai et al. 2001b).

DH has been shown to be the most potent seizurogenic excitatory amino acid isolated from natural sources (Fig. 5a) (Sakai et al. 2001b). The convulsant activity of DH was found to be approximately six-fold more potent than that of DOM (Table 1). Seizure behaviors induced by injection of DH in mice were chiefly distinguished from those elicited by kainoids in the duration of the status epilepticus. Mice receiving DH

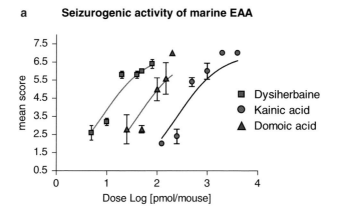

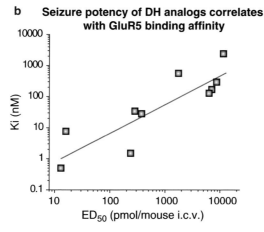

Fig. 5 Seizure activity of marine excitatory amino acids and synthetic analogs of neodysiherbaine A. (**a**) Seizure behaviors induced by i.c.v. injection of dysiherbaine, kainic acid, and domoic acid in mice were graded using a seven-point scale (Sakai et al. 2001b). Values, the mean scores ±S.E.M., were fit on a sigmoidal curve using Prism™ software. (**b**) Binding affinity at GluR5–2a subunits correlates with seizure potency ($r = 0.86$; $p < 0.01$). Linear correlation graph is plotted as Ki (nanomolar) versus ED_{50} (picomoles per mouse) after i.c.v. injection of the following compounds: DH, neoDH, MSVII19, 8-deoxy-neoDH, 9-deoxy-neoDH, 8-epi-neoDH, 9-epi-neoDH, 9-F-8-epineoDH, 2,4-epi-neoDH, and 4-epi-neoDH (Lash et al. 2007)

Table 1 Convulsant activities of marine excitatory amino acids

ED_{50} compound	i.c.v. (pmol/mouse)	i.p. mg/kg
Dysiherbaine[a]	6	0.97
Neodysiherbaine A[b]	15	
Kainic acid[a]	280	
Domoic acid[a]	34	5.7
t-HPIS[c]	20	
Cribronic acid[c]	29	
AMPA[a]	240	
NMDA[a]	430	

[a]Sakai et al. (2001b)
[b]Sakai et al. (2001a)
[c]Sakai et al. (2003)

(40 pmol/mouse, i.c.v. or 1.6 mg/kg, i.p) experienced severe whole body convulsions for more than 3 h, which then stabilized into periodic recurrent seizures. This state, which was not replicated by KA or DOM, lasted for more than 24 h. It is possible that this unique behavior arises from stable binding of the toxin with a subset of kainate receptors similar to that observed with recombinant receptor subunits. Indeed, a recent study showed that the seizurogenic potency of a diverse panel of DH-related molecules was strongly correlated with their affinity for the GluR5 KAR subunit (Fig. 5b), suggesting that activation of receptors comprised of this subunit primarily underlies toxin convulsant activity (Lash et al. 2007). It remains unknown whether the longer-term neuropathological consequences of seizure induction with DH closely resemble the well-characterized pattern of limbic structural reorganization, neuronal loss, and sclerosis produced in the kainate model of mTLE.

The structural determinants that underlie DH and neoDH affinity for kainate receptors have been explored in studies with synthetic analogs (Shoji et al. 2006). The C8 and C9 functional groups, in particular, confer specificity for the GluR5 and GluR6 KAR subunits. Elimination or epimerization of the C9 hydroxyl essentially eliminates binding to GluR6 subunits, as does similar alterations to the C8 group; in contrast, binding to GluR5 subunits is more tolerant to modification at C9 and is unaffected by elimination of the C8 moiety (Lash et al. 2007). This likely arises from the spatially larger binding pocket and the presence of favorable hydrophobic and polar interactions in the GluR5 binding domain as compared to GluR6 (Mayer 2005; Naur et al. 2005; Sanders et al. 2006). Interestingly, removal of both functional groups to produce dideoxy-neoDH (also known as MSVIII-19) fundamentally altered pharmacological activity; this molecule was an antagonist for homomeric GluR5 receptors, rather than an agonist (Sanders et al. 2005). Molecules with similar pharmacological profiles are under active examination for efficacy in a variety of animal models of neuropathologies, including epilepsy and chronic pain. MSVIII-19 is weak convulsant that additionally promotes an reversible unconscious state when injected intracerebroventricularly (i.c.v) in mice (Sasaki et al. 1999) but has relatively modest effects on motor function when introduced via intrathecal or intraperitoneal injection. Finally, epimerization of neoDH at the C4 position, which disrupts the glutamate congener in the molecule, reduced but did not eliminate affinity for GluR5 and GluR6 subunits (Lash et al. 2007). Given these precedents, further modifications to the DH template structure may yet produce molecules with distinct pharmacological profiles or activities on kainate receptor subunits.

5 NMDA Receptor Ligands: Amino Acids

5.1 Natural Sources of Amino Acid Ligands Acting on NMDA Receptors

Several compounds with selectivity for NMDA receptors have been identified from marine organisms. NMDA itself was detected from foot muscle of blood shell, *Scapharca broughtonii* (Sato et al. 1987). More recently, endogenous NMDA was discovered in the tunicate *Ciona intestinalis* (D'Aniello et al. 2003), where it is was biosynthesized from precursor D-aspartate; in the tunicate gonads, NMDA induced synthesis of gonadotropin-releasing hormone (GnRH), which in turn led to production of sex steroid hormones. A survey in marine algae for *N*-methyl aspartic acid (with an unspecified stereochemical configuration) found that eight out of 42 algae collected contained the compound (Sato et al. 1996); the physiological function of these amino acids in the algae is unknown.

Two 4,5-substituted analogs of pipecolic acid with activity on NMDA receptors have been isolated: cribronic acid [(2S,4R,5R)-5-hydroxy-4-sulfooxy-piperidine2-carboxylic acid], from the Palauan sponge *Cribrochalina olemda*, and (2S,4S) 4-sulfooxy-piperidine-2-carboxylic acid (*trans*-4-hydoroxypipecolic acid sulfate, *t*-HPIS) from the Micronesian sponges *Axynella carteri* and *Stylotella aurantium* (Sakai et al. 2003). Cribronic acid was a new compound while *t*-HPIS had been isolated previously from the legume *Peltophorum africanum* and characterized as NMDA agonist (Evans et al. 1985; Moroni et al. 1995).

Lophocladines are alkaloids with 2,7-naphthyridine skeletons isolated from red algae *Lophocladia* sp. collected in the Fijian Islands. One of the isolates, Lophocladine A, was shown to have affinity for the MK-801 binding site in the channel pore of NMDA receptors, suggesting that this compound might represent a novel class of naturally occurring small-molecule NMDA receptor antagonists. Validation of this possibility awaits physiological studies. A closely related analog, Lophocladine B, did not show equivalent affinity for NMDA receptors; rather, it inhibited microtubule formation and was cytotoxic (Gross et al. 2006).

5.2 Biological Activities of Natural NMDA Receptor Agonists

Both cribronic acid and *t*-HPIS were potent convulsants when injected i.c.v. in mice, producing dose-dependent behaviors, from running, jumping, and tonic extension to lethal convulsions, with ED_{50} values of ~20–30 pmol/mouse (Sakai et al. 2003). *t*-HPIS and cribronic acid displaced CGP 39653, a ligand for the glutamate binding site on NMDA receptors, from rat cerebrocortical membrane preparations with IC_{50} values of 214 nM and 83 nM, respectively. Neither compound displaced radiolabeled ligands from AMPA and kainate receptors. The agonist activity of *t*-HPIS on NMDA receptors was confirmed earlier in mouse cortical wedge preparations, in which the molecule caused dose-dependent depolarizations that were reduced by AP-5, an NMDA-receptor antagonist (Moroni et al. 1995). The relative depolarization potency of *t*-HPIS was about 5 times that of NMDA in the cortical preparation. Nothing is known regarding the NMDA receptor selectivity of these compounds. Interestingly, structurally related three- and four-substituted pipecolic acid analogs act as potent NMDA receptor antagonists and have been modified and studied extensively in pursuit of neurotherapeutic drugs (e.g., *cis*-4-phosphonomethyl-2-piperidine carboxylic acid, CGS 19755, Selfotel) (Lehmann et al. 1988). Thus far, however, this pharmacological approach has not proven beneficial in clinical trials for stroke mediation, and instead have tended to exacerbate neurotoxicity associated with ischemia (Davis et al. 2000).

6   NMDA Receptor Ligands: Conantokins

6.1   *Natural Sources of Conantokins*

Venoms from fish hunting snails, a genus of Conus, are a rich source of diverse neuroactive peptides targeting various voltage-gated ion channels, neurotransmitter receptors and transporters (Terlau and Olivera 2004). *Conus* peptides are categorized into a variety of families based on pharmacological targets and structural characteristics. Until very recently, the conantokin family of cone snail peptides contained four members, conantokin-G, -L, -R and –T (con-G, con-L, con-R, and con-T), which were shown to act as NMDA receptor antagonists (reviewed in Prorok and Castellino 2007). Conantokins are relatively unusual because they lack the disulfide bridges critical for structural integrity in most conopeptides and because the conantokins contain four to five γ-carboxyglutamates, a modified amino acid, in their structure. These first four conantokins range in size from 17 (Con-G) to 27 amino acids (Con-R). Con-G, from *Conus geographus* (Fig. 3g), was the first member of the family discovered and was isolated on the basis of its unusual bioactivity in mice: it produced a sleep-like state (McIntosh et al. 1984). This "sleeper peptide" was proposed initially to target NMDA receptors based on indirect biochemical assays (Mena et al. 1990) and later confirmed using physiological recordings from NMDA receptor channels (Hammerland et al. 1992). Con-T was isolated from *Conus tulipa* based on similar behavioral effects ("sleep" induction) in mice and found to have structural similarities to con-G (Haack et al. 1990); namely the initial glycine-glutamate-γ-carboxyglutamate-γ-carboxyglutamate residues were conserved in the two peptides. This sequence is also present in con-R, from *Conus radiatus* (White et al. 2000), and con-L, from *Conus lynceus* (Jimenez et al. 2002). A new, closely related series of conopeptides targeting NMDA receptors, conantokin-Pr1 to -Pr3 (con-Pr1 to -Pr3), was recently discovered from *Conus parius* (Teichert et al. 2007). Notably, this species of cone snail is the first whose venom contains multiple conantokin peptides. They diverge structurally from the four original conantokins, primarily in that the con-Pr toxins have only three γ-carboxyglutamate residues and two of the members (con-Pr2 and con-Pr3) contain another modified amino acid, 4-*trans*-hydroxyproline. Con-Pr peptides also induce a sleep-like state in mice (Teichert et al. 2007).

6.2 Biological Activities of Conantokins

Conantokins are peptide antagonists selective for NMDA receptors (Haack et al. 1990; Mena et al. 1990; Hammerland et al. 1992; Jimenez et al. 2002; Prorok and Castellino 2007). Con-G, the most extensively characterized conantokin, has appeared to have both competitive and noncompetitive antagonist activity in different assays for NMDA receptor function (Prorok and Castellino 2007). Competitive antagonism occurs at the glutamate binding site on the NR2 subunit (Hammerland et al. 1992; Donevan and McCabe 2000; Wittekindt et al. 2001), with an IC_{50} of ~0.5–1 µM for inhibition of NMDA-evoked currents in cultured mouse cortical neurons or for inhibition of synaptic NMDA-EPSCs in CA1 pyramidal neurons in rat hippocampal slice preparations (Donevan and McCabe 2000; Barton et al. 2004). The molecular binding site of con-G could be heterotopic, because the inhibitory activity is enhanced by polyamines such as spermine (Donevan and McCabe 2000), which binds to an extracellular allosteric modulatory site. Con-G also exhibits a high degree of subunit selectivity; inhibition of NR1/NR2B NMDA receptors is potent, whereas NR1/NR2A receptors (or those containing NR2C or NR2D) are relatively unaffected by the peptide (Donevan and McCabe 2000; Klein et al. 2001). This degree of selectivity does not extend to all conantokins, however, as con-R and con-T inhibit both NR2A- and NR2B-containing receptors (White et al. 2000; Klein et al. 2001). Conantokins, and in particular con-G, have attracted significant attention for their potential as therapeutics in a variety of neuropathologies (Layer et al. 2004). Con-G and con-T are effective antinociceptive agents in models of chronic pain (Malmberg et al. 2003), although their peptide structures restrict potential routes of administration. Con-G and con-R, but not con-L, also show anticonvulsant efficacy in a number of mouse seizures models (White et al. 2000; Jimenez et al. 2002; Barton et al. 2003). Con-G, which in its pre-clinical form is known as CGX-1007, also has neuroprotective effects in stroke models (Williams et al. 2000, 2003). While CGX-1007 was found to be safe in Phase I clinical trials, further clinical trials have not been disclosed (Olivera 2006).

Con-Pr peptides also induced the characteristic sleep-like state after injection observed decades earlier with con-G (McIntosh et al. 1984) and inhibited recombinant

NR1/NR2 NMDA receptors expressed in *Xenopus* oocytes, with varying degrees of selectivity for NR2B-containing receptors (Teichert et al. 2007). Their pharmacological profiles were distinct from those of the earlier conantokins and consequently will provide additional tools and clues for understanding how the structure of these unusual peptides determines their functional activity and subunit specificity.

7 Unpurified Bioactive Extracts

Several studies found crude extracts with bioactivity that appeared to target ionotropic glutamate receptors. For example, extracts of cultured bacteria associated with the marine sponge *Halichondria panacea* activate rat cortical NMDA receptors; the active principle(s) were not further isolated, however (Perovic et al. 1998). Garateix and colleagues carried out an ecologically-inspired search for iGluR ligands from marine organisms that prey on crustaceans, leading to the discovery of bioactive peptide-containing fractions from a sea anemone, *Phyllactis flosculifera* (Garateix et al. 1996). Peptide fractions of extracts from the animal diminished both the excitatory and the inhibitory responses to glutamate agonists in neurons of the land snail *Zachrysia guanensisin*. Similarly, a crude extract from the sea anemone *Bunodosoma caissarum* induced convulsions following intracerebroventricular (i.c.v.) injection in mice. The convulsion was suppressed by chlorokynurenic acid, an antagonist of the glycine site on the NMDA receptor (Gondran et al. 2002). Isolation and structures for these sea anemone products have not been reported to date.

8 Conclusion

Marine-derived compounds have played key roles in iGluR research. The recent discovery of the dysiherbaines and the con-Pr peptides, and the variety of bioactive extracts with unknown active principles, suggest that additional and novel molecules await the attention of neuroscience researchers. One of the central challenges for finding new molecules lies in the very early steps of characterizing bioactivity. While stereotypic seizure behavior and convulsions or sleep-inducing activity are obvious behavioral responses that lend insight into potential biological activity, extracts containing bioactive molecules that elicit less dramatic responses, but which have novel pharmacological profiles, could potentially be overlooked. It is clear, however, that synthetic modification of natural analogs can dramatically alter their pharmacological action (see, for example the DH analog MSVIII-19), and thus it is worthwhile to pursue even those active principles lacking obvious clinical application (such as convulsants). A revived appreciation of the potential utility of drugs from the sea and other natural sources is evident in the form of new initiatives, from funding bodies such as the US National Institutes of Health, which hope to spur development of higher-throughput identification and isolation of natural source compounds that could impact human health. It is likely that these efforts will produce molecules with new structures and specificities for iGluRs to supplement those described in this chapter.

Acknowledgments We thank Professor Y. Kotaki (Kitasato University, Japan), Professor K. Koike (Hiroshima University, Japan), and Mr. K. Nomura for images of natural organisms and Dr. Anis Contractor (Northwestern University School of Medicine, US) for helpful comments on the text. G. T. Swanson is supported by a grant from the National Institutes for Health (R01 NS44322). R. Sakai was financially supported in part by a Grant-in-Aid for Scientific Research from the Ministry of Education, Culture, Sports, Science and Technology, Japan (15580183 and 17380125).

References

Aamodt SM, Constantine-Paton M (1999) The role of neural activity in synaptic development and its implications for adult brain function. Adv Neurol 79:133–144.

Agrawal SG, Evans RH (1986) The primary afferent depolarizing action of kainate in the rat. Br J Pharmacol 87:345–355.

Alt A, Weiss B, Ornstein PL, Gleason SD, Bleakman D, Stratford RE, Jr., Witkin JM (2007) Anxiolytic-like effects through a GLU(K5) kainate receptor mechanism. Neuropharmacology 52:1482–1487.

Araki K, Meguro H, Kushiya E, Takayama C, Inoue Y, Mishina M (1993) Selective expression of the glutamate receptor channel delta 2 subunit in cerebellar Purkinje cells. Biochem Biophys Res Commun 197:1267–1276.

Armstrong N, Sun Y, Chen GQ, Gouaux E (1998) Structure of a glutamate-receptor ligand-binding core in complex with kainate. Nature 395:913–917.

Ashworth TS, Brown EG, Roberts FM (1972) Biosynthesis of willardiine and isowillardiine in germinating pea seeds and seedlings. Biochem J 129:897–905.

Awobuluyi M, Yang J, Ye Y, Chatterton JE, Godzik A, Lipton SA, Zhang D (2007) Subunit-specific roles of glycine-binding domains in activation of NR1/NR3 *N*-methyl-D-aspartate receptors. Mol Pharmacol 71:112–122.

Bahn S, Volk B, Wisden W (1994) Kainate receptor gene expression in the developing rat brain. J Neurosci 14:5525–5547.

Bargu S, Powell CL, Coale SL, Busman M, Doucette GJ, Silver MW (2002) Krill: a potential vector for domoic acid in marine food webs. Mar Ecol Prog Ser 237:209–216.

Bargu S, Lefebvre K, Silver MW (2006) Effect of dissolved domoic acid on the grazing rate of krill *Euphausia pacifica*. Mar Ecol Prog Ser 312:169–175.

Barton ME, Peters SC, Shannon HE (2003) Comparison of the effect of glutamate receptor modulators in the 6 Hz and maximal electroshock seizure models. Epilepsy Res 56:17–26.

Barton ME, White HS, Wilcox KS (2004) The effect of CGX-1007 and CI-1041, novel NMDA receptor antagonists, on NMDA receptor-mediated EPSCs. Epilepsy Res 59:13–24.

Bates SS (2000) Domoic-acid-producing diatoms: another genus added! J Phycol 36:978–983.

Bates SS, Bird CJ, Defreitas ASW, Foxall R, Gilgan M, Hanic LA, Johnson GR, McCulloch AW, Odense P, Pocklington R, Quilliam MA, Sim PG, Smith JC, Rao DVS, Todd ECD, Walter JA, Wright JLC (1989) Pennate diatom *Nitzschia-Pungens* as the primary source of domoic acid, a toxin in shellfish from eastern Prince Edward Island, Canada. Can J Fish Aquat Sci 46:1203–1215.

Ben-Ari Y (1985) Limbic seizure and brain damage produced by kainic acid: mechanisms and relevance to human temporal lobe epilepsy. Neuroscience 14:375–403.

Ben-Ari Y, Cossart R (2000) Kainate, a double agent that generates seizures: two decades of progress. Trends Neurosci 23:580–587.

Bettler B, Boulter J, Hermans-Borgmeyer I, O'Shea-Greenfield A, Deneris ES, Moll C, Borgmeyer U, Hollmann M, Heinemann S (1990) Cloning of a novel glutamate receptor subunit, GluR5: expression in the nervous system during development. Neuron 5:583–595.

Bettler B, Egebjerg J, Sharma G, Pecht G, Hermans-Borgmeyer I, Moll C, Stevens CF, Heinemann S (1992) Cloning of a putative glutamate receptor: a low affinity kainate-binding subunit. Neuron 8:257–265.

Bill BD, Cox FH, Horner RA, Borchert JA, Trainer VL (2006) The first closure of shellfish harvesting due to domoic acid in Puget Sound, Washington, USA. Afr J Mar Sci 28:435–440.

Blagbrough IS, Moya E, Taylor S (1994) Polyamines and polyamine amides from wasps and spiders. Biochem Soc Trans 22:888–893.

Bleakman D, Alt A, Nisenbaum ES (2006) Glutamate receptors and pain. Semin Cell Dev Biol 17:592–604.

Bliss TV, Collingridge GL (1993) A synaptic model of memory: long-term potentiation in the hippocampus. Nature 361:31–39.

Bortolotto ZA, Clarke VR, Delany CM, Parry MC, Smolders I, Vignes M, Ho KH, Miu P, Brinton BT, Fantaske R, Ogden A, Gates M, Ornstein PL, Lodge D, Bleakman D, Collingridge GL (1999) Kainate receptors are involved in synaptic plasticity. Nature 402:297–301.

Bruce M, Bukownik R, Eldefrawi AT, Eldefrawi ME, Goodnow R, Jr., Kallimopoulos T, Konno K, Nakanishi K, Niwa M, Usherwood PN (1990) Structure-activity relationships of analogues of the wasp toxin philanthotoxin: non-competitive antagonists of quisqualate receptors. Toxicon 28:1333–1346.

Burnashev N, Zhou Z, Neher E, Sakmann B (1995) Fractional calcium currents through recombinant GluR channels of the NMDA, AMPA and kainate receptor subtypes. J Physiol (Lond) 485:403–418.

Calaf R, Barlatier A, Garçon D, Balansard G, Pellegrini M, Reynaud J (1989) Isolation of an unknown kainic peptide from the red alga *Alsidium helminthocorton*. J Appl Phycol 1:257–266.

Cantrell BE, Zimmerman DM, Monn JA, Kamboj RK, Hoo KH, Tizzano JP, Pullar IA, Farrell LN, Bleakman D (1996) Synthesis of a series of aryl kainic acid analogs and evaluation in cells stably expressing the kainate receptor humGluR6. J Med Chem 39:3617–3624.

Castillo PE, Malenka RC, Nicoll RA (1997) Kainate receptors mediate a slow postsynaptic current in hippocampal CA3 neurons. Nature 388:182–186.

Cendes F, Andermann F, Carpenter S, Zatorre RJ, Cashman NR (1995) Temporal lobe epilepsy caused by domoic acid intoxication: evidence for glutamate receptor-mediated excitotoxicity in humans. Ann Neurol 37:123–126.

Chatterton JE, Awobuluyi M, Premkumar LS, Takahashi H, Talantova M, Shin Y, Cui J, Tu S, Sevarino KA, Nakanishi N, Tong G, Lipton SA, Zhang D (2002) Excitatory glycine receptors containing the NR3 family of NMDA receptor subunits. Nature 415:793–798.

Chiamulera C, Costa S, Valerio E, Reggiani A (1992) Domoic acid toxicity in rats and mice after intracerebroventricular administration: comparison with excitatory amino acid agonists. Pharmacol Toxicol 70:115–120.

Clark RB, Donaldson PL, Gration KA, Lambert JJ, Piek T, Ramsey R, Spanjer W, Usherwood PN (1982) Block of locust muscle glutamate receptors by delta-philanthotoxin occurs after receptor activations. Brain Res 241:105–114.

Clark RF, Williams SR, Nordt SP, Manoguerra AS (1999) A review of selected seafood poisonings. Undersea Hyperb Med 26:175–184.

Clayden J, Read B, Hebditch KR (2005) Chemistry of domoic acid, isodomoic acids, and their analogues. Tetrahedron 61:5713–5724.

Cohen JL, Limon A, Miledi R, Chamberlin AR (2006) Design, synthesis, and biological evaluation of a scaffold for iGluR ligands based on the structure of (−)-dysiherbaine. Bioorg Med Chem Lett 16:2189–2194.

Contractor A, Swanson GT, Sailer A, O'Gorman S, Heinemann SF (2000) Identification of the kainate receptor subunits underlying modulation of excitatory synaptic transmission in the CA3 region of the hippocampus. J Neurosci 20:8269–8278.

Contractor A, Swanson GT, Heinemann SF (2001) Kainate receptors are involved in short and long term plasticity at mossy fiber synapses in the hippocampus. Neuron 29:209–216.

Contractor A, Sailer AW, Darstein M, Maron C, Xu J, Swanson GT, Heinemann SF (2003) Loss of kainate receptor-mediated heterosynaptic facilitation of mossy-fiber synapses in KA2$^{-/-}$ mice. J Neurosci 23:422–429.

Cusack CK, Bates SS, Quilliam MA, Patching JW, Raine R (2002) Confirmation of domoic acid production by *Pseudo-nitzschia australis* (bacillariophyceae) isolated from Irish waters. J Phycol 38:1106–1112.

D'Aniello A, Spinelli P, De Simone A, D'Aniello S, Branno M, Aniello F, Fisher GH, Di Fiore MM, Rastogi RK (2003) Occurrence and neuroendocrine role of D-aspartic acid and N-methyl-D-aspartic acid in *Ciona intestinalis*. FEBS Lett 552:193–198.

Davis SM, Lees KR, Albers GW, Diener HC, Markabi S, Karlsson G, Norris J (2000) Selfotel in acute ischemic stroke: possible neurotoxic effects of an NMDA antagonist. Stroke 31:347–354.

Debonnel G, Beauchesne L, de Montigny C (1989) Domoic acid, the alleged "mussel toxin," might produce its neurotoxic effect through kainate receptor activation: an electrophysiological study in the dorsal hippocampus. Can J Physiol Pharmacol 67:29–33.

Dingledine R, Hume RI, Heinemann SF (1992) Structural determinants of barium permeation and rectification in non-NMDA glutamate receptor channels. J Neurosci 12:4080–4087.

Dingledine R, Borges K, Bowie D, Traynelis SF (1999) The glutamate receptor ion channels. Pharmacol Rev 51:7–62.

Donevan SD, McCabe RT (2000) Conantokin G is an NR2B-selective competitive antagonist of N-methyl-D-aspartate receptors. Mol Pharmacol 58:614–623.

Dudek FE, Clark S, Williams PA, Grabenstatter HL (2006) Kainate-induced Status Epilepticus: a chronic model of acquired epilepsy. In: Models of Seizures and Epilepsy (Pitkänen A, Schwartzkroin PA, Moshé SL, eds.). Burlington, MA: Elsevier Academic Press.

Egebjerg J, Bettler B, Hermans-Borgmeyer I, Heinemann S (1991) Cloning of a cDNA for a glutamate receptor subunit activated by kainate but not AMPA. Nature 351:745–748.

Estrada G, Villegas E, Corzo G (2007) Spider venoms: a rich source of acylpolyamines and peptides as new leads for CNS drugs. Nat Prod Rep 24:145–161.

Evans ML, Usherwood PN (1985) The effect of lectins on desensitisation of locust muscle glutamate receptors. Brain Res 358:34–39.

Evans SV, Shing TKM, Aplin RT, Fellows LE, Fleet GWJ (1985) Sulphate ester of *trans*-4-hydroxypipecolic acid in seeds of Peltophorum. Phytochemistry 24:2593–2596.

Fehling J, Green DH, Davidson K, Bolch CJ, Bates SS (2004) Domoic acid production by *Pseudo-nitzschia seriata* (bacillariophyceae) in Scottish waters. J Phycol 40:622–630.

Fisahn A, Heinemann SF, McBain CJ (2005) The kainate receptor subunit GluR6 mediates metabotropic regulation of the slow and medium AHP currents in mouse hippocampal neurones. J Physiol 562:199–203.

Frerking M, Ohliger-Frerking P (2002) AMPA receptors and kainate receptors encode different features of afferent activity. J Neurosci 22:7434–7443.

Frerking M, Schmitz D, Zhou Q, Johansen J, Nicoll RA (2001) Kainate receptors depress excitatory synaptic transmission at CA3- > CA1 synapses in the hippocampus via a direct presynaptic action. J Neurosci 21:2958–2966.

Fritz L, Quilliam MA, Wright JLC, Beale AM, Work TM (1992) An outbreak of domoic acid and poisoning attributed to the pennate diatom *Pseudonitzschia australis*. J Phycol 28:438–442.

Furukawa H, Singh SK, Mancusso R, Gouaux E (2005) Subunit arrangement and function in NMDA receptors. Nature 438:185–192.

Garateix A, Flores A, Garcia-Andrade JM, Palmero A, Aneiros A, Vega R, Soto E (1996) Antagonism of glutamate receptors by a chromatographic fraction from the exudate of the sea anemone *Phyllactis flosculifera*. Toxicon 34:443–450.

Geiger JR, Melcher T, Koh DS, Sakmann B, Seeburg PH, Jonas P, Monyer H (1995) Relative abundance of subunit mRNAs determines gating and Ca^{2+} permeability of AMPA receptors in principal neurons and interneurons in rat CNS. Neuron 15:193–204.

Gereau RW, Swanson GT, eds. (2008) The Glutamate Receptors. Totawa, NJ: Humana Press.

Gmelin R (1959) The free amino acids in the seeds of *Acacia willardiana* (Mimosaceae). Isolation of willardiin, a new plant amino acid which is probably L-beta-(3-uracil)-alpha-aminopropionic acid. Hoppe Seylers Z Physiol Chem 316:164–169.

Goldstein T, Mazet JA, Zabka TS, Langlois G, Colegrove KM, Silver M, Bargu S, Van Dolah F, Leighfield T, Conrad PA, Barakos J, Williams DC, Dennison S, Haulena M, Gulland FM (2008) Novel symptomatology and changing epidemiology of domoic acid toxicosis in California sea lions (*Zalophus californianus*): an increasing risk to marine mammal health. Proc Biol Sci 275(1632):267–276.

Gondran M, Eckeli AL, Migues PV, Gabilan NH, Rodrigues AL (2002) The crude extract from the sea anemone, *Bunodosoma caissarum* elicits convulsions in mice: possible involvement of the glutamatergic system. Toxicon 40:1667–1674.

Greger IH, Esteban JA (2007) AMPA receptor biogenesis and trafficking. Curr Opin Neurobiol 17:289–297.

Gross H, Goeger DE, Hills P, Mooberry SL, Ballantine DL, Murray TF, Valeriote FA, Gerwick WH (2006) Lophocladines, bioactive alkaloids from the red alga *Lophocladia sp*. J Nat Prod 69:640–644.

Haack JA, Rivier J, Parks TN, Mena EE, Cruz LJ, Olivera BM (1990) Conantokin-T. A gamma-carboxyglutamate containing peptide with *N*-methyl-D-aspartate antagonist activity. J Biol Chem 265:6025–6029.

Hammerland LG, Olivera BM, Yoshikami D (1992) Conantokin-G selectively inhibits *N*-methyl-D-aspartate-induced currents in *Xenopus* oocytes injected with mouse brain mRNA. Eur J Pharmacol 226:239–244.

Hampson DR, Manalo JL (1998) The activation of glutamate receptors by kainic acid and domoic acid. Nat Toxins 6:153–158.

Hampson DR, Huang XP, Wells JW, Walter JA, Wright JL (1992) Interaction of domoic acid and several derivatives with kainic acid and AMPA binding sites in rat brain. Eur J Pharmacol 218:1–8.

Hardy RW, Scott TM, Hatfield CL, Barnett HJ, Gauglitz EJ, Wekell JC, Eklund MW (1995) Domoic acid in rainbow trout (*Oncorhynchus mykiss*) feeds. Aquaculture 131:253–260.

Hart AC, Sims S, Kaplan JM (1995) Synaptic code for sensory modalities revealed by *C. elegans* GLR-1 glutamate receptor. Nature 378:82–85.

Hayashi Y, Shi SH, Esteban JA, Piccini A, Poncer JC, Malinow R (2000) Driving AMPA receptors into synapses by LTP and CaMKII: requirement for GluR1 and PDZ domain interaction. Science 287:2262–2267.

Herb A, Burnashev N, Werner P, Sakmann B, Wisden W, Seeburg PH (1992) The KA-2 subunit of excitatory amino acid receptors shows widespread expression in brain and forms ion channels with distantly related subunits. Neuron 8:775–785.

Hirano T (2006) Cerebellar regulation mechanisms learned from studies on GluRdelta2. Mol Neurobiol 33:1–16.

Holland PT, Selwood AI, Mountfort DO, Wilkins AL, McNabb P, Rhodes LL, Doucette GJ, Mikulski CM, King KL (2005) Isodomoic acid C, an unusual amnesic shellfish poisoning toxin from *Pseudo-nitzschia australis*. Chem Res Toxicol 18:814–816.

Hollmann M, Heinemann S (1994) Cloned glutamate receptors. Annu Rev Neurosci 17:31–108.

Hollmann M, Hartley M, Heinemann S (1991) Ca^{2+} permeability of KA-AMPA gated glutamate receptor channels depends on subunit composition. Science 252:851–853.

Huettner JE (1990) Glutamate receptor channels in rat DRG neurons: activation by kainate and quisqualate and blockade of desensitization by Con A. Neuron 5:255–266.

Huettner JE (2003) Kainate receptors and synaptic transmission. Prog Neurobiol 70:387–407.

Iino M, Koike M, Isa T, Ozawa S (1996) Voltage-dependent blockage of Ca(2+) permeable AMPA receptors by joro spider toxin in cultured rat hippocampal neurones. J Physiol 496:431–437.

Impellizzeri G, Mangiafico S, Oriente G, Piattelli M, Sciuto S, Fattorusso E, Magno S, Santacroce C, Sica D (1975) Amino acids and low-molecularweight carbohydrates of some marine red algae. Phytochemistry 14:1549–1557.

Iverson F, Truelove J (1994) Toxicology and seafood toxins: domoic acid. Nat Toxins 2:334–339.

Jaskolski F, Coussen F, Mulle C (2005) Subcellular localization and trafficking of kainate receptors. Trends Pharmacol Sci 26:20–26.

Jeffery B, Barlow T, Moizer K, Paul S, Boyle C (2004) Amnesic shellfish poison. Food Chem Toxicol 42:545–557.

Jiang L, Xu J, Nedergaard M, Kang J (2001) A kainate receptor increases the efficacy of GABAergic synapses. Neuron 30:503–513.

Jimenez EC, Donevan S, Walker C, Zhou LM, Nielsen J, Cruz LJ, Armstrong H, White HS, Olivera BM (2002) Conantokin-L, a new NMDA receptor antagonist: determinants for anti-convulsant potency. Epilepsy Res 51:73–80.

Jin XT, Pare JF, Raju DV, Smith Y (2006) Localization and function of pre- and postsynaptic kainate receptors in the rat globus pallidus. Eur J Neurosci 23:374–386.

Johnson JW, Ascher P (1987) Glycine potentiates the NMDA response in cultured mouse brain neurons. Nature 325:529–531.

Johnston GA, Kennedy SM, Twitchin B (1979) Action of the neurotoxin kainic acid on high affinity uptake of L-glutamic acid in rat brain slices. J Neurochem 32:121–127.

Kamiya H, Ozawa S (1998) Kainate receptor-mediated inhibition of presynaptic Ca^{2+} influx and EPSP in area CA1 of the rat hippocampus. J Physiol (Lond) 509:833–845.

Kamiya H, Ozawa S (2000) Kainate receptor-mediated presynaptic inhibition at the mouse hippocampal mossy fibre synapse. J Physiol (Lond) 523:653–665.

Klein RC, Prorok M, Galdzicki Z, Castellino FJ (2001) The amino acid residue at sequence position 5 in the conantokin peptides partially governs subunit-selective antagonism of recombinant N-methyl-D-aspartate receptors. J Biol Chem 276:26860–26867.

Konno K, Hashimoto K, Ohfune Y, Shirahama H, Matsumoto T (1988) Acromelic acids A and B. Potent neuroexcitatory amino acids isolated from Clitocybe acromelalga. J Am Chem Soc 110:4807–4815.

Kotaki Y, Koike K, Sato S, Ogata T, Fukuyo Y, Kodama M (1999) Confirmation of domoic acid production of Pseudo-nitzschia multiseries isolated from Ofunato Bay, Japan. Toxicon 37:677–682.

Kotaki Y, Koike K, Yoshida M, Van Thuoc C, Huyen NTM, Hoi NC, Fukuyo Y, Kodama M (2000) Domoic acid production in Nitzschia sp (Bacillariophyceae) isolated from a shrimp-culture pond in Do Son, Vietnam. J Phycol 36:1057–1060.

Kotaki Y, Lundholm N, Onodera H, Kobayashi K, Bajarias FFA, Furio EF, Iwataki M, Fukuyo Y, Kodama M (2004) Wide distribution of Nitzschia navis-varingica, a new domoic acid-producing benthic diatom found in Vietnam. Fish Sci 70:28–32.

Kung SS, Wu YM, Chow WY (1996) Characterization of two fish glutamate receptor cDNA molecules: absence of RNA editing at the Q/R site. Brain Res Mol Brain Res 35:119–130.

Kuryatov A, Laube B, Betz H, Kuhse J (1994) Mutational analysis of the glycine-binding site of the NMDA receptor: structural similarity with bacterial amino acid-binding proteins. Neuron 12:1291–1300.

Lash LL, Sanders JM, Akiyama N, Shoji M, Postila P, Pentikainen OT, Sasaki M, Sakai R, Swanson GT (2007) Novel analogs and stereoisomers of the marine toxin neodysiherbaine with specificity for kainate receptors. J Pharmacol Exp Ther 324:484–496.

Laube B, Hirai H, Sturgess M, Betz H, Kuhse J (1997) Molecular determinants of agonist discrimination by NMDA receptor subunits: analysis of the glutamate binding site on the NR2B subunit. Neuron 18:493–503.

Lauri SE, Bortolotto ZA, Bleakman D, Ornstein PL, Lodge D, Isaac JT, Collingridge GL (2001) A critical role of a facilitatory presynaptic kainate receptor in mossy fiber LTP. Neuron 32:697–709.

Laycock MV, de Freitas ASW, Wright JLC (1989) Glutamate agonists from marine algae. J Appl Phycol 1:113–122.

Layer RT, Wagstaff JD, White HS (2004) Conantokins: peptide antagonists of NMDA receptors. Curr Med Chem 11:3073–3084.

Lefebvre KA, Powell CL, Busman M, Doucette GJ, Moeller PD, Silver JB, Miller PE, Hughes MP, Singaram S, Silver MW, Tjeerdema RS (1999) Detection of domoic acid in northern anchovies and California sea lions associated with an unusual mortality event. Nat Toxins 7:85–92.

Lefebvre KA, Dovel SL, Silver MW (2001) Tissue distribution and neurotoxic effects of domoic acid in a prominent vector species, the northern anchovy *Engraulis mordax*. Marine Biology 138:693–700.

Lefebvre KA, Bargu S, Kieckhefer T, Silver MW (2002a) From sanddabs to blue whales: the pervasiveness of domoic acid. Toxicon 40:971–977.

Lefebvre KA, Silver MW, Coale SL, Tjeerdema RS (2002b) Domoic acid in planktivorous fish in relation to toxic *Pseudo-nitzschia* cell densities. Marine Biology 140:625–631.

Lefebvre KA, Noren DP, Schultz IR, Bogard SM, Wilson J, Eberhart BT (2007) Uptake, tissue distribution and excretion of domoic acid after oral exposure in coho salmon (*Oncorhynchus kisutch*). Aquat Toxicol 81:266–274.

Lehmann J, Hutchison AJ, McPherson SE, Mondadori C, Schmutz M, Sinton CM, Tsai C, Murphy DE, Steel DJ, Williams M, et al. (1988) CGS 19755, a selective and competitive *N*-methyl-D-aspartate-type excitatory amino acid receptor antagonist. J Pharmacol Exp Ther 246:65–75.

Lerma J (2006) Kainate receptor physiology. Curr Opin Pharmacol 6:89–97.

Lomeli H, Wisden W, Köhler M, Keinänen K, Sommer B, Seeburg PH (1992) High-affinity kainate and domoate receptors in rat brain. FEBS Lett 307:139–143.

Lomeli H, Sprengel R, Laurie DJ, Kohr G, Herb A, Seeburg PH, Wisden W (1993) The rat delta-1 and delta-2 subunits extend the excitatory amino acid receptor family. FEBS Lett 315:318–322.

Lomeli H, Mosbacher J, Melcher T, Hoger T, Geiger JR, Kuner T, Monyer H, Higuchi M, Bach A, Seeburg PH (1994) Control of kinetic properties of AMPA receptor channels by nuclear RNA editing. Science 266:1709–1713.

Loscher W (2002) Animal models of epilepsy for the development of antiepileptogenic and disease-modifying drugs. A comparison of the pharmacology of kindling and post-status epilepticus models of temporal lobe epilepsy. Epilepsy Res 50:105–123.

Lundholm N, Moestrup Ø (2000) Morphology of the marine diatom *Nitzschia navis-varingica*, sp. Nov. (bacillariophyceae), another producer of the neurotoxin domoic acid. J Phycol 36:1162–1174.

Lundholm N, Hansen PJ, Kotaki Y (2005) Lack of allelopathic effects of the domoic acid-producing marine diatom *Pseudo-nitzschia multiseries*. Mar Ecol Prog Ser 288:21–33.

Lygo B, Slack D, Wilson C (2005) Synthesis of neodysiherbaine. Tetrahedron Lett 46:6629–6632.

Lynch G, Gall CM (2006) Ampakines and the threefold path to cognitive enhancement. Trends Neurosci 29:554–562.

MacDermott AB, Mayer ML, Westbrook GL, Smith SJ, Barker JL (1986) NMDA-receptor activation increases cytoplasmic calcium concentration in cultured spinal cord neurones. Nature 321:519–522.

Madden DR (2002) The structure and function of glutamate receptor ion channels. Nat Rev Neurosci 3:91–101.

Maeda M, Kodama T, Tanaka T, Yoshizumi H, Takemoto T, Nomoto K, Fujita T (1986) Structures Of isodomoic acid-A, acid-B And acid-C, novel insecticidal amino-acids from the red alga *Chondria armata*. Chem Pharm Bull 34:4892–4895.

Maeda M, Kodama T, Tanaka T, Yoshizumi H, Takemoto T, Nomoto K, Fujita T (1987) Structures of domoilactone A and B, novel amino acids from the red alga. Tetrahedron Lett 28:633–636.

Magee JC, Johnston D (1997) A synaptically controlled, associative signal for Hebbian plasticity in hippocampal neurons. Science 275:209–213.

Malmberg AB, Gilbert H, McCabe RT, Basbaum AI (2003) Powerful antinociceptive effects of the cone snail venom-derived subtype-selective NMDA receptor antagonists conantokins G and T. Pain 101:109–116.

Maricq AV, Peckol E, Driscoll M, Bargmann CI (1995) Mechanosensory signalling in *C. elegans* mediated by the GLR-1 glutamate receptor. Nature 378:78–81.

Masaki H, Maeyama J, Kamada K, Esumi T, Iwabuchi Y, Hatakeyama S (2000) Total synthesis of (-)-dysiherbaine. J Am Chem Soc 122:5216–5217.

Mayer ML (2005) Crystal structures of the GluR5 and GluR6 ligand binding cores: molecular mechanisms underlying Kainate receptor selectivity. Neuron 45:539–552.

Mayer ML, Westbrook GL (1987) Permeation and block of *N*-methyl-D-aspartic acid receptor channels by divalent cations in mouse cultured central neurones. J Physiol 394:501–527.

Mayer ML, Benveniste M, Patneau DK, Vyklicky L, Jr. (1992) Pharmacologic properties of NMDA receptors. Ann N Y Acad Sci 648:194–204.

McCormick J, Li Y, McCormick K, Duynstee HI, van Engen AK, van der Marel GA, Ganem B, van Boom JH, Meinwald J (1999) Structure and total synthesis of HF-7, a neuroactive glyconucleoside disulfate from the funnel-web spider *Hololena curta*. J Am Chem Soc 121:5661–5665.

McIntosh JM, Olivera BM, Cruz LJ, Gray WR (1984) Gamma-carboxyglutamate in a neuroactive toxin. J Biol Chem 259:14343–14346.

Melyan Z, Wheal HV, Lancaster B (2002) Metabotropic-mediated kainate receptor regulation of I$_{\text{sAHP}}$ and excitability in pyramidal cells. Neuron 34:107–114.

Melyan Z, Lancaster B, Wheal HV (2004) Metabotropic regulation of intrinsic excitability by synaptic activation of kainate receptors. J Neurosci 24:4530–4534.

Mena EE, Gullak MF, Pagnozzi MJ, Richter KE, Rivier J, Cruz LJ, Olivera BM (1990) Conantokin-G: a novel peptide antagonist to the *N*-methyl-Daspartic acid (NMDA) receptor. Neurosci Lett 118:241–244.

Moloney MG (1998) Excitatory amino acids. Nat Prod Rep 15:205–219.

Moloney MG (1999) Excitatory amino acids. Nat Prod Rep 16:485–498.

Moloney MG (2002) Excitatory amino acids. Nat Prod Rep 19:597–616.

Moroni F, Galli A, Mannaioni G, Carla V, Cozzi A, Mori F, Marinozzi M, Pellicciari R (1995) NMDA receptor heterogeneity in mammalian tissues: focus on two agonists, (2S,3R,4S) cyclopropylglutamate and the sulfate ester of 4-hydroxy-(S)-pipecolic acid. Naunyn Schmiedebergs Arch Pharmacol 351:371–376.

Mos L (2001) Domoic acid: a fascinating marine toxin. Environ Toxicol Pharmacol 9:79–85.

Mulle C, Sailer A, Pérez-Otaño I, Dickinson-Anson H, Castillo PE, Bureau I, Maron C, Gage FH, Mann JR, Bettler B, Heinemann SF (1998) Altered synaptic physiology and reduced susceptibility to kainate-induced seizures in GluR6-deficient mice. Nature 392:601–605.

Mulle C, Sailer A, Swanson GT, Brana C, O'Gorman S, Bettler B, Heinemann SF (2000) Subunit composition of kainate receptors in hippocampal interneurons. Neuron 28:475–484.

Murakami S, Takemoto T, Shimizu Z (1953) Studies on the effective principles of *Digenea simplex* Aq. I. Separation of the effective fraction by liquid chromatography. J Pharm Soc Jpn 73:1026–1028.

Nadler JV (1979) Kainic acid: neurophysiological and neurotoxic actions. Life Sci 24:289–299.

Nadler JV (1981) Kainic acid as a tool for the study of temporal lobe epilepsy. Life Sci 29:2031–2042.

Naur P, Vestergaard B, Skov LK, Egebjerg J, Gajhede M, Kastrup JS (2005) Crystal structure of the kainate receptor GluR5 ligand-binding core in complex with (S)-glutamate. FEBS Lett 579:1154–1160.

Naur P, Hansen KB, Kristensen AS, Dravid SM, Pickering DS, Olsen L, Vestergaard B, Egebjerg J, Gajhede M, Traynelis SF, Kastrup JS (2007) Ionotropic glutamate-like receptor delta2 binds D-serine and glycine. Proc Natl Acad Sci USA 104:14116–14121.

Nawy S, Copenhagen DR (1987) Multiple classes of glutamate receptor on depolarizing bipolar cells in retina. Nature 325:56–58.

Nicoll RA, Malenka RC (1999) Expression mechanisms underlying NMDA receptor-dependent long-term potentiation. Ann N Y Acad Sci 868:515–525.

Nitta I, Watase H, Tomiie Y (1958) Structure of kainic acid and its isomer, allokainic acid. Nature 181:761–762.

Nowak L, Bregestovski P, Ascher P, Herbet A, Prochiantz A (1984) Magnesium gates glutamate-activated channels in mouse central neurones. Nature 307:462–465.

Olivera BM (2006) Conus peptides: biodiversity-based discovery and exogenomics. J Biol Chem 281:31173–31177.

Olney JW (1994) Excitotoxins in foods. Neurotoxicology 15:535–544.

Paoletti P, Neyton J (2007) NMDA receptor subunits: function and pharmacology. Curr Opin Pharmacol 7:39–47.

Paternain AV, Morales M, Lerma J (1995) Selective antagonism of AMPA receptors unmasks kainate receptor-mediated responses in hippocampal neurons. Neuron 14:185–189.

Paternain AV, Vicente A, Nielsen EO, Lerma J (1996) Comparative antagonism of kainate-activated kainate and AMPA receptors in hippocampal neurons. Eur J Neurosci 8:2129–2136.

Pei-Gen X, Shan-Lin F (1986) Traditional antiparasitic drugs in China. Parasitol Today 2:353–355.

Perez-Otaño I, Ehlers MD (2004) Learning from NMDA receptor trafficking: clues to the development and maturation of glutamatergic synapses. Neurosignals 13:175–189.

Perl TM, Bedard L, Kosatsky T, Hockin JC, Todd EC, McNutt LA, Remis RS (1990a) Amnesic shellfish poisoning: a new clinical syndrome due to domoic acid. Can Dis Wkly Rep 16 (Suppl. 1E):7–8.

Perl TM, Bedard L, Kosatsky T, Hockin JC, Todd EC, Remis RS (1990b) An outbreak of toxic encephalopathy caused by eating mussels contaminated with domoic acid. N Engl J Med 322:1775–1780.

Perovic S, Wickles A, Schutt C, Gerdts G, Pahler S, Steffen R, Muller WEG (1998) Neuroactive compounds produced by bacteria from the marine sponge *Halichondria panicea*: activation of the neuronal NMDA receptor. Environ Toxicol Pharmacol 6:125–133.

Phillips D, Chamberlin AR (2002) Total synthesis of dysiherbaine. J Org Chem 67:3194–3201.

Pinheiro PS, Perrais D, Coussen F, Barhanin J, Bettler B, Mann JR, Malva JO, Heinemann SF, Mulle C (2007) GluR7 is an essential subunit of presynaptic kainate autoreceptors at hippocampal mossy fiber synapses. Proc Natl Acad Sci USA 104:12181–12186.

Planells-Cases R, Lerma J, Ferrer-Montiel A (2006) Pharmacological intervention at ionotropic glutamate receptor complexes. Curr Pharm Des 12:3583–3596.

Prorok M, Castellino FJ (2007) The molecular basis of conantokin antagonism of NMDA receptor function. Curr Drug Targets 8:633–642.

Ramsey UP, Bird CJ, Shacklock PF, Laycock MV, Wright JLC (1994) Kainic acid and 1'-hydrox-ykainic acid from *Palmariales*. Nat Toxins 2:286–292.

Ratte S, Lacaille JC (2006) Selective degeneration and synaptic reorganization of hippocampal interneurons in a chronic model of temporal lobe epilepsy. Adv Neurol 97:69–76.

Ren Z, Riley NJ, Garcia EP, Sanders JM, Swanson GT, Marshall J (2003) Multiple trafficking signals regulate kainate receptor KA2 subunit surface expression. J Neurosci 23:6608–6616.

Rodriguez-Moreno A, Herreras O, Lerma J (1997) Kainate receptors presynaptically downregulate GABAergic inhibition in the rat hippocampus. Neuron 19:893–901.

Sakai R, Kamiya H, Murata M, Shimamoto K (1997) Dysiherbaine: a new neurotoxic amino acid from the Micronesian marine sponge *Dysidea herbacea*. J Am Chem Soc 119:4112–4116.

Sakai R, Koike T, Sasaki M, Shimamoto K, Oiwa C, Yano A, Suzuki K, Tachibana K, Kamiya H (2001a) Isolation, structure determination, and synthesis of neodysiherbaine A, a new excitatory amino acid from a marine sponge. Org Lett 3:1479–1482.

Sakai R, Swanson GT, Shimamoto K, Green T, Contractor A, Ghetti A, Tamura-Horikawa Y, Oiwa C, Kamiya H (2001b) Pharmacological properties of the potent epileptogenic amino acid dysiherbaine, a novel glutamate receptor agonist isolated from the marine sponge *Dysidea herbacea*. J Pharmacol Exp Ther 296:650–658.

Sakai R, Matsubara H, Shimamoto K, Jimbo M, Kamiya H, Namikoshi M (2003) Isolations of N-methyl-D-aspartic acid-type glutamate receptor ligands from Micronesian sponges. J Nat Prod 66:784–787.

Sakai R, Suzuki K, Shimamoto K, Kamiya H (2004) Novel betaines from a Micronesian sponge *Dysidea herbacea*. J Org Chem 69:1180–1185.

Sakai R, Minato S, Koike K, Koike K, Jimbo M, Kamiya H (2005) Cellular and subcellular localization of kainic acid in the marine red alga *Digenea simplex*. Cell Tissue Res 322:491–502.

Sakai R, Swanson GT, Sasaki M, Shimamoto K, Kamiya H (2006) Dysiherbaine: a new generation of excitatory amino acids of marine origin. CNS Agents Med Chem 6:83–108.

Sakai R, Yoshida K, Kimura A, Koike K, Jimbo M, Koike K, Kobiyama A, Kamiya H (2008) Cellular origin of dysiherbaine, a marine sponge-derived excitatory amino acid. ChemBioChem 9(4):543–551.

Sanders JM, Ito K, Settimo L, Pentikainen OT, Shoji M, Sasaki M, Johnson MS, Sakai R, Swanson GT (2005) Divergent pharmacological activity of novel marine-derived excitatory amino acids on glutamate receptors. J Pharmacol Exp Ther 314:1068–1078.

Sanders JM, Pentikainen OT, Settimo L, Pentikainen U, Shoji M, Sasaki M, Sakai R, Johnson MS, Swanson GT (2006) Determination of binding site residues responsible for the subunit selectivity of novel marine-derived compounds on kainate receptors. Mol Pharmacol 69:1849–1860.

Sasaki M, Maruyama T, Sakai R, Tachibana K (1999) Synthesis and biological activity of dysiherbaine model compound. Tetrahedron Lett 40:3195–3198.

Sasaki M, Koike T, Sakai R, Tachibana K (2000) Total synthesis of (−)-dysiherbaine, a novel neuroexcitotoxic amino acid. Tetrahedron Lett 41:3923–3926.

Sasaki M, Tsubone K, Shoji M, Oikawa M, Shimamoto K, Sakai R (2006) Design, total synthesis, and biological evaluation of neodysiherbaine A derivative as potential probes. Bioorg Med Chem Lett 16:5784–5787.

Sasaki M, Akiyama N, Tsubone K, Shoji M, Oikawa M, Sakai R (2007) Total synthesis of dysiherbaine. Tetrahedron Lett 48:5697–5700.

Sato M, Inoue F, Kanno N, Sato Y (1987) The occurrence of *N*-methyl-D-aspartic acid in muscle extracts of the blood shell, *Scapharca broughtonii*. Biochem J 241:309–311.

Sato M, Nakano T, Takeuchi M, Kanno N, Nagahisa E, Sato Y (1996) Distribution of neuroexcitatory amino acids in marine algae. Phytochemistry 42:1595–1597.

Sawant PM, Weare BA, Holland PT, Selwood AI, King KL, Mikulski CM, Doucette GJ, Mountfort DO, Kerr DS (2007) Isodomoic acids A and C exhibit low KA receptor affinity and reduced in vitro potency relative to domoic acid in region CA1 of rat hippocampus. Toxicon 50:627–638.

Schiffer HH, Swanson GT, Heinemann SF (1997) Rat GluR7 and a carboxyterminal splice variant, GluR7b, are functional kainate receptor subunits with a low sensitivity to glutamate. Neuron 19:1141–1146.

Schmitz D, Frerking M, Nicoll RA (2000) Synaptic activation of presynaptic kainate receptors on hippocampal mossy fiber synapses. Neuron 27:327–338.

Schmitz D, Mellor J, Breustedt J, Nicoll RA (2003) Presynaptic kainate receptors impart an associative property to hippocampal mossy fiber long-term potentiation. Nat Neurosci 6:1058–1063.

Scholin CA, Gulland F, Doucette GJ, Benson S, Busman M, Chavez FP, Cordaro J, DeLong R, De Vogelaere A, Harvey J, Haulena M, Lefebvre K, Lipscomb T, Loscutoff S, Lowenstine LJ, Marin R, III, Miller PE, McLellan WA, Moeller PDR, Powell CL, Rowles T, Silvagni P, Silver M, Spraker T, Trainer V, Van Dolah FM (2000) Mortality of sea lions along the central California coast linked to a toxic diatom bloom. Nature 403:80–84.

Schuster CM, Ultsch A, Schloss P, Cox JA, Schmitt B, Betz H (1991) Molecular cloning of an invertebrate glutamate receptor subunit expressed in *Drosophila* muscle. Science 254:112–114.

Shi S, Hayashi Y, Esteban JA, Malinow R (2001) Subunit-specific rules governing AMPA receptor trafficking to synapses in hippocampal pyramidal neurons. Cell 105:331–343.

Shin-ya K, Kim J-S, Furihata K, Hayakawa Y, Seto H (1997a) Structure of kaitocephalin, a novel glutamate receptor antagonist produced by *Eupenicillium shearii*. Tetrahedron Lett 38:7079–7082.

Shin-ya K, Kim JS, Hayakawa Y, Seto H (1997b) Protective effect of a novel AMPA and NMDA antagonist kaitocephalin against glutamate neurotoxicity. J Neurochem 73:S190.

Shinozaki H (1988) Pharmacology of the glutamate receptor. Prog Neurobiol 30:399–435.

Shoji M, Akiyama N, Tsubone K, Lash LL, Sanders JM, Swanson GT, Sakai R, Shimamoto K, Oikawa M, Sasaki M (2006) Total synthesis and biological evaluation of neodysiherbaine A and analogues. J Org Chem 71:5208–5220.

Sierra Beltran A, Palafox-Uribe M, Grajales-Montiel J, Cruz-Villacorta A, Ochoa JL (1997) Sea bird mortality at Cabo San Lucas, Mexico: evidence that toxic diatom blooms are spreading. Toxicon 35:447–453.

Simmons RM, Li DL, Hoo KH, Deverill M, Ornstein PL, Iyengar S (1998) Kainate GluR5 receptor subtype mediates the nociceptive response to formalin in the rat. Neuropharmacology 37:25–36.

Smolders I, Bortolotto ZA, Clarke VR, Warre R, Khan GM, O'Neill MJ, Ornstein PL, Bleakman D, Ogden A, Weiss B, Stables JP, Ho KH, Ebinger G, Collingridge GL, Lodge D, Michotte Y (2002) Antagonists of GLU(K5) containing kainate receptors prevent pilocarpine-induced limbic seizures. Nat Neurosci 5:796–804.

Smothers CT, Woodward JJ (2007) Pharmacological characterization of glycine-activated currents in HEK 293 cells expressing *N*-methyl-Daspartate NR1 and NR3 subunits. J Pharmacol Exp Ther 322:739–748.

Snider BB, Hawryluk NA (2000) Synthesis of (−)-dysiherbaine. Org Lett 2:635–638.

Sommer B, Keinanen K, Verdoorn TA, Wisden W, Burnashev N, Herb A, Köhler M, Takagi T, Sakmann B, Seeburg PH (1990) Flip and flop: a cell-specific functional switch in glutamate-operated channels of the CNS. Science 249:1580–1585.

Sommer B, Köhler M, Sprengel R, Seeburg PH (1991) RNA editing in brain controls a determinant of ion flow in glutamate-gated channels. Cell 67:11–19.

Sommer B, Burnashev N, Verdoorn TA, Keinänen K, Sakmann B, Seeburg PH (1992) A glutamate receptor channel with high affinity for domoate and kainate. EMBO J 11:1651–1656.

Sperk G (1994) Kainic acid seizures in the rat. Prog Neurobiol 42:1–32.

Spruston N, Schiller Y, Stuart G, Sakmann B (1995) Activity-dependent action potential invasion and calcium influx into hippocampal CA1 dendrites. Science 268:297–300.

Stern-Bach Y, Bettler B, Hartley M, Sheppard PO, O'Hara PJ, Heinemann SF (1994) Agonist selectivity of glutamate receptors is specified by two domains structurally related to bacterial amino acid-binding proteins. Neuron 13:1345–1357.

Swanson GT, Gereau RW, IV, Green T, Heinemann SF (1997) Identification of amino acid residues that control functional behavior in GluR5 and GluR6 kainate receptors. Neuron 19:913–926.

Swanson GT, Green T, Sakai R, Contractor A, Che W, Kamiya H, Heinemann SF (2002) Differential activation of individual subunits in heteromeric kainate receptors. Neuron 34:589–598.

Takahashi K, Matsumura T, Corbin GRM, Ishihara J, Hatakeyama S (2006) A highly stereocontrolled total synthesis of (−)-neodysiherbaine A. J Org Chem 71:4227–4231.

Takemoto T (1978) Isolation and structural identification of naturally occurring excitatory amino acids. In: Kainic Acid as a Tool in Neurobiology (McGreer EG, ed.), New York pp. 1–15: Raven Press.

Takemoto T, Daigo K (1958) Constituents of *Chondria armata*. Chem Pharm Bull 6:578–580.

Takemoto T, Nakajima T, Sakuma R (1964) Isolation of a flycidal constituent "Ibotenic Acid" from Amanita muscaria and A. pantherina. Yakugaku Zasshi 84:1233–1234.

Takemoto T, Nakajima T, Arihara S, Koike K (1975) Studies on the constituents of Quisqualis Fructus. II. Structure of quisqualic acid. Yakugaku Zasshi 95:326–332.

Teichert RW, Jimenez EC, Twede V, Watkins M, Hollmann M, Bulaj G, Olivera BM (2007) Novel conantokins from conus parius venom are specific antagonists of NMDA receptors. J Biol Chem 282(51):36905–36913.

Teitelbaum JS, Zatorre RJ, Carpenter S, Gendron D, Evans AC, Gjedde A, Cashman NR (1990) Neurologic sequelae of domoic acid intoxication due to the ingestion of contaminated mussels. N Engl J Med 322:1781–1787.

Terlau H, Olivera BM (2004) Conus venoms: a rich source of novel ion channel-targeted peptides. Physiol Rev 84:41–68.

Tiedeken JA, Ramsdell JS, Ramsdell AF (2005) Developmental toxicity of domoic acid in zebrafish (*Danio rerio*). Neurotoxicol Teratol 27:711–717.

Trainer VL, Adams NG, Bill BD, Anulacion BF, Wekell JC (1998) Concentration and dispersal of a *Pseudo-nitzschia* bloom in Penn Cove, Washington, USA. Nat Toxins 6:113–126.

Tremblay J-F (2000) Shortage of kainic acid hampers neuroscience research. Chem Eng News 78:14–15.

Tsai C, Schneider JA, Lehmann J (1988) *Trans*-2-carboxy-3-pyrrolidineacetic acid (CPAA), a novel agonist at NMDA-type receptors. Neurosci Lett 92:298–302.

Usherwood PN (2000) Natural and synthetic polyamines: modulators of signalling proteins. Farmaco 55:202–205.

Vignes M, Collingridge GL (1997) The synaptic activation of kainate receptors. Nature 388:179–182.

Washburn MS, Dingledine R (1996) Block of alpha-amino-3-hydroxy-5-methyl4-isoxazolepropionic acid (AMPA) receptors by polyamines and polyamine toxins. J Pharmacol Exp Ther 278:669–678.

Watanabe H, Kitahara T (2007) Revision of the stereochemistries of natural products through the synthetic study: Synthesis of fudecalone and kaitocephalin. J Synth Org Chem Jpn 65:511–519.

Watters MR (1995) Organic neurotoxins in seafoods. Clin Neurol Neurosurg 97:119–124.

Wekell JC, Gauglitz Jr EJ, Barnett HJ, Hatfield CL, Simons D, Ayres D (1994) Occurrence of domoic acid in Washington State razor clams (*Siliqua patula*) during 1991–1993. Nat Toxins 2:197–205.

Werner P, Voigt M, Keinänen K, Wisden W, Seeburg PH (1991) Cloning of a putative high-affinity kainate receptor expressed predominantly in hippocampal CA3 cells. Nature 351:742–744.

White HS, McCabe RT, Armstrong H, Donevan SD, Cruz LJ, Abogadie FC, Torres J, Rivier JE, Paarmann I, Hollmann M, Olivera BM (2000) In vitro and in vivo characterization of conantokin-R, a selective NMDA receptor antagonist isolated from the venom of the fish-hunting snail *Conus radiatus*. J Pharmacol Exp Ther 292:425–432.

Wilding TJ, Huettner JE (1997) Activation and desensitization of hippocampal kainate receptors. J Neurosci 17:2713–2721.

Williams AJ, Dave JR, Phillips JB, Lin Y, McCabe RT, Tortella FC (2000) Neuroprotective efficacy and therapeutic window of the high-affinity *N*-methyl-D-aspartate antagonist conantokin-G: in vitro (primary cerebellar neurons) and in vivo (rat model of transient focal brain ischemia) studies. J Pharmacol Exp Ther 294:378–386.

Williams AJ, Ling G, Berti R, Moffett JR, Yao C, Lu XM, Dave JR, Tortella FC (2003) Treatment with the snail peptide CGX-1007 reduces DNA damage and alters gene expression of c-fos and bcl-2 following focal ischemic brain injury in rats. Exp Brain Res 153:16–26.

Wisden W, Seeburg PH (1993) A complex mosaic of high-affinity kainate receptors in rat brain. J Neurosci 13:3582–3598.

Wittekindt B, Malany S, Schemm R, Otvos L, Maccecchini ML, Laube B, Betz H (2001) Point mutations identify the glutamate binding pocket of the *N*-methyl-D-aspartate receptor as major site of conantokin-G inhibition. Neuropharmacology 41:753–761.

Work TM, Barr BB, Beale AM, Fritz L, Quilliam MA, Wright JLC (1993) Epidemiology of domoic acid poisoning in brown pelicans (*Pelicanus occidentalis*) and Brandt's cormorants (*Phalacrocorax pencillatus*) in California. J Zool Wildlife Med 24:54–62.

Wright JLC, Boyd RK, Defreitas ASW, Falk M, Foxall RA, Jamieson WD, Laycock MV, McCulloch AW, McInnes AG, Odense P, Pathak V, Quilliam MA, Ragan M, Sim PG, Thibault P, Walter JA, Gilgan M, Richard DJA, Dewar D (1989) Identification of domoic acid, a neuroexcitatory amino acid, in toxic mussels from eastern Prince Edward Island. Can J Chem 67:481–490.

Xia Z, Storm DR (2005) The role of calmodulin as a signal integrator for synaptic plasticity. Nat Rev Neurosci 6:267–276.

Yuzaki M (2004) The delta2 glutamate receptor: a key molecule controlling synaptic plasticity and structure in Purkinje cells. Cerebellum 3:89–93.

Zaman L, Arakawa O, Shimosu A, Onoue Y, Nishio S, Shida Y, Noguchi T (1997) Two new isomers of domoic acid from a red alga, *Chondria armata*. Toxicon 35:205–212.

Zhu JJ, Esteban JA, Hayashi Y, Malinow R (2000) Postnatal synaptic potentiation: delivery of GluR4-containing AMPA receptors by spontaneous activity. Nat Neurosci 3:1098–1106.

Marine Toxins Potently Affecting Neurotransmitter Release[1]

Frédéric A. Meunier, César Mattei, and Jordi Molgó

F.A. Meunier (✉)
Queensland Brain Institute and School of Biomedical Sciences, The University of Queensland, St. Lucia, Queensland 4061, Australia

C. Mattei
CNRS, Institut de Neurobiologie Alfred Fessard, FRC2118, Laboratoire de Neurobiologie Cellulaire et Moléculaire, UPR 9040, 1, avenue de la Terrasse, Gif sur Yvette, F-91198, France and Délégation Générale pour l'Armement 7-9 rue des Mathurins, Bagneux, F-92220, France

J. Molgó
CNRS, Institut de Neurobiologie Alfred Fessard, FRC2118, Laboratoire de Neurobiologie Cellulaire et Moléculaire, UPR 9040, 1, avenue de la Terrasse, Gif sur Yvette, F-91198, France

[1]This chapter is dedicated to honour the memory of Professor André Ménez (1943–2008)

N. Fusetani and W. Kem (eds.), *Marine Toxins as Research Tools*,
Progress in Molecular and Subcellular Biology, Marine Molecular Biotechnology 46,
DOI: 10.1007/978-3-540-87895-7, © Springer-Verlag Berlin Heidelberg 2009

Abstract Synapses are specialised structures where interneuronal communication takes place. Not only brain function is absolutely dependent on synaptic activity, but also most of our organs are intimately controlled by synaptic activity. Synapses are therefore an ideal target to act upon and poisonous species have evolved fascinating neurotoxins capable of shutting down neuronal communication by blocking or activating essential components of the synapse. By hijacking key proteins of the communication machinery, neurotoxins are therefore extremely valuable tools that have, in turn, greatly helped our understanding of synaptic biology. Moreover, analysis and understanding of the molecular strategy used by certain neurotoxins has allowed the design of entirely new classes of drugs acting on specific targets with high selectivity and efficacy. This chapter will discuss the different classes of marine neurotoxins, their effects on neurotransmitter release and how they act to incapacitate key steps in the process leading to synaptic vesicle fusion.

1   Introduction

Neurotoxins have paved the way of modern neurobiology by providing indispensable tools to unravel pre- and post-synaptic mechanisms. In particular, marine neurotoxins have been very useful in determining the role played by calcium, sodium and potassium channels in mediating neurotransmitter release and controlling excitability in various synaptic preparations. The exquisite specificity of certain neurotoxins facilitated the identification of various channel subtypes and their functional roles. The mode of action of neurotoxins has not been only centred on presynaptic mechanisms: the post-synaptic machinery has also greatly benefited from their use. Neurotoxins targeting muscular nicotinic receptor were the first to be characterised (see review by Terlau and Olivera 2004) but the interest soon shifted to CNS transmission and novel categories of marine neurotoxins blocking nicotinic (Cartier et al. 1996), glutamate (Haack et al. 1990) receptors and other important synaptic targets were discovered (Terlau and Olivera 2004). In this chapter, we will first summarise the mechanism of quantal neurotransmitter release in order to set the scene for the examination of marine neurotoxins and their presynaptic actions.

Quantal transmitter release depends on the controlled fusion of neurotransmitter-containing synaptic vesicles with the presynaptic plasma membrane at specialised sites called active zones. There are two major types of quantal acetylcholine (ACh) release taking place at motor nerve terminals:

Spontaneous neurotransmitter release which is asynchronous in nature and result from the random fusion of synaptic vesicles normally occurring at a low frequency. This can be detected post-synaptically as small amplitude depolarisations called miniature endplate potential (MEPP).

Ca^{2+}-dependent phasic neurotransmitter release which results from the mobilisation and synchronised fusion of 40–200 synaptic vesicles. The postsynaptic depolarisation resulting from this ACh release called endplate potential (EPP) is of much larger amplitude and normally capable of triggering a post-synaptic action potential (Van der Kloot and Molgó 1994). Both types of quantal release can be affected by marine neurotoxins. The fusion event also called exocytosis occurs via the formation of a lipidic hemifusion intermediate corresponding to the merging of the proximal leaflets of the synaptic vesicle and plasma membrane bilayers preceding the merging of the two distal leaflets to form the fusion pore. Both fusion and fission require large deformations in membrane curvature. This deformation would be facilitated by the formation of high local concentrations of lipids with altered shapes that can be provided by the action of certain marine neurotoxins. Following the opening of the fusion pore, the neurotransmitter is released by simple diffusion and then binds to postsynaptic receptors, which promote either depolarisation or hyperpolarisation according to the nature of the neurotransmitter released and the post-synaptic receptor composition. There are a wide variety of post-synaptic receptors which are affected by marine neurotoxins (as seen in Chap. 2). There are two major ways of clearing the neurotransmitter from the synaptic cleft. At the neuromuscular junction, ACh is degraded by specialised enzymes called cholinesterases. One of the products of this enzymatic reaction, choline, is actively re-uptaken by a choline transport mechanism back into the presynaptic nerve terminal. In central synapses, the most common way of clearing the transmitter is through the action of specialised transporters located both on the presynaptic terminal and the peri-synaptic glial cell. Marine neurotoxins affecting transporters have recently been discovered (see Chap. 2). Some neurotransmitter receptors are also located presynaptically and function as modulators.

Phasic transmitter release is initiated by the opening of voltage-gated Ca^{2+} channels following arrival of an action potential at the nerve terminal. The depolarisation phase of the action potential is driven by Na^+ channels and a variety of marine neurotoxin producers have used this strategy to promote fast paralysis of their prey or predators. Following depolarisation of the nerve terminal, voltage-gated Ca^{2+} channels open, promoting an influx of extracellular Ca^{2+}. The latter is directly responsible for triggering synaptic vesicle fusion by acting on synaptotagmin, the Ca^{2+} sensor for exocytosis (Südhof 2004). Perhaps not surprisingly, a large variety of marine toxins target voltage-gated Ca^{2+} channels – a key component of the exocytosis machinery. In many of the previously critical components of the communication apparatus, marine neurotoxins have allowed accurate dissection of physiological mechanisms and thus have contributed greatly to our basic knowledge of synaptic function – which in many cases was confirmed with modern techniques of molecular biology. Because of their often-exquisite specificities, marine neurotoxins are also now being successfully used for therapeutic purposes (Lewis and Garcia 2003; Terlau and Olivera 2004) (see Chap. 2).

2   Neurotoxins with Pore-Forming Activity

2.1   *Trachynilysin*

Few marine neurotoxins have been found to promote neurotransmitter release by promoting the formation of non-specific ionic pores on presynaptic nerve terminals. The venom of the stonefish is considered one of the most dangerous of the

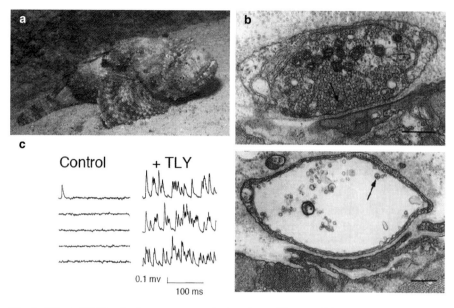

Fig. 1 Effect of TLY at the frog neuromuscular junction. **a** The stonefish *Synanceia trachynis* (©
D. Scarr). **(b)** Effect of TLY on nerve terminal ultrastructure. Electron micrographs of motor nerve
terminals before (top) and 3 h after (bottom) exposure to TLY. Scale bars, 0.5 mm. **c** Effect of
TLY (10 pg/ml) on spontaneous acetylcholine release at the frog neuromuscular junction. MEPPs
recorded intracellularly before (control) and 10 min after addition of TLY (+TLY) to the nerve-
muscle preparation. Adapted from Colasante et al. (1996)

scorpionfish genus (*Synanceia horrida, S. verucossa, S. trachynis, S. nana* and
S. alula) as envenomation from stepping on one of the dorsal spines can be lethal.
One of the best characterised is trachynilysin (TLY), a 159 kDa di-chain toxic protein isolated from the venom of the stonefish *Synanceia trachynis* (Fig. 1a). The
α- and β-subunits (76 and 83 kDa, respectively) linked by disulfide bridges were
originally found to have a haemolytic activity at high concentrations (Kreger 1991).
The neurotoxin is believed to be glycosylated since pre-treatment with concanavalin A completely prevent TLY's stimulatory effect (Ouanounou et al. 2002). TLY is
structurally related to other haemolytic neurotoxins from stonefish such as stonustoxin (*S. horrida*), verrucotoxin and neoverrucotoxin (*S. verrucosa*) (Khoo 2002).

At the neuromuscular junction, nanomolar concentrations of TLY promote a
long-lasting increase in asynchronous transmitter release (Fig. 1b) (Kreger et al.
1993) leading to a specific depletion of small clear synaptic vesicles and subsequent
neuromuscular paralysis (Colasante et al. 1996). This effect is reminiscent of the
stimulatory effect observed on the same preparation when treated with α-latrotoxin
from the venom of the black widow spider *Latrodectus mactans tredecimguttatus*
(see review by Ceccarelli and Hurlbut 1980). The similarity in effect between the
two neurotoxins is even greater as both seem to selectively act on small synaptic
vesicles without affecting the release of large dense core vesicles at the frog presynaptic motor nerve terminals (Fig. 1c) (Colasante et al. 1996). However, TLY does
not interact with at least one α-latrotoxin receptor, CIRL-latrophilin (Meunier et al.

2002b). More work is needed to investigate whether TLY acts through other known receptors for α-latrotoxin such as neurexins.

TLY stimulatory effect is not restricted to motor nerve terminals. In bovine chromaffin cells, TLY triggers a in Ca^{2+} influx that is responsible for promoting catecholamine release. TLY-induced Ca^{2+} rise is not dependent on activation of L, N, nor P/Q voltage-dependent Ca^{2+} channels since their respective blockade did not affect the Ca^{2+} signal (Meunier et al. 2000). However, removal of extracellular Ca^{2+} and addition of EGTA or La^{3+} completely abolished both secretion and Ca^{2+} signals (Meunier et al. 2000). This stimulation was shown to be highly dependent on caffeine-sensitive Ca^{2+} stores suggesting that catecholamine secretion from adrenal chromaffin cells is initiated by pore-formation and sustained by subsequent mobilisation of intracellular Ca^{2+} stores (Meunier et al. 2000). The pore-forming activity of TLY was demonstrated with TLY spontaneously forming cationic pores in neuroblastoma cells (Ouanounou et al. 2002).

2.2   Sea Anemone Pore-Forming Toxins (Actinoporins)

Sea anemones are providing an ever-increasing array of toxins besides modifiers of Na^+ and K^+ channels, that have found some utility in the treatment of multiple sclerosis (Lewis and Garcia 2003; Norton et al. 2004) including cytolytic pore-forming toxins (Fig. 2). The most commonly described toxins are equinatoxin from *Actinia equina* and sticholysins from *Stichodactyla helianthus*, homologous 20kDa polypeptides known to produce cation-permeant pores in sphingomyelin-containing membranes.

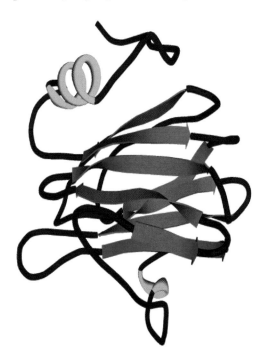

Fig. 2 Structure of equinatoxin II. Structure generated with protein workshop (green: α-helice, red: β-sheet), PDB: 1KD6 (Hinds et al. 2002)

Although the structures and mechanisms of pore-formation by these toxins are well documented, little information is currently available on their effects on neurons. Equinatoxins and related sticholysins are cytolysins that can produce pores with an estimated diameter of 1 nm (Anderluh et al. 2000; de los Rios et al. 1999; Malovrh et al. 2000; Tejuca et al. 1996). One hallmark of these cytolysins is that their effect can be potently inhibited by the membrane phospholipid sphingomyelin. The pore formed is likely to allow diffusion of small molecules ranging from 400 to 900 Da and cations such as Ca^{2+}. Bc2 (20 kDa), isolated from the sea anemone *Bunodosoma caissarum*, was shown to promote glutamate release from rat cortical synaptosomes (Migues et al. 1999) and catecholamine release from bovine chromaffin cells (Ales et al. 2000). Most interestingly, the effect of this toxin was reversible by simple washing, suggesting the specific and reversible nature of the pore formed.

3 Neurotoxins Acting on Na$^+$ Channels

Voltage-sensitive sodium channels (VSSC) play a crucial role in the generation and propagation of the action potential and thus the communication between excitable cells (Catterall 2000). VSSC are transmembrane proteins that control the Na$^+$ ion flux through the cytoplasmic membrane. The channel is formed of a 260 kDa α-subunit associated with one or multiple b subunits of 33–36 kDa. The a-subunit is composed of four homologous domains (I–IV). Each of them holds six trans-membrane α-helices (S1–S6) and a pore loop between the S5 and S6 segments.

A standardised nomenclature is now accepted to classify mammalian VSSC: the Na_{v1} family comprises nine different a subunits ranging from $Na_v1.1$ to $Na_v1.9$ (Goldin et al. 2000; Yu and Catterall 2003). Another isoform, termed Na_{vx}, may be related to this family. These ten different VSSC are structurally and functionally similar but their tissue distribution, pharmacological properties and physiological regulation vary greatly. VSSC are responsible for initiating action potential, their opening allowing an influx of Na$^+$ promotes a fast depolarisation that correspond to the rising phase of the action potential. Interestingly, numerous marine neurotoxins have been found to bind with a high affinity to the α-subunit of VSSC and to affect their related physiological functions such as neurotransmitter release. These toxins act by altering specific biophysical properties of VSSC (Catterall 2000).

The use of such neurotoxins as tools to study the Na$^+$ channels has allowed the discovery of most of their molecular, physiological and functional properties over the past 30 years. Remarkably, some Na$^+$ channel-binding toxins can even mimic channelopathies by modifying the channel biophysical properties (Ogata and Ohishi 2002). It is interesting to note that VSSC are the target of neurotoxins produced by a variety of marine living organisms: predators, putative preys, microorganisms such as dinoflagellates or others. Na$^+$ channels toxins are therefore targets of choice for both predation and defensive strategy of some animals (fish, cones, sea anemones). The advantage conferred (if any) to other toxin producers (such as dinoflagellates) is not fully understood. Six different binding sites have been characterised all on the α-subunits of VSSC (Catterall et al. 2007). These toxins modify the activation and inactivation properties of VSSC, or block the pore of the channels (Ogata and Ohishi 2002). As a consequence these neurotoxins affect the action potential initiation and propagation, and downstream excitability and neurotransmission.

3.1 Tetrodotoxins and Saxitoxins

Among the different receptor sites of VSSC, the site 1 binds only Na^+ channel inhibitors. Three different types of marine neurotoxins are known to occlude the channel pore, which leads to the inhibition of the Na^+ conductance: the nonpeptide tetrodotoxin (TTX) and saxitoxin (STX), and the peptide μ-conotoxins (discussed in Chap. 2). Site 1 for the binding of these neurotoxins comprises amino acid from the pore loops and on the extracellular side of the pore loops at the outer end of the pore (Catterall 2000; Catterall et al. 2005, 2007). TTXs and STXs have been extensively studied and used for decades as potent Na^+ channel inhibitors. Their pore-blocking activities were exploited to purify the Na^+ channel proteins and to identify the residues involved in the outer pore and in the selectivity filter (Yu and Catterall 2003).

Both toxins are implicated in human food intoxications resulting from the consumption of fish and shellfish: TTXs are the causative agents of puffer fish poisoning (PFP), whereas STXs are responsible for paralytic shellfish poisoning (PSP). The main symptoms of human intoxications associated with PFP and PSP include neurological, cardiac and gastrological complications. Numbness of the mouth, face and fingertips, nausea and vomiting, paralysis of the arms and legs, difficulty in breathing and eventually – for extreme cases – death associated with respiratory paralysis have been described (Landsberg et al. 2006; Schantz and Johnson 1992). For more information see Chap. 3. TTX and STX exert similar effects: they are both potent and selective inhibitors of the VSSC (Kao and Nishiyama 1965; Narahashi et al. 1960, 1964, 1967; Nishiyama 1967). As TTX had been shown to block the skeletal muscle action potential without any effect on the resting potential, it was believed that its effect was due to an inhibiting action on Na^+ channels (Narahashi et al. 1960). These data were confirmed by voltage-clamp experiments on lobster giant axons (Narahashi et al. 1964). TTX and STX exert an inhibitory action on neurotransmission through their blocking effect on VSSC: the inhibition of neuromuscular transmission affects the motor axon and muscle membrane, while the end plate receptor is not affected (Nishiyama 1967). Indeed, the toxins inhibit the nerve impulse, resulting in a blocking effect of the action potential-evoked neurotransmission without membrane depolarisation. This implies that at the physiological target of the toxins, the neuromuscular junction and action potential-induced endplate potentials are totally inhibited while spontaneous quantal release of neurotransmitter is unaffected (Kao and Nishiyama 1965; Miledi 1967). As TTX was shown to act as a selective VSSC inhibitor and was commercially available, it became a precious tool to study the allosteric properties of these channels, kinetics of other, mainly K^+ channels expressed in the same cell, that also are involved in synaptic transmission (Catterall 2000). See Chap. 3 for more details.

3.2 Ciguatoxins and Brevetoxins

"Ciguatera fish poisoning" is one of the five phycotoxin-induced human intoxications. Several tens of thousands people experience ciguatera-related troubles every year, especially in coral-reef ecosystems from the Caribbean area and the Pacific and Indian oceans (Anderson and Lobel 1987; Levine 1995; Swift and Swift 1993). The main agent of this human intoxication is the dinoflagellate *Gambierdiscus toxicus* (Adachi and Fukuyo 1979; Yasumoto et al. 1986). Its subsequent intoxication in human is mainly characterised by neurological, gastrointestinal and cardiac disturbances. Neurological troubles usually appear a few hours after consumption of intoxicated but fresh fish (Allsop et al. 1986; Cameron et al. 1991; Poli et al. 1997) due to heat-resistant ciguatoxins that accumulate in muscles, liver and flesh (Tosteson et al. 1988). For more information see Chap. 3.

Ciguatoxins (CTX) are lipid-soluble cyclic polyether compounds produced by the benthic dinoflagellate *Gambierdiscus toxicus*. They target the VSSC, at site 5 of their α-subunit (Lombet et al. 1987; Poli et al. 1986). In neuronal cells, this interaction produces a shift in the voltage dependence of the channel activation to more negative (depolarised) potentials and inhibits VSSC inactivation, both causing resting membrane depolarisation by increasing Na^+ influx (Benoit et al. 1996; Bidard et al. 1984; Catterall et al. 2007; Sharkey et al. 1987). At the neuromuscular junction, CTX produces transient neuronal discharges and trains of repetitive action potentials. Eventually, the toxins reduce and block irreversibly nerve-evoked transmitter release (Huang et al. 1984; Molgó et al. 1990). CTX also enhances spontaneous quantal acetylcholine release from neuromuscular junctions in the presence or the absence of external Ca^{2+}. This effect leads to a complete blockade of neurotransmitter release resulting from a total depletion of synaptic vesicles. The recycling of synaptic vesicles is also inhibited. CTX increases synaptic transmission by promoting both spontaneous and repetitive synchronous release of neurotransmitter evoked by a single presynaptic stimulation. Due to a membrane depolarisation and impairment of action potential generation, this toxin-evoked neurotransmitter release decreases and is eventually totally inhibited (Molgó et al. 1990).

Brevetoxins (PbTx-1 to PbTx-10) are potent polyether neurotoxins synthesized by the marine dinoflagellate *Karenia brevis* previously known as *Gymnodinium breve* and *Ptychodiscus brevis*. The microalgae proliferates during red outbreaks mainly in the gulf of Mexico (Baden 1989) and in New Zealand (Ishida et al. 2004). *Karenia brevis* blooms have been implicated in massive fish kills, bird deaths, and recently demonstrated to be directly responsible for marine mammal mortalities such as manatees and dolphins (Flewelling et al. 2005). The toxins can be transferred to humans through shellfish consumption causing a syndrome named neurotoxic shellfish poisoning

(NSP) generally accompanied with the development of asthma. The symptoms are inversion in temperature sensation, fever, tingling feeling in various area of the body such as the face and digits, dizziness, muscle pain, nausea, diarrhea, pupil dilation and bradycardia. So far no fatalities were reported from NSP. Brevetoxins consist of 9–11 structurally-related, lipid soluble cyclic polyethers derivatives with molecular mass of about 900 Da. There are two main classes of brevetoxins depending on the number of polyether rings (Baden 1989). PbTx, like CTX, interact with site 5 of the α-subunit of the VSSC thereby augmenting Na^+ influx by (1) increasing the mean open time of the channels, (2) inhibiting VSSC inactivation and (3) shifting the activation potential to more hyperpolarised potentials (Jeglitsch et al. 1998; Lombet et al. 1987). As a consequence a population of VSSC remains permanently opened at resting membrane potential triggering spontaneous and/or repetitive train of action potentials in neuronal cells (Lombet et al. 1987). Interestingly, PbTx have recently been shown to allosterically modify TRPV channel – a result that could explain some of alterations in temperature sensation occasioned by NSP (Cuypers et al. 2007). Some of the central effects could also result from indirect effects of PbTx on NMDA receptors and phosphorylation-dependent long-term synaptic effects (Dravid et al. 2005).

At the neuromuscular junction, PbTx promotes an increase in spontaneous quantal transmitter release in the presence of external Ca^{2+}. This effect was completely blocked by TTX pre-treatment of the preparation, or removal of extracellular Na^+ (Atchison et al. 1986). In the absence of external Ca^{2+}, PbTx were later shown to promote a delayed albeit dramatic increase in miniature endplate potential (MEPP) frequency associated with depletion of small clear synaptic vesicles and subsequent blockade of neurotransmission (Meunier et al. 1997). This effect was also prevented by TTX treatment and Na^+ removal. Interestingly, the role of Ca^{2+} in mediating PbTx-induced asynchronous release has been questioned since strong Ca^{2+} buffers such as BAPTA/AM did not prevent PbTx's stimulatory effect (Meunier et al. 1997). However, the difficulty in adequately delivering and controlling BAPTA concentrations in motor nerve terminals renders this particular result difficult to interpret.

Recently, Brevenal, a novel nontoxic natural polyether product from the dinoflagellate *K. brevis*, was shown to compete with tritiated PbTx for site 5 of VSSC. The therapeutic potential of Brevenal is highlighted by the protection of fish from PbTx exposure (LePage et al. 2007; Potera 2007). In view of the similarity of CTX1B and PbTx targeting site 5 on VSSC, it is perhaps not surprising that brevenal was revealed to block the neurosecretory effect of CTX1B in bovine adrenal chromaffin cells (Mattei et al. 2008).

4   Neurotoxins Acting on K^+ Channels

Potassium channels are transmembrane proteins that have been found in all the known organisms and exhibit a high level of heterogeneity (Jan and Jan 1997). These channels play an important role in the control of the resting membrane potential, or in restoring the membrane potential following a depolarising stimulus and therefore are important for setting membrane excitability. In addition, K$^+$ channels control the repolarisation phase of action potentials in some nerve terminals and this action is of foremost importance for depolarisation-neurotransmitter release coupling at motor nerve endings (Meir et al. 1999; Van der Kloot and Molgó 1994). Furthermore, K$^+$ channels have a variety of specialised functions in a wide range of cell types (Hille 2001). Molecular biology approaches have identified more than 70 different genes that assemble to form diverse functional classes of K$^+$ channels. Although functional K$^+$ channels are present within presynaptic nerve endings, direct studies of their precise identity and function have been generally limited to large and specialised presynaptic terminals, such as the ones in the squid giant synapse, basket cell terminals, and Calyx of Held.

Several families of voltage-gated K$^+$ channels are known, each with an α-subunit containing six transmembrane domains. The region spanning the fifth and the sixth transmembrane segments (the pore loop) forms the ion conduction pathway, with four subunits assembled in a functional channel, which may be either homomeric (four identical subunits) or heteromeric (two or more different α-subunits) (for more details on potassium channels see Chap. 4).

4.1   Gambierol and Polyether Toxins

Gambierol is a marine polycyclic ether toxin (Fig. 3) characterised by a transfused octacyclic polyether core containing 18 stereogenic centers, and a partially skipped triene side chain including a conjugated (Z,Z)-diene system (Fuwa et al. 2002; Johnson et al. 2006; Morohashi et al. 1999; Satake et al. 1993).

Gambierol is considered as one of the possible toxins involved in ciguatera fish poisoning caused by the consumption of fish contaminated with toxins produced by the dinoflagellate *Gambierdiscus toxicus* (Fuwa et al. 2002; Johnson et al. 2006; Morohashi et al. 1999; Satake et al. 1993). Gambierol causes toxicity to mice with associated symptoms resembling those produced by ciguatoxin (Fuwa et al. 2004; Ito et al. 2003; Satake et al. 1993). This toxin is naturally produced in very low

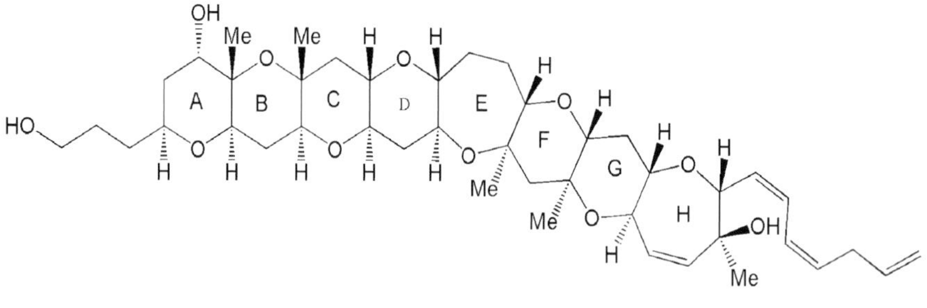

Fig. 3 Chemical structure of gambierol

amounts by the dinoflagellate, but its total chemical synthesis performed in two different laboratories (Fuwa et al. 2002, 2003, 2004; Johnson et al. 2006) allowed a detailed pharmacological analysis of its cellular actions.

Gambierol has been reported to irreversibly inhibit K^+ currents in the nanomolar range in excitable taste cells (Ghiaroni et al. 2005), with a maximum efficiency of 60%. In addition to affecting the K^+ current amplitude, gambierol significantly altered both activation and inactivation processes of K^+ currents, whereas it showed no significant effect on Na^+ or Cl^- currents even at micromolar concentrations. Voltage-gated K^+ currents play an important role in the generation of the firing pattern during chemotransduction. Thus, gambierol may alter action potential discharge in taste cells and this could be associated with the taste alterations reported in the clinical symptoms of ciguatera.

Gambierol, in nanomolar concentrations, also blocks voltage-dependent K^+ channels in primary cultures of rat embryonic chromaffin cells, and in vertebrate motor nerve terminals. As a consequence of this latter action, gambierol greatly increases evoked- and delayed-quantal ACh release from isolated frog and mouse neuromuscular preparations (Schlumberger et al. 2007). Furthermore, in contrast to ciguatoxins (Molgó et al. 1990), sub-micromolar concentrations of gambierol did not affect spontaneous quantal ACh release, detected as MEPPs-frequency.

Micromolar concentrations of gambierol were reported to inhibit the binding of radiolabeled brevetoxin-3 ([^{3}H]PbTx-3) to rat brain synaptosomes (Inoue et al. 2003; LePage et al. 2007). Also, high concentrations of gambierol (30 µM) were reported to produced Na^+-dependent membrane depolarisation, potentiated the effect of veratridine, decreased P-CTX-3C-induced depolarisation and increased cytosolic Ca^{2+} concentration in human neuroblastoma cells (Louzao et al. 2006). Interestingly, the membrane depolarisation caused by gambierol was partially blocked by neosaxitoxin, a well known Na^+-channel blocker. The increase in intracellular Ca^{2+} was attributed to membrane depolarisation, calcium channels opening, and subsequent activation of the Na^+-Ca^{2+} exchanger in the reversed mode operated by the increase in intracellular Na^+. In cerebellar granule neurons, gambierol was shown to be a potent inhibitor of the elevation of intracellular Ca^{2+} triggered by the activation of VSSC by brevetoxin-2 (PbTx-2) (LePage et al. 2007), in a similar manner to that of brevenal, a natural inhibitor of PbTx-toxic effects isolated from cultures of the dinoflagellate *Karenia brevis* (Bourdelais et al. 2004). Available evidence indicates that gambierol acts primarily with a high affinity on voltage-gated K^+ channels, and secondarily with a lower affinity on neuronal VSSC.

Until recently it was well accepted that the voltage-gated Na^+ channel was the primary target of Pacific ciguatoxin-1 (P-CTX-1 or P-CTX-1B). However, an increase in the duration of spontaneous action potentials has been reported with gambiertoxin (P-CTX-4B) in single frog myelinated axons (Benoit and Legrand 1994), suggesting that this toxin could have an action on K^+ channels. P-CTX-4B at higher concentrations than those needed to activate Na^+ channels also targets K^+ channels of single frog myelinated nerve fibres (E. Benoit, personal communication). In rat myotubes, low concentrations of P-CTX-1 (2–5 nM) were reported to block macroscopic K$^+$ currents without significantly changing their activation kinetics, and in a manner that was dependent on the membrane holding potential. Higher P-CTX-1 concentrations (2–12 nM) caused a small membrane depolarisation (3–5 mV), and an increase in the frequency of spontaneous muscle action potential discharges (Hidalgo et al. 2002). Further studies in rat dorsal root ganglion neurons showed that P-CTX-1 inhibited, in a dose-dependent manner, both delayed-rectifier and 'A-type' K$^+$ currents. These effects would most likely contribute to membrane depolarisation and to the prolongation of the action potential and after hyperpolarisation duration. They could therefore contribute to the increase in neuronal excitability observed in sensory neurons subjected to P-CTX-1 (Birinyi-Strachan et al. 2005). Altogether, these results provide a further understanding of the predominant sensory and muscular neurological symptoms associated with ciguatera fish poisoning in the Pacific region.

4.2 Peptide Toxins

Small peptide toxins from distinct origins bind with high specificity to the external mouth of K$^+$-channel. Interestingly, toxins from marine snail (κ-conotoxin PVIIA) and sea anemone (ShK, from *Stichodactyla helianthus* and BgK from *Bunodosoma granulifera*) have functionally converged towards a similar mechanism of inhibition including a lysine and an aromatic residue ~7 Å apart form a functional dyad that is critical for binding (Dauplais et al. 1997; Menez 1998); but see also Mouhat et al. (2005). The snail κ-conotoxins bind to K$^+$ channels with a 1:1 stoichiometry, compete with external tetraethyl ammonium, and are sensitive to the degree of occupancy of the pore, indicating that they occlude the ion conduction pathway (Boccaccio et al. 2004).

5 Neurotoxins Acting on Ca^{2+} Channels

Voltage-sensitive calcium channels (VSCC) have been classified according to their activation range: the low voltage activation threshold (LVA channels called T channels) and the high voltage activation threshold (HVA channels named L, N, P/Q and R channels). HVA are the main VSCC involved in mediating phasic neurotransmitter release. The presynaptic VSCC are composed of three subunits: the main α1 subunit forms the pore of the protein through which Ca^{2+} ions flow. The β subunit is an associated protein that act both as a chaperone promoting surface expression of the α1 subunit, and as modulator altering the inactivation of the channel. The 2α-δ subunit also modifies the biophysical properties of the channel. Marine neurotoxins such as ω-conotoxins have been instrumental in our understanding of the function and distribution of VSCC. More recently, the purification and characterisation of glycerotoxin from the annelid *Glycera convoluta* added a selective stimulatory neurotoxin to the arsenal of tools available to decipher VSCC function (Meunier et al. 2002a).

5.1 *ω-Conotoxins*

Predatory molluscs from the *Conus* genus produce venoms containing a wide variety of neuroactive cysteine-rich peptides ranging from 10 to 40 amino acid residues. For more details see Chap. 2. These peptides are *bona fide* mini-proteins with very well defined tertiary structures stabilised by disulfide bridges. The cocktail of peptides present in the venom is highly efficient in stunning and immobilising the conus prey by acting on key ion channels and transporters present in axons and their synapses. One of the milestones in the pharmacological characterisation of the N-type VSCCs ($Ca_v2.2$) came with the discovery of the selective blocking effect of ω-conotoxin GVIA extracted from *Conus geographus* (Cruz et al. 1987; McCleskey et al. 1987; Reynolds et al. 1986). A number of other ω-conotoxins has been found to block several subtypes of VSCC with various degrees of specificity. This topic has been extensively reviewed (see Terlau and Olivera 2004). Notably, ω-conotoxin GVIA, MVIIA and CVID (Lewis et al. 2000) have good selectivity for $Ca_v2.2$ channels. Other ω-conotoxins such as MVIIC and SVIB have less specificity and block N-, P/Q- ($Ca_v2.1$) and -($Ca_v2.2$) type VSCC (Terlau and Olivera 2004). The modality of ω-conotoxin GVIA inhibition is through binding to the outer vestibule of the $α_{1b}$ subunit and more precisely in the S5–S6 region of domain III. This interaction is ideally located to promote a physical occlusion of the channel pore (Ellinor et al. 1994).

The effect of ω-conotoxins on neurotransmission have been assessed on a number of models (Terlau and Olivera 2004) including synaptosomes (Schenning et al. 2006), neuromuscular junction (Bowersox et al. 1995; Rosato Siri and Uchitel 1999) and the auditory synapse formed by the Calyx of Held (Iwasaki et al. 2000). A number of studies concur to assign a major role for $Ca_v2.1$ and $Ca_v2.2$ in controlling presynaptic neurotransmitter release in cortical neurons (Schenning et al. 2006; Timmermann et al. 2001). R-type ($Ca_v2.3$) is found in the soma and L-type (Ca_v1) predominantly located in dendrites (Timmermann et al. 2001).

More recently, toxins targeting the $Ca_v2.2$ channels have been shown to suppress pain with no development of tolerance when injected intrathecally in rats (Bowersox et al. 1996; Miljanich 2004). The prospect of discovering novel potent antinociceptive drugs has promoted investigations into neurotransmitter release from c-fibres in the spinal cord. One of the most challenging features of the use of peptides as drugs is their high sensitivity to protease degradation – a problem that can be overcome by strategies such as tail to tail cyclisation (Clark et al. 2005).

5.2 *Glycerotoxin*

Glycerotoxin (GLTx) is a potent neurotoxin extracted from the venom of the marine annelid *Glycera convoluta*. This species of bloodworm inhabits sandy burrows and can detect its prey using a set of antennae located near the mouth (Fig. 4a). *Glycera*

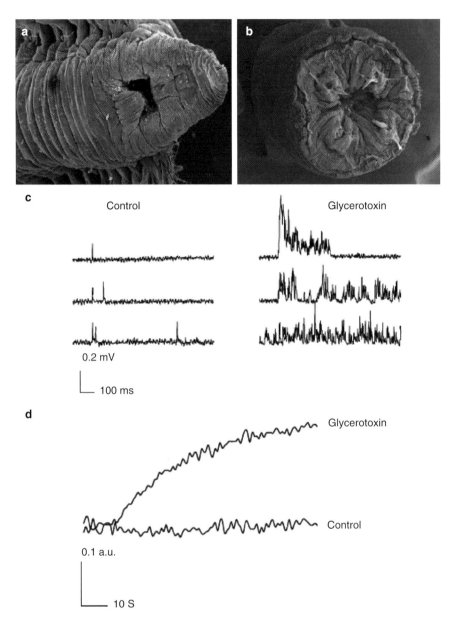

Fig. 4 Effect of GLTx on neurotransmitter release. (**a**) Scanning electron micrograph of the head of *Glycera convoluta* and (**b**) its eversible pharynx fitted with four fangs used to inject prey. (**c**) MEPPs were recorded intracellularly at the frog neuromuscular junction; samples of MEPPs recorded before (control) and after the addition of GLTx (150 nM). (**d**) Endogenous glutamate release from rat brain synaptosomes was recorded in the presence or the absence of GLTx (150 nM), as indicated. Adapted from Meunier et al. (2002a) and Schenning et al. (2006)

is equipped with an eversible pharynx. Upon detection of a prey (shrimp or crab), the worm promptly everts its pharynx towards the prey and uses four fangs to grab its victim and simultaneously inject a venom stored in a gland located at the base of each fang (Fig. 4b).

A crude *Glycera* venom extract was initially shown to cause spontaneous release of acetylcholine at the frog neuromuscular junction for extended periods of time without apparent synaptic vesicle depletion (Morel et al. 1983). The stimulatory effect of both the crude extract and the purified toxin required the presence of external Ca^{2+} (Fig. 4c). The neurotoxin responsible for this effect was later purified and attributed to a 320 kDa single chain glycoprotein. Importantly this stimulatory effect of purified GLTx at the frog neuromuscular junction was prevented by pre-treatment with N-type Ca^{2+} channel ($Ca_V2.2$) blockers (ω-conotoxin GVIA and MVIIA) (Meunier et al. 2002a). GLTx was also shown to stimulate catecholamine secretion from chromaffin cells, an effect also blocked by $Ca_V2.2$ blockers, but not L- or P/Q-type channel blockers (D600 and agatoxin-IVA). The same pharmacological profile was found in rat cortical synaptosomes where sub-nanomolar concentrations of GLTx were shown to promote glutamate release, suggesting that GLTx might selectively act via $Ca_V2.2$ (Fig. 4d) (Schenning et al. 2006). This was confirmed by two pieces of evidence: GLTx directly interacts with $Ca_V2.2$ in vitro and promotes Ca^{2+} influx in HEK293 cells expressing $Ca_V2.2$, but not any other VGCCs (Meunier et al. 2002a; Schenning et al. 2006).

An important point to consider is that GLTx not only increases spontaneous release but also phasic quantal transmitter release for sustained periods of time (Meunier et al. 2002a; Schenning et al. 2006). In view of the latter property and the selective pharmacology of GLTx towards $Ca_V2.2$ including human channel, it is anticipated that GLTx or active fragments of the neurotoxin could be useful as a general strategy to enhance neurotransmitter in $Ca_V2.2$ expressing neurons in the brain. Similar general strategy has been used with some success to ameliorate the cognitive dysfunction of Alzheimer's disease with anticholinesterase drugs (Schenning et al. 2006).

5.3   *Prymnesium Extracts*

Although the neurotoxin from the marine flagellate *Prymnesium patelliferum* has not been fully characterised to date, this extract has been shown to stimulate acetylcholine release from bronchial cholinergic nerves. It was suggested that this extract contains a neurotoxin capable of activating non-specifically HVA VSCC as the potentiating effect could be prevented through the blockade of various VSCC (Meldahl et al. 1996). It is worth noting that the chemical structure of prymnesin 1 and 2 isolated from related *Prymnesium parvum* has been solved and that they possess high haemolytic activity (see review by Yasumoto, 2001). A more recent study has dissected further the neurotoxin effect of a partially purified extract from *Prymnesium parvum* and found that it prevents glutamate uptake in rat synaptosomes and stimulated Ca^{2+}-dependent glutamate release. This effect was blocked by vera-pamil, an L-type Ca^{2+} channel inhibitor, suggesting that the active compound has an activity similar to that of maitotoxin (Mariussen et al. 2005).

6   Neurotoxins Acting on Lipid Metabolism

Pardaxins are a potent family of peptide neurotoxins isolated from the venom of the flatfishes *Pardachirus marmoratus* and *P. pavoninous*. The venomous glands are located on the dorsal fins and have a cytotoxic and haemolytic activity at high con-centrations. It is believed that the venom has evolved as a defence mechanism used by those flatfishes to repel large predators such as sharks. Five pardaxins have been isolated so far (Adermann et al. 1998). They contain 33 amino acids and differ by no more than 3 residues (Fig. 5).

Like a few other pore-forming peptides, pardaxins have a strong antibacterial activity. They actively lyse the bacterial wall and could therefore be useful as an alternative to the presently used antibiotics (Oren and Shai 1996). They are struc-turally related to other peptide families of pore-forming toxins such as mellitin and have a tendency to aggregate both in solution and in lipidic membrane and to create voltage-gated ion channels (Lazarovici et al. 1986). The pore-forming activity is highly dependent on the ability on both the peptide to form α helices and an intramolecular interaction between the N- and C-termini of the peptide suggesting that cyclisation might play an important role in this process, as suggested for a number of other neurotoxins (Clark et al. 2005). A molecular view of membrane inserted pardaxin was recently obtained by NMR, which highlighted the impor-tance of non-covalent interactions in promoting a stable secondary structure (Porcelli et al. 2004). Of interest, pardaxins have recently been shown to promote a redistribution of cholesterol in membranes (Epand et al. 2006).

At the frog neuromuscular junction, *Pardachirus* venom and purified pardaxin enhance both spontaneous and phasic transmitter release, leading to paralysis due to depletion of small synaptic vesicles (Renner et al. 1987). In neurosecretory cells, pardaxins have been shown to promote both secretion, by stimulating intracellular

```
                    10         20         30
Pa1   GFFALIPKII SSPLFKTLLS AVGSALSSSG EQE
Pa2   GFFALIPKII SSPIFKTLLS AVGSALSSSG GQE
Pa3   GFFAFIPKII SSPLFKTLLS AVGSALSSSG EQE
Pa4   GFFALIPKII SSPLFKTLLS AVGSALSSSG GQE
Pa5   GFFAFIPKII SSPLFKTLLS AVGSALSSSG DQE
```

Fig. 5 Primary amino acid sequences of pardaxins (Pa1-Pa5) (Porcelli et al. 2004)

phospholipase A2, and production of arachidonic acid and eicosanoids (Abu-Raya et al. 1998). Arachidonic acid potentiates exocytosis by favouring vesicle fusion competence (Renner et al. 1987) and has recently been shown to act directly on the release machinery by stimulating Soluble N-ethylmaleimide sensitive factor attachment protein receptor (SNARE) protein complex production by altering Munc18 function (Latham and Meunier 2007; Latham et al. 2007). SNAREs are a family of proteins implicated in membrane fusion. Munc18a is an important regulator of the SNARE protein function. Importantly, pardaxin induces cell death in neurosecretory cells by activating a cascade of mitogen-activated protein kinases (MAPKs) (Bloch-Shilderman et al. 2002).

7 Neurotoxin Acting on Non-Specific Channels

7.1 Palytoxins

Palytoxin (PTX) is a large, water-soluble, polyalcohol (Fig. 6). It was first isolated from marine coelenterates (soft corals) of the genus *Palythoa* notably *Palythoa tuberculosa* (Moore and Bartolini 1981; Moore and Scheuer 1971), and subsequently

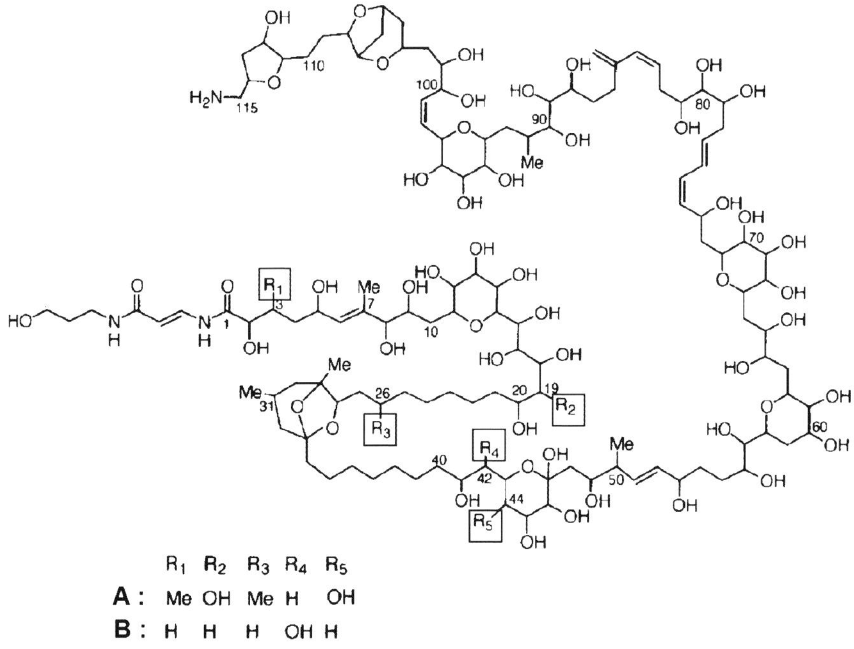

Fig. 6 The structure of palytoxin (**a**) from soft coral species and ostreocin-D (**b**), a palytoxin analogue isolated from the marine dinoflagellate *Ostreopsis siamensis*. The differences (squares) between ostreocin-D and palytoxin are reported in the inset

from the benthic dinoflagellate *Ostreopsis siamensis* which also produce PTX and ostreocin D (42-hydroxy-3, 26-didemethyl-19, 44-dideoxypalytoxin) (Fig. 4), a PTX analog (Rhodes et al. 2002; Taniyama et al. 2003; Ukena et al. 2001; Usami et al. 1995). Recently, PTX and a new analogue, Ovatoxin-a have been identified in the Mediterranean Sea during a massive bloom of the dinoflagellate *Ostreopsis ovata* (Ciminiello et al. 2006; Ciminiello et al. 2008). Thus, at present at least two distinct dinoflagellates belonging to the same genus are known to produce PTX and analogues.

PTX is one of the most potent toxins known to mankind and is also highly lethal when injected to mice. Its lethality is mainly due to a rapid disruption of cardiac function together with severe vasoconstriction. This toxin constitutes a serious threat to human health, as observed in human intoxication caused by PTX after the consumption of crabs (Alcala et al. 1988), mackerel (Kodama et al. 1989), trigger-fish and sardines (reviewed by Molgo et al., 1999). Human symptoms of PTX poisoning are characterised by abdominal cramps, nausea, diarrhoea, paraesthesia, severe muscle spasms and respiratory distress (reviewed by Molgó et al., 1999). Clupeotoxism is a type of seafood poisoning associated with PTX and character-ised by a high mortality rate (Onuma et al. 1999).

PTX is a potent toxin that impairs ATPase activity of the Na^+/K^+ pump and simultaneously increases cation permeability of mammalian cells (Habermann 1989). The following lines of evidence established that PTX interacts with the Na^+/K^+ pump itself: (1) PTX induced ouabain-sensitive cation flux in yeast cells (which lack endogenous Na^+/K^+ pumps) just after the expression of both Na^+/K^+-ATPase α- and β-subunits, and not of either subunit alone (Scheiner-Bobis et al. 1994) (2) incorporation of *in vitro*-translated Na^+/K^+-ATPase α- and β-subunits into lipid bilayers allowed PTX to open cation channels with a unitary conductance of about 7 pS (Hirsh and Wu 1997) similar to that of the relatively nonselective channels opened by PTX in various excitable tissues (Artigas and Gadsby 2004; Ikeda et al. 1988; Muramatsu et al. 1988; Rakowski et al. 2007).

Stimulation of ion flux appears to be a critical component of PTX action that leads to membrane depolarisation (via an increase in Na^+ permeability), triggering a secondary activation of voltage-dependent Ca^{2+} channels, and the activation of both Na^+/Ca^{2+} and Na^+/H^+ exchangers and the Ca^{2+}-ATPase. Altogether, these effects of PTX lead to increases of intracellular Na^+, Ca^{2+} and H^+ that have great impact on cell homeostasis and various physiological processes including the release of neurotransmitters. Despite considerable investigation, the neurotoxic effect of palytoxin is still poorly understood.

PTX has also been reported to induce the release of [³H]norepinephrine in a concentration-dependent manner from clonal rat pheochromocytoma cells (Tatsumi et al. 1984), rabbit aorta (Nagase and Karaki 1987), rat isolated tail artery (Karaki et al. 1988), and catecholamine release from cultured bovine adrenal chromaffin cells (Nakanishi et al. 1991; Yoshizumi et al. 1991).

The release of [³H]norepinephrine induced by low concentrations of PTX is abolished in Na^+-free medium, but the release induced by high PTX concentration (1μM) was unaffected by external Na^+ concentration. PTX caused a concentration-dependent increase in $^{22}Na^+$ and $^{45}Ca^{2+}$ influxes into pheochromocytoma cells (Tatsumi et al. 1984). The PTX-induced $^{45}Ca^{2+}$ influx was markedly inhibited by Co^{2+}, whereas the PTX-induced $^{22}Na^+$ influx was not affected by TTX. Similarly, PTX-induced catecholamine release was dependent on both extracellular Na^+ and Ca^{2+} and was inhibited by organic and inorganic Ca^{2+} channel blockers, but not by TTX.

The PTX-induced increase in intracellular Ca^{2+} can be prevented and reversed by ouabain. This effect is due to the interaction of both PTX and ouabain with the Na^+-K^+ ATPase. Blockade of the normal exchange of Na^+ and K^+ causes the Na^+-Ca^{2+} exchanger to work in reverse mode and this causes a large elevation in the cytosolic Ca^{2+} concentration. This creates an intracellular acidification generated by an influx of protons promoted by the activation of the plasma Ca^{2+}-ATPase (Vale-Gonzalez et al. 2007). PTX (10^{-8} M) has also been reported to increased the release of prostaglandins E2, F2 alpha and 6-keto-prostaglandin F1 alpha from rat aorta possessing an intact endothelium (Nagase and Karaki 1987).

PTX in sub-nanomolar concentrations greatly increases spontaneous quantal ACh release from motor nerve terminals in isolated frog and mouse neuromuscular preparations (Shimahara and Molgó 1990). The increase in miniature endplate frequency induced by PTX is irreversible, long-lasting and occurs even when the external medium contains no Ca^{2+} and is supplemented with a Ca^{2+}-chelating agent (EGTA). Blockade of sodium channels with TTX (see Sect. 3.1) neither prevents nor antagonizes the enhancement of quantal ACh release caused by PTX. Since the concentrations of PTX that increased ACh release also depolarised the muscle fibres, the amplitude of spontaneous and evoked endplate potentials was reduced.

Nerve-evoked quantal transmitter release is transiently enhanced by PTX, but thereafter is blocked probably due to nerve terminal depolarisation and failure of action potentials to invade motor nerve terminals. In rat cerebrocortical synaptosomes, PTX also causes in sub-nano molar concentrations an important increase in [^{3}H]ACh release, together with an increase of cytosolic free Ca^{2+}, an uptake of $^{22}Na^+$ and a decrease in membrane potential (Satoh and Nakazato 1991). Therefore, it is likely that low concentrations of PTX cause membrane depolarisation as a result of an increased Na^+ permeability which, in turn, promotes Ca^{2+} influx and leads to an increase in ACh release. Also, PTX may mobilise Ca^{2+} from intracellular stores, as it has been shown in neuronal cells, in calcium-free medium containing EGTA (Shimahara and Molgó 1990). Also, PTX induces the release of glutamate in primary cultures of rat cerebellar granule cells (Vale-Gonzalez et al. 2007). In conclusion, an important action of PTX at nerve terminals is to promote the release of endogenous transmitters, leading to their eventual depletion.

7.2 Maitotoxin

Maitotoxin (MTX), is an extremely potent highly polar, water soluble, membrane-impermeant polycyclic ether, and the largest nonpolymeric natural product

(3,422 Da) so far known (Murata and Yasumoto 2000; Nicolaou and Frederick 2007). It was first discovered in the surgeon fish *Ctenochaetus striatus* and was named according to its Tahitian name ("maito") for this fish (Yasumoto et al. 1976). MTX is produced by the benthic dinoflagellate *Gambierdiscus toxicus* which also elaborates a number of ciguatoxin precursors known as gambiertoxins that are involved in ciguatera fish poisoning. The exact role of MTX in ciguatera has not been established, but it has been suggested that it contributes to diversifying ciguatera symptoms.

Although attempts have been made to understand the effects of MTX at the cellular and molecular level, its mechanism of action is not yet completely elucidated. It is currently accepted that the main effect of MTX is to activate a non-selective cation channel (NSCC) (Bielfeld-Ackermann et al. 1998; Dietl and Volkl 1994; Estacion et al. 1996; Musgrave et al. 1994; Schilling et al. 1999; Wisnoskey et al. 2004; Worley et al. 1994). Thus, the initial MTX-induced increase in intracellular Ca^{2+} concentration reflects the activation of a Ca^{2+}-permeable non-selective cation channel, which has a reported conductance in the range of 12–40 pS, depending on ionic conditions (Dietl and Volkl 1994; Kobayashi et al. 1987; Nishio et al. 1996). In addition, this NSCC channel poorly discriminates between Na^+ and K^+ (Bielfeld-Ackermann et al. 1998; Dietl and Volkl 1994; Escobar et al. 1998; Estacion et al. 1996; Gusovsky et al. 1990) and causes rapid membrane depolarisation, which in excitable cells, leads to activation of voltage-sensitive channels. Several cations like La^{3+}, Gd^{3+}, Ni^{2+} and Ca^{2+} have been shown to block MTX-elicited currents. Furthermore, several reports have suggested that MTX targets members of the transient receptor potential (TRP) family of channels in several cell types, notably TRPC1 (Brereton et al. 2001; Chen and Barritt 2003; Trevino et al. 2006). In addition, NSCC inhibitor of TRP channels, including di- and tri-valent cations that block MTX currents also block TRP channels (Clapham 2007). The TRP channel families are a large and growing class of channels, encoded by at least 28 genes, displaying an extraordinary selectivity assortment and activation mechanisms (Flockerzi 2007; Owsianik et al. 2006).

At nanomolar concentrations MTX has been reported to produce many responses in a wide variety of mammalian cells, including the stimulation of hormone release from secretory cells and neurotransmitter release from nerve terminals, positive inotropic effect in cardiac muscle, contraction of smooth and striated muscles (Escobar et al. 1998; Gusovsky and Daly 1990), and the stimulation of inositol phosphate production (Sladeczek et al. 1988). All of these diverse actions are dependent on the stimulation of Ca^{2+} influx into the cells. Thus, MTX-induced Ca^{2+} influx is accompanied by a transient increase in catecholamine release in PC12 cells (Takahashi et al. 1983), and norepinephrine release from adrenergic nerve terminals of the guinea pig vas deferens (Kobayashi et al. 1985). MTX applied to isolated diaphragm neuromuscular preparations dramatically enhances spontaneous quantal ACh release, recorded as an increase in MEPPs frequency. This effect requires external Ca^{2+} and is accompanied by muscle fibre depolarisation. In contrast to the enhancement of spontaneous quantal ACh release, MTX causes only a modest increase in the quantal content of endplate potentials representing evoked ACh release (Kim et al. 1985). These actions are probably related to an important rise of intraterminal background Ca^{2+} levels that mainly enhances spontaneous quantal ACh release, rather than evoked transmitter release that is mainly dependent on phasic Ca^{2+} entry during the presynaptic action potential.

8   Conclusions

Marine neurotoxins have been instrumental in achieving our current understanding of synaptic biology. The genomic and proteomic era has provided us with a listing of proteins comparable to the most challenging puzzle and the daunting task to assign function to each individual synaptic protein. In this context, neurotoxins will continue to shed new light on the specific functions of the most essential proteins. Conotoxins are amongst the most well characterised marine toxins and it is somehow reassuring that it has been estimated that only 0.1% of these neuroactive peptides have been characterised (see Chap. 3). No doubt, more exciting milestones in neurobiology will be achieved using marine and other naturally occurring toxins. Moreover, the generation in the laboratory of custom-made neurotoxins that target specific receptor sites is leading the way of modern molecular pharmacology, where rational design is slowly replacing screening strategies.

References

Abu-Raya S, Bloch-Shilderman E, Shohami E, Trembovler V, Shai Y, Weidenfeld J, Yedgar S, Gutman Y, Lazarovici P (1998) Pardaxin, a new pharmacological tool to stimulate the arachidonic acid cascade in PC12 cells. J Pharmacol Exp Ther 287:889–896.

Adachi R, Fukuyo Y (1979) The thecal structure of a marine toxic dinoflagellate *Gambierdiscus toxicus* gen et sp nov collected in a ciguatera-endemic area. Bull Jnp Soc Sci Fish 45:67–71.

Adermann K, Raida M, Paul Y, Abu-Raya S, Bloch-Shilderman E, Lazarovici P, Hochman J, Wellhoner H (1998) Isolation, characterization and synthesis of a novel pardaxin isoform. FEBS Lett 435:173–177.

Alcala AC, Alcala LC, Garth JS, Yasumura D, Yasumoto T (1988) Human fatality due to ingestion of the crab *Demania reynaudii* that contained a palytoxin-like toxin. Toxicon 26:105–107.

Ales E, Gabilan NH, Cano-Abad MF, Garcia AG, Lopez MG (2000) The sea anemone toxin Bc2 induces continuous or transient exocytosis, in the presence of sustained levels of high cytosolic Ca^{2+} in chromaffin cells. J Biol Chem 275:37488–37495.

Allsop JL, Martini L, Lebris H, Pollard J, Walsh J, Hodgkinson S (1986) [Neurologic manifestations of ciguatera 3 cases with a neurophysiologic study and examination of one nerve biopsy]. Rev Neurol (Paris) 142:590–597.

Anderluh G, Barlic A, Potrich C, Macek P, Menestrina G (2000) Lysine 77 is a key residue in aggregation of equinatoxin II, a pore-forming toxin from sea anemone *Actinia equina*. J Membr Biol 173:47–55.

Anderson DM, Lobel PS (1987) The continuing enigma of ciguatera. Biol Bill 172:89–107.

Artigas P, Gadsby DC (2004) Large diameter of palytoxin-induced Na/K pump channels and modulation of palytoxin interaction by Na/K pump ligands. J Gen Physiol 123:357–376.

Atchison W, Luke V, Narahashi T, Vogel S (1986) Nerve membrane sodium channels as the target site of brevetoxins at neuromuscular junctions. Br J Pharmacol 89:731–738.

Baden DG (1989) Brevetoxins: unique polyether dinoflagellate toxins. FASEB J 3:1807–1817.

Benoit E, Legrand A-M (1994) Gambiertoxin-induced modifications of the membrane potential of myelinated nerve fibres. Mem Queensland Mus 34:461–464.

Benoit E, Juzans P, Legrand AM, Molgó J (1996) Nodal swelling produced by ciguatoxin-induced selective activation of sodium channels in myelinated nerve fibers. Neuroscience 71:1121–1131.

Bidard JN, Vijverberg HP, Frelin C, Chungue E, Legrand A-M, Bagnis R, Lazdunski M (1984) Ciguatoxin is a novel type of Na$^+$ channel toxin. J Biol Chem 259:8353–8357.

Bielfeld-Ackermann A, Range C, Korbmacher C (1998) Maitotoxin (MTX) activates a nonselective cation channel in Xenopus laevis oocytes. Pflügers Arch 436:329–337.

Birinyi-Strachan LC, Gunning SJ, Lewis RJ, Nicholson GM (2005) Block of voltage-gated potassium channels by Pacific ciguatoxin-1 contributes to increased neuronal excitability in rat sensory neurons. Toxicol Appl Pharmacol 204:175–186.

Bloch-Shilderman E, Jiang H, Lazarovici P (2002) Pardaxin, an ionophore neurotoxin, induces PC12 cell death: activation of stress kinases and production of reactive oxygen species. J Nat Toxins 11:71–85.

Boccaccio A, Conti F, Olivera BM, Terlau H (2004) Binding of kappa-conotoxin PVIIA to Shaker K$^+$ channels reveals different K$^+$ and Rb$^+$ occupancies within the ion channel pore. J Gen Physiol 124:71–81.

Bourdelais AJ, Campbell S, Jacocks H, Naar J, Wright JL, Carsi J, Baden DG (2004) Brevenal is a natural inhibitor of brevetoxin action in sodium channel receptor binding assays. Cell Mol Neurobiol 24:553–563.

Bowersox SS, Miljanich GP, Sugiura Y, Li C, Nadasdi L, Hoffman BB, Ramachandran J, Ko CP (1995) Differential blockade of voltage-sensitive calcium channels at the mouse neuromuscular junction by novel ω–conopeptides and ω-agatoxin-IVA. J Pharmacol Exp Ther 273:248–256.

Bowersox SS, Gadbois T, Singh T, Pettus M, Wang YX, Luther RR (1996) Selective N-type neuronal voltage-sensitive calcium channel blocker, SNX-111, produces spinal antinociception in rat models of acute, persistent and neuropathic pain. J Pharmacol Exp Ther 279:1243–1249.

Brereton HM, Chen J, Rychkov G, Harland ML, Barritt GJ (2001) Maitotoxin activates an endogenous non-selective cation channel and is an effective initiator of the activation of the heterologously expressed hTRPC-1 (transient receptor potential) non-selective cation channel in H4-IIE liver cells. Biochim Biophys Acta 1540:107–126.

Cameron J, Flowers AE, Capra MF (1991) Electrophysiological studies on ciguatera poisoning in man (Part II). J Neurol Sci 101:93–97.

Cartier GE, Yoshikami D, Gray WR, Luo S, Olivera BM, McIntosh JM (1996) A new α-conotoxin which targets α3β2 nicotinic acetylcholine receptors. J Biol Chem 271:7522–7528.

Catterall WA (2000) From ionic currents to molecular mechanisms: the structure and function of voltage-gated sodium channels. Neuron 26:13–25.

Catterall WA, Goldin AL, Waxman SG (2005) International Union of Pharmacology XLVII Nomenclature and structure-function relationships of voltage-gated sodium channels. Pharmacol Rev 57:397–409.

Catterall WA, Cestele S, Yarov-Yarovoy V, Yu FH, Konoki K, Scheuer T (2007) Voltage-gated ion channels and gating modifier toxins. Toxicon 49:124–141.

Ceccarelli B, Hurlbut WP (1980) Vesicle hypothesis of the release of quanta of acetylcholine. Physiol Rev 60:396–441.

Chen J, Barritt GJ (2003) Evidence that TRPC1 (transient receptor potential canonical 1) forms a Ca^{2+}-permeable channel linked to the regulation of cell volume in liver cells obtained using small interfering RNA targeted against. TRPC1 Biochem J 373:327–36.

Ciminiello P, Dell'Aversano C, Fattorusso E, Forino M, Magno GS, Tartaglione L, Grillo C, Melchiorre N (2006) The Genoa 2005 outbreak. Determination of putative palytoxin in Mediterranean Ostreopsis ovata by a new liquid chromatography tandem mass spectrometry method. Anal Chem 78:6153–6159.

Ciminiello P, Dell'aversano C, Fattorusso E, Forino M, Tartaglione L, Grillo C, Melchiorre N (2008) Putative palytoxin and its new analogue, ovatoxin-a, in *Ostreopsis ovata* collected along the Ligurian coasts during the 2006 toxic outbreak. J Am Soc Mass Spectrom 19:111–210.

Clapham DE (2007) SnapShot: mammalian TRP channels. Cell 129:220.

Clark RJ, Fischer H, Dempster L, Daly NL, Rosengren KJ, Nevin ST, Meunier FA, Adams DJ, Craik DJ (2005) Engineering stable peptide toxins by means of backbone cyclization: stabilization of the α-conotoxin MII. Proc Natl Acad Sci USA 102:13767–13772.

Colasante C, Meunier FA, Kreger AS, Molgó J (1996) Selective depletion of clear synaptic vesicles and enhanced quantal transmitter release at frog motor nerve endings produced by trachynilysin, a protein toxin isolated from stonefish (*Synanceia trachynis*) venom. Eur J Neurosci 8:2149–2156.

Cruz LJ, Johnson DS, Olivera BM (1987) Characterization of the ω-conotoxin target. Evidence for tissue-specific heterogeneity in calcium channel types. Biochemistry 26:820–824.

Cuypers E, Yanagihara A, Rainier JD, Tytgat J (2007) TRPV1 as a key determinant in ciguatera and neurotoxic shellfish poisoning. Biochem Biophys Res Commun 361:214–217.

Dauplais M, Lecoq A, Song J, Cotton J, Jamin N, Gilquin B, Roumestand C, Vita C, de Medeiros CL, Rowan EG, Harvey AL, Menez A (1997) On the convergent evolution of animal toxins. Conservation of a diad of functional residues in potassium channel-blocking toxins with unrelated structures. J Biol Chem 272:4302–4309.

de los Rios V, Mancheno JM, Martinez del Pozo A, Alfonso C, Rivas G, Onaderra M, Gavilanes JG (1999) Sticholysin II, a cytolysin from the sea anemone *Stichodactyla helianthus*, is a monomer-tetramer associating protein. FEBS Lett 455:27–30.

Dietl P, Volkl H (1994) Maitotoxin activates a nonselective cation channel and stimulates Ca^{2+} entry in MDCK renal epithelial cells. Mol Pharmacol 45:300–305.

Dravid SM, Baden DG, Murray TF (2005) Brevetoxin augments NMDA receptor signaling in murine neocortical neurons. Brain Res 1031:30–38.

Ellinor PT, Zhang JF, Horne WA, Tsien RW (1994) Structural determinants of the blockade of N-type calcium channels by a peptide neurotoxin. Nature 372:272–275.

Epand RF, Ramamoorthy A, Epand RM (2006) Membrane lipid composition and the interaction of pardaxin: the role of cholesterol. Protein Pep Lett 13:1–5.

Escobar LI, Salvador C, Martinez M, Vaca L (1998) Maitotoxin, a cationic channel activator. Neurobiology 6:59–74.

Estacion M, Nguyen HB, Gargus JJ (1996) Calcium is permeable through a maitotoxin-activated nonselective cation channel in mouse L cells. Am J Physiol 270:C1145–1152.

Flewelling LJ, Naar JP, Abbott JP, Baden DG, Barros NB, Bossart GD, Bottein MYD, Hammond DG, Haubold EM, Heil CA, Henry MS, Jacocks HM, Leighfield TA, Pierce RH, Pitchford TD, Rommel SA, Scott PS, Steidinger KA, Truby EW, Van Dolah FM, Landsberg JH (2005) Brevetoxicosis: Red tides and marine mammal mortalities. Nature 435:755–756.

Flockerzi V (2007) An introduction on TRP channels. Handb Exp Pharmacol 179:1–19.

Fuwa H, Kainuma N, Satake M, Sasaki M (2003) Synthesis and biological evaluation of gambierol analogues. Bioorg Med Chem Lett 13:2519–2522.

Fuwa H, Kainuma N, Tachibana K, Sasaki M (2002) Total synthesis of (-)-gambierol. J Am Chem Soc 124:14983–14992.

Fuwa H, Kainuma N, Tachibana K, Tsukano C, Satake M, Sasaki M (2004) Diverted total synthesis and biological evaluation of gambierol analogues: elucidation of crucial structural elements for potent toxicity. Chem Eur J 10:4894–4909.

Ghiaroni V, Sasaki M, Fuwa H, Rossini G P, Scalera G, Yasumoto T, Pietra P, Bigiani A (2005). Inhibition of voltage-gated potassium currents by gambierol in mouse taste cells. Toxicol Sci 85:657–665.

Goldin AL, Barchi RL, Caldwell JH, Hofmann F, Howe JR, Hunter JC, Kallen RG, Mandel G, Meisler MH, Netter YB, Noda M, Tamkun MM, Waxman SG, Wood JN, Catterall WA (2000) Nomenclature of voltage-gated sodium channels. Neuron 28:365–368.

Gusovsky F, Daly JW (1990) Maitotoxin: a unique pharmacological tool for research on calcium-dependent mechanisms. Biochem Pharmacol 39:1633–1639.

Gusovsky F, Bitran JA, Yasumoto T, Daly JW (1990) Mechanism of maitotoxin-stimulated phosphoinositide breakdown in HL-60 cells. J Pharmacol Exp Ther 252:466–473.

Haack JA, Rivier J, Parks TN, Mena EE, Cruz LJ, Olivera BM (1990) Conantokin-T A gamma-carboxyglutamate containing peptide with N-methyl-d-aspartate antagonist activity. J Biol Chem 265:6025–6029.

Habermann E (1989) Palytoxin acts through Na$^+$, K$^+$-ATPase. Toxicon 27:1171–1187.

Hidalgo J, Liberona JL, Molgó J, Jaimovich E (2002) Pacific ciguatoxin-1b effect over Na$^+$ and K$^+$ currents, inositol 1,4,5-triphosphate content and intracellular Ca^{2+} signals in cultured rat myotubes. Br J Pharmacol 137:1055–62.

Hille B (2001) Ion channels of excitable membranes. 3rd Ed, Sinauer Assoc, Sunderland, MA.

Hinds MG, Zhang W, Anderluh G, Hansen PE, Norton RS (2002) Solution structure of the eukaryotic pore-forming cytolysin equinatoxin II: implications for pore formation. J Mol Biol 315:1219–1229.

Hirsh JK, Wu CH (1997) Palytoxin-induced single-channel currents from the sodium pump synthesized by in vitro expression. Toxicon 35:169–176.

Huang JM, Wu CH, Baden DG (1984) Depolarizing action of a red-tide dinoflagellate brevetoxin on axonal membranes. J Pharmacol Exp Ther 229:615–621.

Ikeda M, Mitani K, Ito K (1988) Palytoxin induces a nonselective cation channel in single ventricular cells of rat. Naunyn Schmiedebergs Arch Pharmacol 337:591–593.

Inoue M, Hirama M, Satake M, Sugiyama K, Yasumoto T (2003) Inhibition of brevetoxin binding to the voltage-gated sodium channel by gambierol and gambieric acid-A. Toxicon 41:469–474.

Ishida H, Nozawa A, Nukaya H, Tsuji K (2004) Comparative concentrations of brevetoxins PbTx-2, PbTx-3, BTX-B1 and BTX-B5 in cockle, *Austrovenus stutchburyi*, greenshell mussel, *Perna canaliculus*, and Pacific oyster, *Crassostrea gigas*, involved neurotoxic shellfish poisoning in New Zealand. Toxicon 43:779–789.

Ito E, Suzuki-Toyota F, Toshimori K, Fuwa H, Tachibana K, Satake M, Sasaki M (2003) Pathological effects on mice by gambierol, possibly one of the ciguatera toxins. Toxicon 42:733–740.

Iwasaki S, Momiyama A, Uchitel OD, Takahashi T (2000) Developmental changes in calcium channel types mediating central synaptic transmission. J Neurosci 20:59–65.

Jan LY, Jan YN (1997) Voltage-gated and inwardly rectifying potassium channels. J Physiol 505:267–282.

Jeglitsch G, Rein K, Baden DG, Adams DJ (1998) Brevetoxin-3 (PbTx-3) and its derivatives modulate single tetrodotoxin-sensitive sodium channels in rat sensory neurons. J Pharmacol Exp Ther 284:516–525.

Johnson HW, Majumder U, Rainier JD (2006) Total synthesis of gambierol: subunit coupling and completion. Chem Eur J 12:1747–1753.

Kao CY, Nishiyama A (1965) Actions of saxitoxin on peripheral neuromuscular systems. J Physiol 180:50–66.

Karaki H, Nagase H, Ohizumi Y, Satake N, Shibata S (1988) Palytoxin-induced contraction and release of endogenous noradrenaline in rat tail artery. Br J Pharmacol 95:183–188.

Khoo HE (2002) Bioactive proteins from stonefish venom. Clin Exp Pharmacol Physiol 29: 802–806.

Kim YI, Login IS, Yasumoto T (1985) Maitotoxin activates quantal transmitter release at the neuromuscular junction: evidence for elevated intraterminal Ca^{2+} in the motor nerve terminal. Brain Res 346:357–362.

Kobayashi M, Ohizumi Y, Yasumoto T (1985) The mechanism of action of maitotoxin in relation to Ca^{2+} movements in guinea-pig and rat cardiac muscles. Br J Pharmacol 86:385–391.

Kobayashi M, Ochi R, Ohizumi Y (1987) Maitotoxin-activated single calcium channels in guinea-pig cardiac cells. Br J Pharmacol 92:665–671.

Kodama AM, Hokama Y, Yasumoto T, Fukui M, Manea SJ, Sutherland N (1989) Clinical and laboratory findings implicating palytoxin as cause of ciguatera poisoning due to *Decapterus macrosoma* (mackerel). Toxicon 27:1051–1053.

Kreger AS (1991) Detection of a cytolytic toxin in the venom of the stonefish (*Synanceia trachynis*). Toxicon 29:733–743.

Kreger AS, Molgó J, Comella JX, Hansson B, Thesleff S (1993) Effects of stonefish (*Synanceia trachynis*) venom on murine and frog neuromuscular junctions. Toxicon 31:307–317.

Landsberg JH, Hall S, Johannessen JN, White KD, Conrad SM, Abbott JP, Flewelling LJ, Richardson RW, Dickey RW, Jester EL, Etheridge S, Deeds J, Van Dolah F, Leighfield T, Zou Y, Beaudry C, Benner R, Rogers P, Scott P, Kawabata K, Wolny J, Steidinger K (2006) Saxitoxin puffer fish poisoning in the United States, with the first report of *Pyrodinium bahamense* as the putative toxin source. Environ Health Perspect 114:1502–1507.

Latham CF, Meunier FA (2007) Munc18a: Munc-y business in mediating exocytosis. Int J Biochem Cell Biol 39:1576–1581.

Latham CF, Osborne SL, Cryle MJ, Meunier FA (2007) Arachidonic acid potentiates exocytosis and allows neuronal SNARE complex to interact with Munc18a. J Neurochem 100:1543–1554.

Lazarovici P, Primor N, Loew LM (1986) Purification and pore-forming activity of two hydrophobic polypeptides from the secretion of the Red Sea Moses sole (*Pardachirus marmoratus*). J Biol Chem 261:16704–16713.

LePage KT, Rainier JD, Johnson HW, Baden DG, Murray TF (2007) Gambierol acts as a functional antagonist of neurotoxin site 5 on voltage-gated sodium channels in cerebellar granule neurons. J Pharmacol Exp Ther 323:174–179.

Levine DZ (1995) Ciguatera: current concepts. J Am Osteopath Assoc 95:193–198.

Lewis RJ, Garcia ML (2003) Therapeutic potential of venom peptides. Nat Rev Drug Discov 2:790–802.

Lewis RJ, Nielsen KJ, Craik DJ, Loughnan ML, Adams DA, Sharpe IA, Luchian T, Adams DJ, Bond T, Thomas L, Jones A, Matheson JL, Drinkwater R, Andrews PR, Alewood PF (2000) Novel ω-conotoxins from *Conus catus* discriminate among neuronal calcium channel subtypes. J Biol Chem 275:35335–35344.

Lombet A, Bidard J-N, Lazdunski M (1987) Ciguatoxin and brevetoxins share a common receptor site on the neuronal voltage-dependent Na^+ channel. FEBS Lett 219:355–359.

Louzao MC, Cagide E, Vieytes MR, Sasaki M, Fuwa H, Yasumoto T, Botana LM (2006) The sodium channel of human excitable cells is a target for gambierol. Cell Physiol Biochem 17:257–268.

Malovrh P, Barlic A, Podlesek Z, MaCek P, Menestrina G, Anderluh G (2000) Structure-function studies of tryptophan mutants of equinatoxin II, a sea anemone pore-forming protein. Biochem J 346:223–232.

Mariussen E, Nelson G N, Fonnum F (2005) A toxic extract of the marine phytoflagellate Prymnesium parvum induces calcium-dependent release of glutamate from rat brain synaptosomes. J Toxicol Environ Health A 68:67–79.

Mattei C, Wen PJ, Nguyen-Huu TD, Alvarez M, Benoit E, Bourdelais AJ, Lewis RJ, Baden DG, Molgo J, Meunier, FA (2008) Brevenal inhibits pacific ciguatoxin-1B-induced neurosecretion from bovine chromaffin cells. PLoS ONE 3:e3448.

McCleskey EW, Fox AP, Feldman DH, Cruz LJ, Olivera BM, Tsien RW, Yoshikami D (1987) ω-Conotoxin: direct and persistent blockade of specific types of calcium channels in neurons but not muscle. Proc Natl Acad Sci USA 84:4327–4331.

Meir A, Ginsburg S, Butkevich A, Kachalsky SG, Kaiserman I, Ahdut R, Demirgoren S, Rahamimoff R (1999) Ion channels in presynaptic nerve terminals and control of transmitter release. Physiol Rev 79:1019–1088.

Meldahl AS, Aas P, Fonnum F (1996) Extract of the marine alga *Prymnesium patelliferum* induces release of acetylcholine from cholinergic nerves in the rat bronchial smooth muscle. Acta Physiol Scand 156:99–107.

Menez A (1998) Functional architectures of animal toxins: a clue to drug design? Toxicon 36: 1557–1572.

Meunier FA, Colasante C, Molgó J (1997) Sodium-dependent increase in quantal secretion induced by brevetoxin-3 in Ca^{2+}-free medium is associated with depletion of synaptic vesicles and swelling of motor nerve terminals in situ. Neuroscience 78:883–893.

Meunier FA, Mattei C, Chameau P, Lawrence G, Colasante C, Kreger AS, Dolly JO, Molgó J (2000) Trachynilysin mediates SNARE-dependent release of catecholamines from chromaffin cells via external and stored Ca^{2+}. J Cell Sci 113:1119–1125.

Meunier FA, Feng ZP, Molgó J, Zamponi GW, Schiavo G (2002a) Glycerotoxin from *Glycera convoluta* stimulates neurosecretion by up-regulating N-type Ca^{2+} channel activity. EMBO J 21:6733–6743.

Meunier FA, Ouanounou G, Mattei C, Chameau P, Colasante C, Ushkaryov YA, Dolly JO, Kreger AS, Molgó J (2002b) Secretagogue activity of trachynilysin, a neurotoxic protein isolated from stonefish (*Synanceia trachynis*) venom. In: Massaro EJ (ed) Handbook of neurotoxicology. Humana Press, Inc., Totowa, NJ, Vol. 1, pp. 595–616.

Migues PV, Leal RB, Mantovani M, Nicolau M, Gabilan NH (1999) Synaptosomal glutamate release induced by the fraction Bc2 from the venom of the sea anemone *Bunodosoma caissarum*. Neuroreport 10:67–70.

Miledi R (1967) Spontaneous synaptic potentials and quantal release of transmitter in the stellate ganglion of the squid. J Physiol (Lond) 192:379–406.

Miljanich GP (2004) Ziconotide: neuronal calcium channel blocker for treating severe chronic pain. Curr Med Chem 11:3029–3040.

Molgó J, Comella JX, Legrand A-M (1990) Ciguatoxin enhances quantal transmitter release from frog motor nerve terminals. Br J Pharmacol 99:695–700.

Molgo J, Benoit E, Legrand AM, Kreger AS (1999) Bioactive agents involved in fish poisoning: an overview. Proceedings of the 5th Indo-Pacific Fish Conference, pp. 721–738.

Moore RE, Bartolini G (1981) Structure of palytoxin. J Am Chem Soc 103:2491–2494.

Moore RE, Scheuer PJ (1971) Palytoxin: a new marine toxin from a coelenterate. Science 172:495–498.

Morel N, Thieffry M, Manaranche R (1983) Binding of a *Glycera convoluta* neurotoxin to cholinergic nerve terminal plasma membranes. J Cell Biol 97:1737–1744.

Morohashi A, Satake M, Naoki H, Kaspar HF, Oshima Y, Yasumoto T (1999) Brevetoxin B4 isolated from greenshell mussels *Perna canaliculus*, the major toxin involved in neurotoxic shellfish poisoning in New Zealand. Nat Toxins 7:45–48.

Mouhat S, De Waard M, Sabatier JM (2005) Contribution of the functional dyad of animal toxins acting on voltage-gated Kv1-type channels. J Pept Sci 11:65–68.

Muramatsu I, Nishio M, Kigoshi S, Uemura D (1988) Single ionic channels induced by palytoxin in guinea-pig ventricular myocytes. Br J Pharmacol 93:811–816.

Murata M, Yasumoto T (2000) The structure elucidation and biological activities of high molecular weight algal toxins: maitotoxin, prymnesins and zooxanthellatoxins. Nat Prod Rep 17: 293–314.

Musgrave IF, Seifert R, Schultz G (1994) Maitotoxin activates cation channels distinct from the receptor-activated non-selective cation channels of HL-60 cells. Biochem J 301:437–441.

Nagase H, Karaki H (1987) Palytoxin-induced contraction and release of prostaglandins and norepinephrine in the aorta. J Pharmacol Exp Ther 242:1120–1125.

Narahashi T, Deguchi T, Urakawa N, Ohkubo Y (1960) Stabilization and rectification of muscle fiber membrane by tetrodotoxin. Am J Physiol 198:934–938.

Nakanishi A, Yoshizumi M, Morita K, Murakumo Y, Houchi H, Oka M (1991) Palytoxin: a potent stimulator of catecholamine release from cultured bovine adrenal chromaffin cells. Neurosci Lett 121:163–165.

Narahashi T, Moore JW, Scott WR (1964) Tetrodotoxin blockage of sodium conductance increase in lobster giant axons. J Gen Physiol 47:965–974.

Narahashi T, Haas HG, Therrien EF (1967) Saxitoxin and tetrodotoxin: comparison of nerve blocking mechanism. Science 157:1441–1442.

Nicolaou KC, Frederick MO (2007) On the structure of maitotoxin. Angew Chem Int Ed Engl 46:5278–5282.

Nishio M, Muramatsu I, Yasumoto T (1996) Na$^+$-permeable channels induced by maitotoxin in guinea-pig single ventricular cells. Eur J Pharmacol 297:293–298.

Nishiyama A (1967) Effect of saxitoxin on the end plate of frog muscle. Nature 215:201–202.

Norton RS, Pennington MW, Wulff H (2004) Potassium channel blockade by the sea anemone toxin ShK for the treatment of multiple sclerosis and other autoimmune diseases. Curr Med Chem 11:3041–3052.

Ogata N, Ohishi Y (2002) Molecular diversity of structure and function of the voltage-gated Na⁺ channels. Jpn J Pharmacol 88:365–377.

Onuma Y, Satake M, Ukena T, Roux J, Chanteau S, Rasolofonirina N, Ratsimaloto M, Naoki H, Yasumoto T (1999) Identification of putative palytoxin as the cause of clupeotoxism. Toxicon 37:55–65.

Oren Z, Shai Y (1996) A class of highly potent antibacterial peptides derived from pardaxin, a pore-forming peptide isolated from Moses sole fish *Pardachirus marmoratus*. Eur J Biochem 237:303–310.

Ouanounou G, Malo M, Stinnakre J, Kreger AS, Molgó J (2002) Trachynilysin, a neurosecretory protein isolated from stonefish (*Synanceia trachynis*) venom, forms nonselective pores in the membrane of NG108–15 cells. J Biol Chem 277:39119–39127.

Owsianik G, Talavera K, Voets T, Nilius B (2006) Permeation and selectivity of TRP channels. Annu Rev Physiol 68:685–717.

Poli MA, Lewis RJ, Dickey RW, Musser SM, Buckner CA, Carpenter LG (1997) Identification of Caribbean ciguatoxins as the cause of an outbreak of fish poisoning among US soldiers in Haiti. Toxicon 35:733–741.

Poli MA, Mende TJ, Baden DG (1986) Brevetoxins, unique activators of voltage-sensitive sodium channels, bind to specific sites in rat brain synaptosomes. Mol Pharmacol 30:129–135.

Porcelli F, Buck B, Lee DK, Hallock KJ, Ramamoorthy A, Veglia G (2004) Structure and orientation of pardaxin determined by NMR experiments in model membranes. J Biol Chem 279:45815–45823.

Potera C (2007) Florida red tide brews up drug lead for cystic fibrosis. Science 316:1561–1562.

Rakowski RF, Artigas P, Palma F, Holmgren M, De Weer P, Gadsby DC (2007) Sodium flux ratio in Na/K pump-channels opened by palytoxin. J Gen Physiol 130:41–54.

Renner P, Caratsch CG, Waser PG, Lazarovici P, Primor N (1987) Presynaptic effects of the pardaxins, polypeptides isolated from the gland secretion of the flatfish *Pardachirus marmoratus*. Neuroscience 23:319–325.

Reynolds IJ, Wagner JA, Snyder SH, Thayer SA, Olivera BM, Miller RJ (1986) Brain voltage-sensitive calcium channel subtypes differentiated by ω-conotoxin fraction GVIA. Proc Natl Acad Sci USA 83:8804–8807.

Rhodes L, Towers N, Briggs L, Munday R, Adamson J (2002) Uptake of palytoxin-like compounds by shellfish fed *Ostreopsis siamensis* (*Dinophyceae*). NZ J Mar Freshwater Res 36:631–636.

Rosato Siri MD, Uchitel OD (1999) Calcium channels coupled to neurotransmitter release at neonatal rat neuromuscular junctions. J Physiol 514:533–540.

Satake M, Murata M, Yasumoto T (1993) Gambierol: a new toxic polyether compound isolated from the marine dinoflagellate *Gambierdiscus toxicus*. J Am Chem Soc 115:361–362.

Satoh E, Nakazato Y (1991) Mode of action of palytoxin on the release of acetylcholine from rat cerebrocortical synaptosomes. J Neurochem 57:1276–1280.

Schantz EJ, Johnson EA (1992) Properties and use of botulinum toxin and other microbial neurotoxins in medicine. Microbiol Rev 56:80–99.

Scheiner-Bobis G, Meyer zu Heringdorf D, Christ M, Habermann E (1994) Palytoxin induces K⁺ efflux from yeast cells expressing the mammalian sodium pump. Mol Pharmacol 45:1132–1136.

Schenning M, Proctor DT, Ragnarsson L, Barbier J, Lavidis NA, Molgó J, Zamponi GW, Schiavo G, Meunier FA (2006) Glycerotoxin stimulates neurotransmitter release from N-type Ca²⁺ channel expressing neurons. J Neurochem 98:894–904.

Schilling WP, Sinkins WG, Estacion M (1999) Maitotoxin activates a nonselective cation channel and a P2Z/P2X(7)-like cytolytic pore in human skin fibroblasts. Am J Physiol 277:C755–C765.

Schlumberger S, Girard E, Bournaud R, Sasaki M, Fuwa H, Cagide E, Louzao MC, Botana LM, Molgó J, Benoit E (2007) Effets du gambierol sur les canaux potassium et la libération de neurotransmetteurs. In: Goudey-Perrière F, Benoit E, Marchot P, Popoff MR (eds) Toxines émergentes: nouveaux risques. Collection rencontres en toxinologie. Librairie Lavoisier, Paris, pp. 157–158.

Sharkey RG, Jover E, Couraud F, Baden DG, Catterall WA (1987) Allosteric modulation of neurotoxin binding to voltage-sensitive sodium channels by *Ptychodiscus brevis* toxin 2. Mol Pharmacol 31:273–278.

Shimahara T, Molgó J (1990) Palytoxin enhances quantal acetylcholine release from motor nerve terminals and increases cytoplasmic calcium levels in a neuronal hybrid cell line. Life Sci Adv Pharmacol 9:785–792.

Sladeczek F, Schmidt BH, Alonso R, Vian L, Tep A, Yasumoto T, Cory RN, Bockaert J (1988) New insights into maitotoxin action. Eur J Biochem 174:663–670

Südhof TC (2004) The synaptic vesicle cycle. Annu Rev Neurosci 27:509–547.

Swift AE, Swift TR (1993) Ciguatera. J Toxicol Clin Toxicol 31:1–29.

Takahashi M, Tatsumi M, Ohizumi Y, Yasumoto T (1983) Ca^{2+} channel activating function of maitotoxin, the most potent marine toxin known, in clonal rat pheochromocytoma cells. J Biol Chem 258:10944–10949.

Taniyama S, Arakawa O, Terada M, Nishio S, Takatani T, Mahmud Y, Noguchi T (2003) *Ostreopsis sp*, a possible origin of palytoxin (PTX) in parrotfish *Scarus ovifrons*.Toxicon 42:29–33.

Tatsumi M, Takahashi M, Ohizumi Y (1984) Mechanism of palytoxin-induced [3H]norepine-phrine release from a rat pheochromocytoma cell line. Mol Pharmacol 25:379–83.

Tejuca M, Serra MD, Ferreras M, Lanio ME, Menestrina G (1996) Mechanism of membrane permeabilization by sticholysin I, a cytolysin isolated from the venom of the sea anemone *Stichodactyla helianthus*. Biochemistry 35:14947–14957.

Terlau H, Olivera BM (2004) Conus venoms: a rich source of novel ion channel-targeted peptides. Physiol Rev 84:41–68.

Timmermann DB, Lund TM, Belhage B, Schousboe A (2001) Localization and pharmacological characterization of voltage dependent calcium channels in cultured neocortical neurons. Int J Dev Neurosci 19:1–10.

Tosteson TR, Ballantine DL, Durst HD (1988) Seasonal frequency of ciguatoxic barracuda in southwest Puerto Rico. Toxicon 26:795–801.

Trevino CL, De la Vega-Beltran JL, Nishigaki T, Felix R, Darszon A (2006) Maitotoxin potently promotes Ca^{2+} influx in mouse spermatogenic cells and sperm, and induces the acrosome reaction. J Cell Physiol 206:449–456.

Ukena T, Satake M, Usami M, Oshima Y, Naoki H, Fujita T, Kan Y, Yasumoto T (2001) Structure elucidation of ostreocin D, a palytoxin analog isolated from the dinoflagellate *Ostreopsis siamensis*. Biosci Biotechnol Biochem 65:2585–2588.

Usami M, Satake M, Ishida S, Inoue A, Kan Y, Yasumoto T (1995) Palytoxin analogs from the dinoflagellate *Ostreopsis siamensis*. J Am Chem Soc 177:5389–5390.

Vale-Gonzalez C, Gomez-Limia B, Vieytes MR, Botana LM (2007) Effects of the marine phyco-toxin palytoxin on neuronal pH in primary cultures of cerebellar granule cells. J Neurosci Res 85:90–98.

Van der Kloot W, Molgó J (1994) Quantal acetylcholine release at the vertebrate neuromuscular junction. Physiol Rev 74:899–991.

Wisnoskey BJ, Estacion M, Schilling WP (2004) Maitotoxin-induced cell death cascade in bovine aortic endothelial cells: divalent cation specificity and selectivity. Am J Physiol Cell Physiol 287:C345–C356.

Worley JF 3rd, McIntyre MS, Spencer B, Dukes ID (1994) Depletion of intracellular Ca^{2+} stores activates a maitotoxin-sensitive nonselective cationic current in β-cells. J Biol Chem 269:32055–32058.

Yasumoto T (2001) The chemistry and biological function of natural marine toxins. Chem Record 1:228–242.

Yasumoto T, Bagnis R, Venoux JP (1976) Toxicity study of the surgeon fishes-II: Properties of the principal water-soluble toxin. Bull Jpn Soc Sci Fish 42:359–336.

Yasumoto T, Nagai H, Yasumura D, Michishita T, Endo A, Yotsu M, Kotaki Y (1986) Interspecies distribution and possible origin of tetrodotoxin. Ann NY Acad Sci 479:44–51.

Yoshizumi M, Nakanishi A, Houchi H, Morita K, Katoh I, Oka M (1991) Characterization of palytoxin-induced catecholamine secretion from cultured bovine adrenal chromaffin cells. Effects of Na^+- and Ca^{2+}-channel blockers. Biochem Pharmacol 42:17–23.

Yu FH, Catterall WA (2003) Overview of the voltage-gated sodium channel family. Genome Biol 4:207.

Toxins Affecting Actin Filaments and Microtubules

Shin-ya Saito

Abstract Actin and tubulin are the two major proteins of the cytoskeleton in eukaryotic cells and both display a common property to reversibly assemble into long and flexible polymers, actin filaments and microtubules, respectively. These proteins play important roles in a variety of cellular functions and are also involved in numbers of diseases. An emerging number of marine-derived cytotoxins have been found to bind either actin or tublin, resulting in either inhibition or enhancement of polymerization. Thus, these toxins are valuable molecular probes for solving complex mechanisms of biological processes. This chapter describes actin- and tubulin-targeting marine natural products and their modes of action, with reference to their use as research tools and their clinical applications.

Shin-ya Saito (✉)

Department of Pharmacology, School of Pharmaceutical Sciences, University of Shizuoka, Yada 52-1, Suruga-ku, Shizuoka 422-8526, Japan
e-mail: synsaito@u-shizuoka-ken.ac.jp

N. Fusetani and W. Kem (eds.), *Marine Toxins as Research Tools*,
Progress in Molecular and Subcellular Biology, Marine Molecular Biotechnology 46,
DOI: 10.1007/978-3-540-87895-7, © Springer-Verlag Berlin Heidelberg 2009

1 Introduction

The cytoskeleton is an important component regulating cell functions. It maintains cell shape, enables cellular motion and plays important roles in both intracellular transport and cellular division. Eukaryotic cells contain three main kinds of cytoskeletal filaments, e.g. actin filaments (microfilaments), intermediate filaments and microtubules. In addition, a large number of proteins are known to regulate cytoskeletal dynamics. Historically, a limited number of terrestrial natural products targeting the cytoskeleton have been used in cellular biology research, e.g. cytochalasins, a family of fungal toxins, which inhibit actin polymerization, and colchicine, isolated from Liliaceae plant, which inhibits tubulin polymerization. Actin and tubulin cytoskeletons are structured by their polymerization, which is also controlled by regulatory proteins. While there are now many available small molecules that directly target the actin cytoskeleton, there are only a few specific inhibitors of actin-binding proteins and other immediate regulators of actin dynamics and cell movement. Similarly, while there are numerous tubulin-targeted natural products used in basic or clinical sciences, fewer molecules are identified as targeting tubulin-binding proteins. This makes the study of actin and tubulin regulatory molecules of exceptional interest to chemical biologists and cell biologists (Fenteany and Zhu 2003).

During the last few decades significant efforts have been made to discover potential anticancer agents from marine organisms by both industry and academia, and this resulted in isolation of a large number of cytotoxic compounds. Modes of action studies of these cytotoxins led to discover considerable numbers of compounds targeting the cytoskeleton. This chapter describes marine-derived inhibitors of the cytoskeleton, their mechanisms of action and application as research tools.

2 Toxins Affecting Actin Filaments

2.1 Actin

Actin, a 42 kDa single polypeptide, composes thin filaments in muscle fibers, a discovery by Straub more than 60 years ago (Straub 1942). In its square-shape, actin molecule is divided into four subdomains (I–IV) in each corner and nucleotide binding site is located in its center (Fig. 1a). Straub found that globular actin (G-actin) transforms to fibrous polymers (F-actin) in the presence of salts. F-actin has a polarity which can be visualized as arrow-like shape by decorating it with myosin heads, creating pointed (−) and barbed (+) ends (Fig. 1b).

In the presence of salts, actin molecules associate and dissociate under Brownian motion; whether the local aggregate grows or dissolves depends on its stability. Although a dimer-form is unstable, higher order aggregates such as a trimer or a tetramer are stable. Therefore, the initial phase of filament formation is the period during which the stable "nuclei" form. This is the first and rate-limiting stage of actin polymerization. Once the nuclei are formed, G-actin molecules bind to nuclei

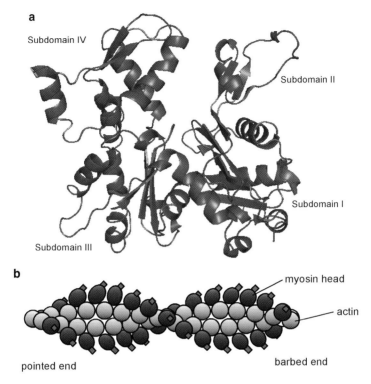

Fig. 1 The structure of actin monomer (G-actin) and actin fiber (F-actin). **a** Molecular structure of G-actin (PDB ID: 1-ATN). **b** A cartoon showing F-actin complexed with myosin head ("rigor complex"). Without nucleotide, myosin head tightly binds to F-actin ("rigor state") with an angle of 45°. Decoration with myosin head pieces reveals the polarity of F-actin

and filaments are formed (elongation state). The population of G-actin still exists in an equilibrium state which is called critical concentration (Oosawa et al. 1959; Oosawa and Asakura 1975) (Fig. 2).

Actin is the key protein not only in the contractile ring which divides cells in mitosis, but also in various microfilaments forming stress fibers, filopodia, lamellipodia or pseudopodia. Since in the living cell salt concentrations are high enough to prevent actin-depolymerization, actin dynamics are controlled by numerous actin binding proteins. Dynamic remodeling of actin filaments in the cytosol fascilitates shape changes, cell locomotion, and chemotactic migration that are vital to tissue development, wound healing, neuron migration, immune responses, and maintenance of homeostasis. Performing these functions demands strict regulation of the spatial and temporal organization of the actin cytoskeleton. This is facilitated on many levels by a myriad of accessory proteins (Korn 1982; Cooper 1991; Dos Remedios et al. 2003). These proteins influence every aspect of actin filament dynamics: the rate of actin polymerization and depolymerization, the balance between the concentration of monomeric and filamentous actin forms within the cell, interfilament interactions and filament branching.

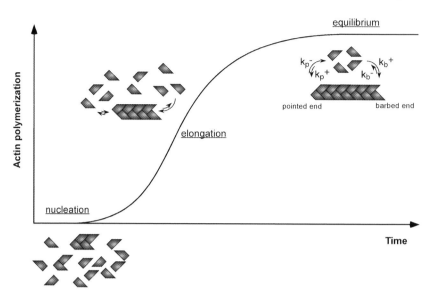

Fig. 2 Three phases in actin polymerization

Since spatiotemporal regulation of actin filament dynamics is required for essential cell processes, an obvious approach to study the relationships between the actin cytoskeleton and cellular functions is the specific perturbation of actin structures using pharmacological tools. Until a decade or so ago only a few agents interfering with cytoskeleton by binding to actin were available. Indeed, most of our initial knowledge concerning the involvement of actin in basic cellular processes was based on the extensive use of cytochalasins, which are derived from fungi. Unfortunately, the mode of actin-binding and cellular effects of cytochalasins are complex; and their binding to actin causes multiple effects on actin dynamics (Cooper 1987; Sampath and Pollard 1991). Therefore, small molecules whose modes of action on actin are different from cytochalasins were required. During the last 20 years a large number of highly cytotoxic compounds were isolated from marine organisms (König et al. 2006; Nagle et al. 2006), and quite a few of these have been shown to affect actin filaments (Spector et al. 1999; Fenteany and Zhu 2003). Interestingly, the majority of them mimic endogenous actin-binding proteins (McGough et al. 2003; Silacci et al. 2004).

2.2 Actin Polymerization

Actin polymerization can be measured by measuring viscosity, flow birefringence, light scattering or fluorescent probe. Flow birefringence reflects the length of polymer but needs special equipments to measure. Viscosity reflects polymer length but is affected by gelation, and orientation of the filaments, which decreases the sensitivity of measurement. Although having these faults, viscometry using Ostwald's viscometer or the falling ball apparatus is a good choice for just checking the effect of toxins on actin polymerization since these instruments are simple and easy to use (MacLean-Fletcher and Pollard 1980). Light scattering can be also used for estimating the size of aggregates, by measuring actin solution turbidity.

Labeling actin with pyrene was first described by Kouyama and Mihashi (1981), which strongly reflects the conformation of actin. Fluorescence (Ex. 365 nm and Em. 407 nm) is enhanced 25-times during polymerization and reflects the fraction of actin molecules in either G- or F-states. The fluorescence is quenched by interaction of actin with myosin heads. Obviously pyrenyl-actin is the most powerful method to measure polymerization state of pure actin.

Another fluorophore, 6-propionyl-2-(dimethylamino)naphthalene (Prodan), is also used as probe (Marriott et al. 1988) for detecting binding of protein or drugs to G-actin (Tanaka et al. 2003).

2.3 *Marine Toxins That Inhibit Actin Polymerization*

As mentioned above, cytochalasins, especially cytochalasin D (Fig. 3), were for a long time the only natural toxins known to inhibit actin polymerization. Cytochalasin D caps barbed ends to inhibit elongation of F-actin. Interestingly, cytochalasin D stabilizes the actin dimer state, instead of the monomer, and this enhances nucleation. These features are similar to those of barbed-end capping and nucleating protein CapZ (Cooper 1987; Sampath and Pollard 1991).

2.3.1 Latrunculins

Coué et al. (1987) first showed inhibition of actin polymerization by marine toxins, latrunculin A and B (Fig. 3), isolated from the Red Sea sponge *Latrunculia*

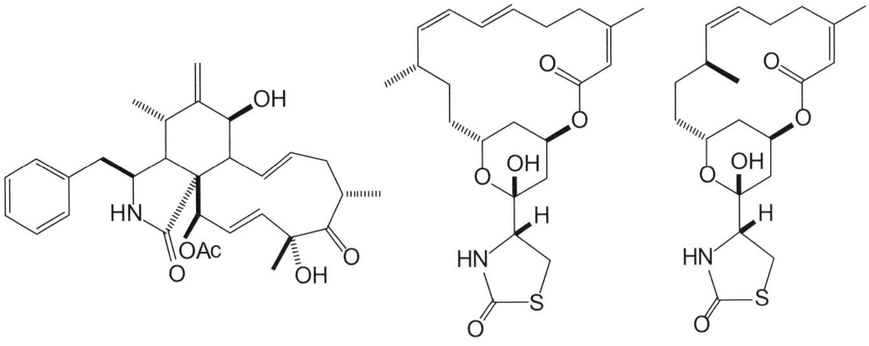

Cytochalasin D (**1**) Latrunculin A (**2**) Latrunculin B (**3**)

Fig. 3 Structures of cytochalasin D, latrunculin A, and latrunculin B *magnifica* (Kashman et al. 1980). Actually, they reversibly disrupt microfilaments of cultured cells at submicromolar concentrations (Spector et al. 1983). Latrunculin A binds to G-actin in a molar ratio of 1:1 with a K_d value of 0.24 µM and sequesters G-actin from actin polymerization (Coué et al. 1987). Since latrunculin A has no severing effect (Saito et al. unpublished observation), its action seems similar to that of monomer-binding proteins such as thymosin β_4 or DNase I. This was a new mechanism of natural toxins to inhibit actin polymerization. It should be noted that latrunculin A is the only known toxin which binds to the cleft between subdomains II and IV near the nucleotide binding pocket of actin monomer (Morton et al. 2000).

2.3.2   Trisoxazole Macrolides and Related Macrolides

Tolytoxin

Tolytoxin (6-hydroxy-7-O-methyl-scytophycin B) (Fig. 4), isolated from marine cyanobacteria (Moore 1981), was shown to disrupt actin cytoskeleton in KB cells with IC_{50} values of ~10 nM and inhibit actin polymerization (Patterson et al. 1993). Although the precise mechanism of action of tolytoxin is not clear, it was the first to be characterized of a family of actin-depolymerizing macrolides discussed below.

Trisoxazole Macrolides

A number of trisoxazole-bearing macrolides (Fig. 4) have been isolated from nudibranches and sponges, including ulapualides (Roesener and Scheuer 1986), kabiramides (Matsunaga et al. 1986; 1989), and jaspisamides (Kobayashi et al. 1993). Saito et al. (1994) analyzed the depolymerizing kinetics of mycalolide B isolated from a marine sponge *Mycale* sp. (Fusetani et al. 1989); it was shown that it not only binds to G-actin in a molar ratio of 1:1, but also severs F-actin. This was the first report that a small molecule can sever actin filaments, which is reminiscent of gelsolin and ADF/cofilin.

Similarly, halichondramide isolated from a *Halichondria* sp. sponge (Kernan and Faulkner 1987) and halishigamide A from a *Halichondria* sp. Okinawan sponge (Kobayashi et al. 1997) sequester G-actin. Halishigamide A and mycalolide B potently sever F-actin, while halichondramide weakly severs it (Saito et al. 1997). However, Allingham et al. (2005), who investigated this by a different method, classified halichondramide as a severing toxin (Allingham et al. 2005). Figure 5 shows the severing effects of trisoxazole macrolides using two methods: seeding and dilution assays. In the seeding assay, small amounts of F-actin are introduced at the same time as initiation of polymerization. In this assay, mycalolide B and halishigamide A but not halichondramide increased filament number. Note that decreases of initial rates relative to control samples without toxins suggest that some filaments are inhibited from elongation by capping with toxin-actin complex

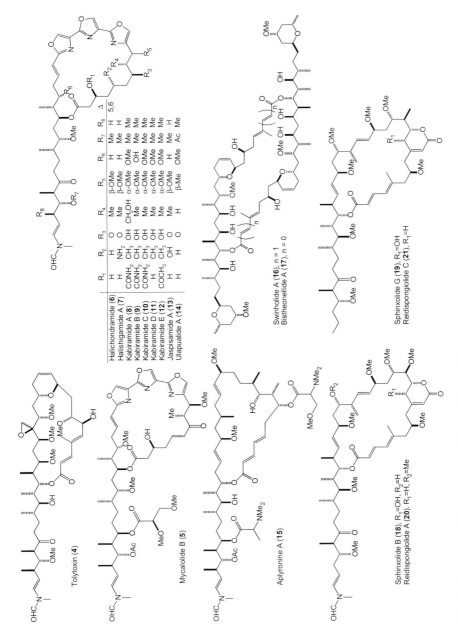

Fig. 4 Structures of trisoxazole macrolides and their relatives

Fig. 5 Measurement of severing activity of trisoxazole macrolides using seeding **a** or dilution **b** assays. **a** 20 μM F-actin was introduced to 20 volumes of 5 μM G-actin (final concentration of actin is 6 μM) simultaneously with 50 mM KCl and 2 mM MgCl$_2$ in the absence (□) or presence of mycalolide B (▲, △), halichondramide (■, □) or halishigamide A (■, □). Each macrolide was applied to F-actin at a concentration of 10 μM (open symbol) or applied to G-actin at a concentration of 0.5 μM (closed symbol). (●): 6 μM G-actin was polymerized by 50 mM KCl and 2 mM MgCl$_2$ without applying F-actin. **b** 5-μM F-actin was diluted 250 times with G-actin buffer without (□) or with each concentration of kabiramide A (●), kabiramide B (●), kabiramide E (●), halichondramide (■), halishigamide A (■) or mycalolide B (▲)

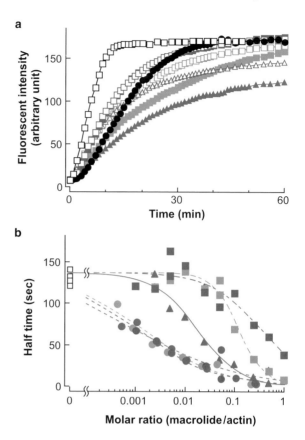

(Tanaka et al. 2003). In the dilution assay, F-actin is diluted by buffer without KCl or MgCl$_2$ so that depolymerization rates reflect the number of filaments. Using this method, kabiramides were shown to be the most powerful severing toxins while halichondramide and halishigamide A were weak severing toxins.

Crystal structures of complexes between G-actin and kabiramide C, jaspisamide A, and halichondramide have been solved (Klenchin et al. 2003; for review see Allingham et al. 2006). These toxins interact with residues between subdomains I and III in actin monomer as does gelsolin (Klenchin et al. 2003).

Aplyronine A

Aplyronine A (Fig. 4) isolated from the sea hare *Aplysia kurodai* (Yamada et al. 1993) exhibited similar effects to those of mycalolide B; it severs F-actin and binds to G-actin with a 1:1 ratio (Saito et al. 1996). Although tropomyosin did not prevent the depolymerizing effect of aplyronine A, a stoichiometric (to actin) concentration of phalloidin completely blocked it. Tropomyosin is known to stabilize actin filaments in muscle cells. The result explains why aplyronine A inhibits smooth muscle contraction, even in the presence of tropomyosin (Saito et al. unpublished observation).

Macrodiolides

Two macrodiolides (Fig. 4) isolated from marine sponges of the genus *Theonella*, swinholide A (Carmeli and Kashman 1985) and bistheonellide A (miaskinolide A) (Kato et al. 1987) bind G-actin with a molar ratio of 1:2, respectively (Bubb et al. 1995; Terry et al. 1997). Strangely, only the former can sever F-actin (Terry et al. 1997).

Klenchin et al. (2005) clearly demonstrated that the dimeric macrolide swinholide A co-crystallizes with two molecules of actins. Since swinholide A binds two actins facing opposite directions, the structure of this complex is different from the dimer precursor in actin polymerization.

Sphinxolides and Reidispongiolides

Sphinxolides and reidispongiolides (Fig. 4) are highly cytotoxic metabolites with IC_{50} values of 0.03–160 nM isolated from the New Caledonian sponges *Neosiphonia supersetes* and *Reidispongia coerulea* (D'Auria et al. 1993; 1994; Carbonelli et al. 1999). Sphinxolide B was reported to inhibit actin polymerization, although its mechanism was not determined (Zhang et al. 1997). Its related macrolides, sphinxolide G, reidispongiolide A and C, showed similar activity (Allingham et al. 2005).

Crystal structures of complexes between reidispongiolides A and C and sphinxolide B with G-actin showed that the positions and directions of side chains in these toxins binding to G-actin are almost identical to those of kabiramide C (Allingham et al. 2005).

Effects of some macrolides mentioned above on actin and cell proliferation are summarized in Table 1. A number of macrolides share essentially the same aliphatic side chain, the presumed major determinant of their binding (reviewed in Yeung and Paterson 2002). In fact, aplyronine A without the side chain lost its activity, whereas the side chain retained the activity (Kigoshi et al. 2002). It was also demonstrated that the FITC-labeled side chain is sufficient to bind with an actin molecule (Kuroda et al. 2006) and the side chain of mycalolide B is responsible for its actin-depolymerizing activity (Suenaga et al. 2004). Recently, Tanaka et al. (2003) proposed a two-step reaction model of kabiramide C interaction with G-actin where it first binds to actin with its ring portion, followed by binding of the side chain which causes its effect.

Table 1 Effects of actin depolymerizing macrolides on actin and cell proliferation

Macrolides	Dissociation constant (μM)	Severing activity	Cytotoxicity IC$_{50}$ value (Cell line)
Tolytoxin	n.d.	n.d.	0.5–8 nM (COLO-201,L1210, KB, HL-60, LoVo)[a]
Mycalolide B	0.013–0.02[b]	Yes[b]	0.5–1 nM (B-16)[c]
Halichondramide	≤0.1[d]	Yes[d] or weak	0.2 μM (K562)[e]
Halishigamide A	n.d.	Yes	4 nM (L1210)[f], 14 nM (KB)[f]
Kabiramide A	n.d.	Yes	30 nM (L1210)[g]
Kabiramide B	n.d.	Yes	30 nM (L1210)[g]
Kabiramide C	0.1[d]	Yes[d]	10 nM (L1210)[g]
Kabiramide D	n.d.	n.d.	20 nM (L1210)[g]
Kabiramide E	n.d.	Yes	20 nM (L1210)[g]
Jaspisamide A	n.d.	n.d.	0.3 μM (K562)[e], 18 nM (KB)[h], <1 nM (L1210)[h]
Ulapualide A	≤0.1[d]	n.d.	10–30 nM (L1210)[i]
Aplyronine A	0.1[j]	Yes[j]	0.4 nM (HeLa-S$_3$)[k]
Swinholide A	0.1[l]	Yes[l]	22 nM (L1210)[m], 29 nM (KB)[n]
Bistheonellide A	0.05[o], 0.15–0.35[p]	No	0.4–7 nM[q]

[a]Patterson and Carmeli (1992); [b]Saito et al. (1994); [c]Fusetani et al. (1989); [d] Tanaka et al. (2003); [e]Shin et al. (2004); [f]Kobayashi et al. (1997); [g]Matsunaga et al. (1989); [h]Kobayashi et al. (1993); [i]Roesener and Scheuer (1986); [j]Saito et al. (1996); [k]Ojika et al. (2007); [l]Bubb et al. (1995); [m]Doi et al. (1991); [n]Kobayashi et al. (1989); [o]Terry et al. (1997); [p]Saito et al. (1998); [q]Sakai et al. (1986)

Like the enzyme hexokinase, actin is expected to have two different conformations (open and closed states) with different affinities for ATP. Chik et al. (1996) demonstrated using β-actin that actin can exist in both an open and a closed state, and that the conformational changes can be much larger than a 5° hinge motion. Crystallographic studies revealed that trisoxazole macrolides (Klenchin et al. 2003), swinholide A (Klenchin et al. 2005) and aplyronine A bind to G-actin (Hirata et al. 2006) in a closed conformation, which explains the inhibitory effect of toxins on exchange of the nucleotide bound in G-actin (Klenchin et al. 2003). It is interesting to note that, although mycalolide B, aplyronine A (Saito et al. 1996), swinholide A (Saito et al. 1998) and latrunculin A (Belmont et al. 1999) inhibit the nucleotide exchange rate in G-actin, only bistheonellide A increases it (Saito et al. 1998). The effect of bistheonellide A is reminiscent of the facilitating effect of profilin on G-actin (Didry et al. 1998). However, in contrast with profilin bistheonellide A inhibits actin polymerization irreversibly, which is similar to mycalolide B (Saito et al. 1998).

2.3.3   Pectenotoxins

Pectenotoxin-2 (Fig. 6) is a diarrheic shellfish toxin produced by the dinoflagellate *Dinophysis fortii* (Yasumoto et al. 1985) and is also cytotoxic with IC$_{50}$ values ranging from 7.8 to 800 nM (Jung et al. 1995). It inhibits actin polymerization (Hori et al. 1999; Spector et al. 1999). Pectenotoxin-2 binds to the cleft between subdomains

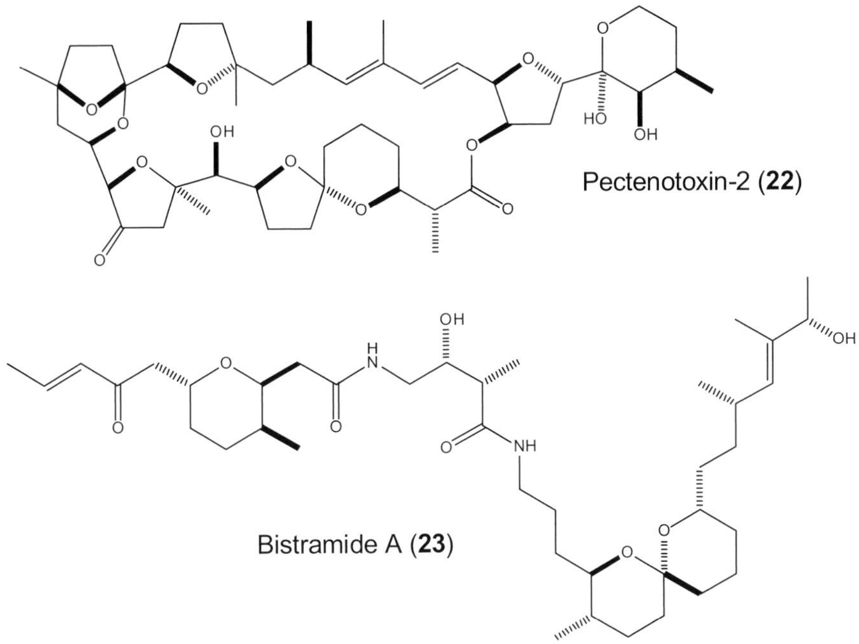

Fig. 6 Structures of pectenotoxin-2 and bistramide A

I and III of the "closed" conformation of actin monomer. However, its binding site is on the opposite face of actin relative to the site occupied by other actin depolymerizing macrolides (Allingham et al. 2007). The binding site of pectenotoxin-2 cannot be accessible in F-actin, although it freely binds to G-actin with 1:1 stoichiometry (Allingham et al. 2007), thus indicating its mode of action on F-actin is different from that of mycalolide B (Fig. 7).

2.3.4 Bistramide A

Bistramide A (bistratene A) (Fig. 6) is a cytotoxic polyether isolated from the ascidian *Lissoclinum bistratum* (Gouiffes et al. 1988; Degnan et al. 1989). Although initially reported as a selective inhibitor of protein kinase C$_\delta$ (Griffiths et al. 1996), later it was identified as a novel actin depolymerizing compound. It binds to G-actin in a 1:1 ratio, but its severing activity against F-actin has not been shown (Statsuk et al. 2005). Bistramide A binds to the cleft between subdomains I and III, and stabilizes actin conformation in the closed state, although its structure is not related to the other toxins mentioned above (Rizvi et al. 2006).

These investigations, demonstrating that all actin depolymerizing macrolides excepting latrunculins bind to the cleft between subdomains I and III, strongly emphasize the importance of this cleft for actin polymerization (Dominguez 2004).

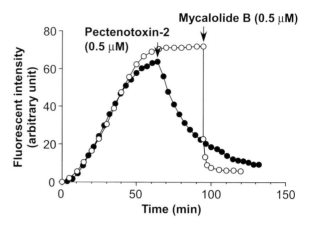

Fig. 7 Effects of mycalolide B and pectenotoxin-2 on F-actin. 1.5 μM G-actin was polymerized by 50 mM KCl and 2 mM MgCl₂; subsequently 0.5 μM of mycalolide B (○) or pectenotoxin-2 (●) was applied

2.4   Marine Toxins That Stabilize Actin Filaments

Phalloidin (Fig. 8) is a toxic cyclopeptide isolated from a mushroom that specifically binds to F-actin but not to G-actin (Estes et al. 1981) and strongly stabilizes the structure of the assembled filaments (summarized by Wieland 1977). This toxin was the only an actin stabilizing compound until jasplakinolide (jaspamide) (Fig. 8), a cyclic peptide isolated from marine sponge *Jaspis* sp. (Crews et al. 1986; Zabriskie et al. 1986), was found to share stabilizing activity and binding site with phalloidin (Bubb et al. 1994). Jaspakinolide is a superior tool, since its cell permeability is much higher than that of phalloidin. However, its strong activity induces not only stabilization of preexisting actin filaments but also polymerization of short filaments in amorphous masses (Bubb et al. 2000).

The sea hare *Dolabella auricularia* often contains highly cytotoxic peptides, some of which are potential anticancer chemicals (Pettit 1997; Ishiwata et al. 1994a; 1994b). Dolastatin 11 and (−)-doliculide (Fig. 8) were reported to enhance actin polymerization (Bai et al. 2001; 2002). Although precise kinetic parameters are not available, (−)-doliculide has similar EC_{50} values to those of jasplakinolide on actin polymerization and on inhibition of phalloidin-binding to F-actin, whereas dolastatin 11 is twice as active as jasplakinolide but has no effect on the binding of phalloidin to actin (Bai et al. 2001; 2002). It was also demonstrated that dolastatin 11 binds to F-actin at a site different from the phalloidin-binding site, though their modes of binding to and stabilization of actin filaments are similar (Oda et al. 2003).

Hectochlorin (Fig. 8), a lipopeptide isolated from the cyanobacterium *Lyngbya majuscula*, promotes actin polymerization with a similar efficiency to that of jasplakinolide, but it does not compete with the binding of FITC-phalloidin to F-actin (Marquez et al. 2002).

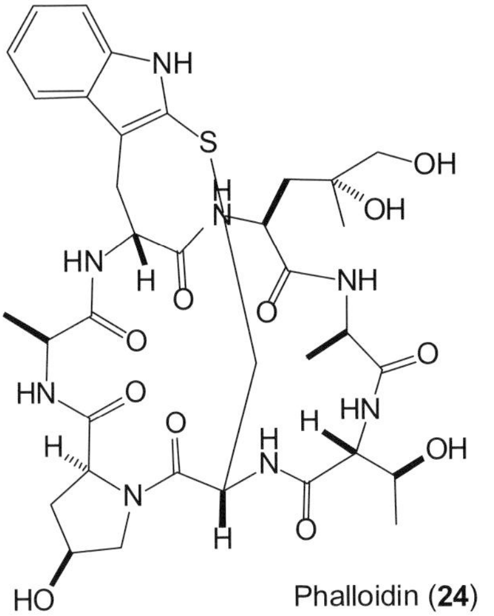

Phalloidin (**24**)

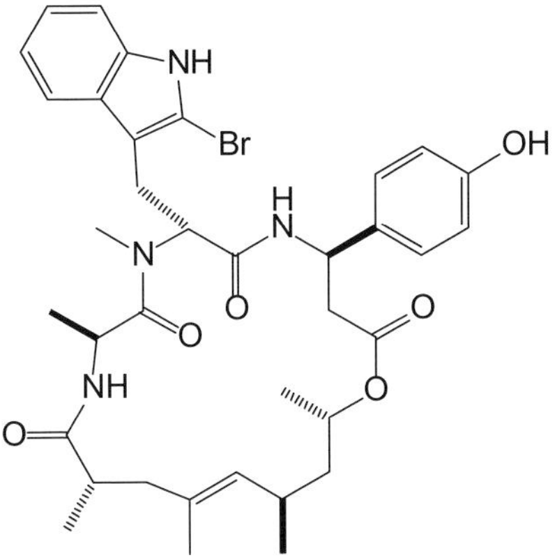

Jasplakinolide (**25**)

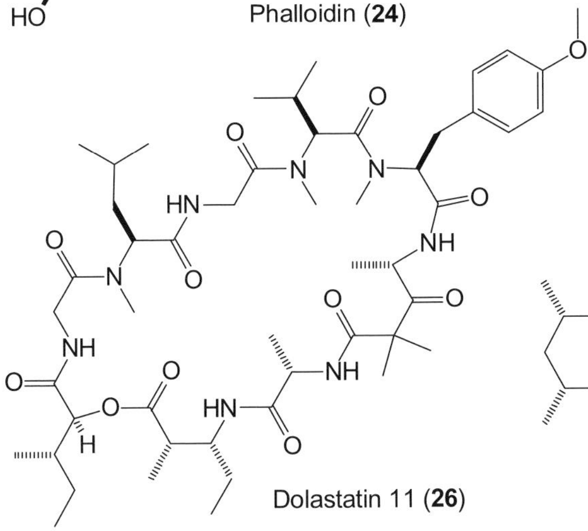

Dolastatin 11 (**26**)

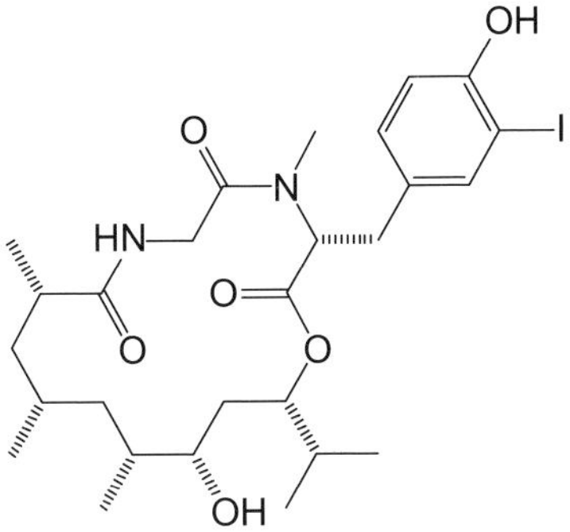

(-)-Doliculide (**27**)

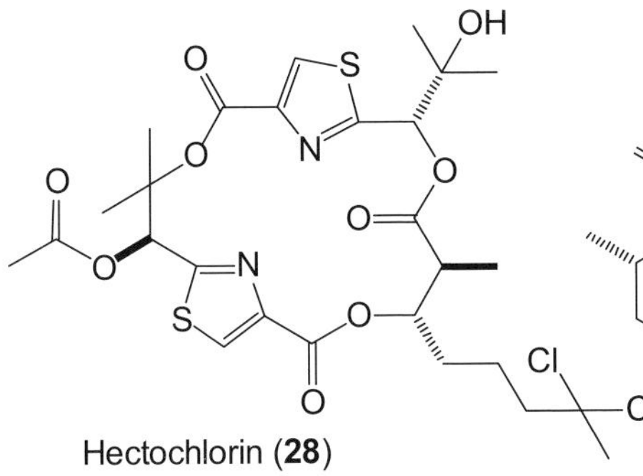

Hectochlorin (**28**)

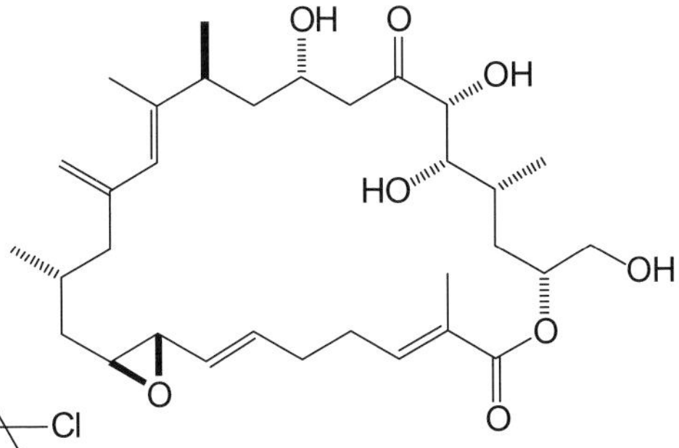

Amphidinolide H (**29**)

Fig. 8 Structures of marine toxins which stabilize actin filaments

Amphidinolide H (Fig. 8), isolated from a dinoflagellate *Amphidinium* sp. (Kobayashi et al. 1991), is the sole nonpeptidic actin stabilizing toxin (Saito et al. 2004). Interestingly it not only enhances phalloidin-binding to F-actin, but also binds to both F- and G-actins with similar affinities of 0.21 μM and 0.43 μM, respectively (Saito et al. 2004). Amphidinolide H covalently binds to Tyr-198 of the actin monomer (Usui et al. 2004). Tyr-198 is in subdomain IV, which faces toward the phalloidin-binding gap (Oda et al. 2005).

2.5   *Actin-Targeted Marine Toxins in Basic and Preclinical Studies*

The high degree of structural and molecular complexity of the actin-based cytoskeletons has made it difficult to assess how they regulate cell functions. An obvious approach to study the roles of actin is the specific perturbation of actin structures using pharmacological tools. During the last 5 years marine-derived cytoskeletal inhibitors were used as research tools in nearly 600 articles. Latrunculin A was used to prove the microtubular involvement of actin in regulating caveolar membrane traffic (Mundy et al. 2002), to prove the association of actin dynamics in regulating synaptic plasticity (Fukazawa et al. 2003), and to study the role of cortical actin as a barrier to the trafficking of secretory vesicles (Toonen et al. 2006). Furthermore, mycalolide B was used to study the roles of F-actin in intracellular transport of HIV-1 in host cells (Sasaki et al. 2004).

Natural product inhibitors of actin cytoskeleton dynamics are not only recognized as valuable molecular probes for dissecting complex mechanisms of cellular function, but also their potential use as chemotherapeutics has become a focus of scientific investigation. Cytoskeletal inhibitors are lethal to tumor cells and several of them have been included in the National Cancer Institute's Molecular Targets Development Program (Fenteany and Zhu 2003; Spector et al. 1999). So far, however, no actin-binding drugs have entered clinical trials due to their lack of selectivity for diseased cells.

3   Toxins Affecting Microtubules

3.1   *Tubulin*

Tubulins are 55 kDa single polypeptides consisting five tubulin families named α-ε. While α- and β-tubulins are known to form cytoskeletal filaments, the functions and distributions of the δ- and ε-tubulins have not yet been determined. Like actin, tubulin also assembles to form filaments with fast growing plus ends and dissociating minus ends by incorporating the αβ heterodimer as protomers where α-tubulin faces to the minus end and β-tubulin faces to the plus end. This filament is called microtubule, because its tube structure is composed of 13 protofibers (Fig. 9). Both α- and β-tubulins possess single nucleotide binding sites where GTP binds. In the αβ heterodimer, GTP bound to α-tubulin cannot be exchanged or hydrolyzed, while GTP bound to β-tubulin (GTP-tubulin) can be hydrolyzed to GDP (GDP-tubulin) which is exchangeable to GTP. Since only GTP-tubulin can be polymerized and GTP is hydrolyzed in microtubule, the major part of the microtubule is composed of GDP-tubulin, while both ends are capped by several molecules of GTP-tubulin (GTP-cap). Once GDP-tubulins are exposed at one end by hydrolysis or depolymerization, the end become unstable enabling acute depolymerization to occur ("catastrophe"). This depolymerization lasts for awhile and then switches to polymerization ("rescue") (Fig. 9). According to this dynamic instability, each single microtubule persists in prolonged states of polymerization and depolymerization that interconvert to each other frequently, in a population of microtubules exhibiting a steady state condition.

Since the protomeric microtubule is an αβ-tubulin dimer whose molecular weight is about 100 kDa, turbidimetric measurement (measuring absorbance or light scattering) is a useful way of monitoring tubulin polymerization (Shelanski et al. 1973; Lee and Timasheff 1977). Tubulin assembly is promoted by microtubule-associated proteins (MAPs), GTP, Mg^{2+} and elevated temperature (37°C). Disassembly is promoted by Ca^{2+}, GDP, and low temperature (4°C).

Microtubules are important in the process of mitosis, during which the duplicated chromosomes of a cell are separated into two identical sets. The word mitosis is derived from Greek word *Mitos* which means thread, and the thread is composed of microtubules.

In the centrosome, γ-tubulin forms a ring structure with associated proteins, providing a nucleating site for tubulin polymerization. The microdubules radiating

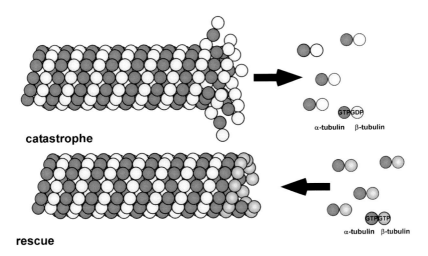

Fig. 9 Dynamic instability of microtubules

from the centrosome are in a state of dynamic instability (Fig. 9), and with a number of lengthening and shortening cycles, they probe a target such as the kinetochore ("search-capture model"): if the filament cannot find the target before transition to catastrophe, the filament may shorten (Fig. 10a), whereas the filament that reaches the kinetochore is stabilized by its interaction with this structure (Fig. 10b). As a result of the search-capture dynamics of microtubules, each aster is positioned properly and forms a spindle body (Fig. 10c). The spindle body is a structure with microtubules connecting two centrosomes and chromosomes are distributed at the equator. As mitosis proceeds, each copy of a duplicated chromosome is separated and assembled by the centrosome to which it is attached.

The cell cycle is strictly monitored at multiple checkpoints, so that damaged cells will not proliferate. There are four checkpoints in the normal cell cycle, three

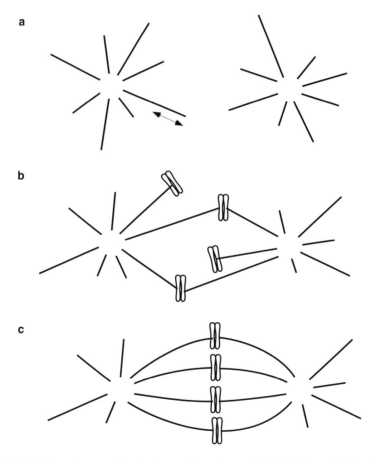

Fig. 10 Spindle formation according to the search–capture model. Each microtubule in the aster-like structures dynamically repeats the rescue and catastrophe processes (**a**). Once a microtubule reaches the target protein, it becomes stable (**b**). By linking both centrosomes with their respective kinetochores, two aster-like structures form the spindle (**c**)

of which monitor whether DNA is copied properly: DNA repair prior to DNA replication, completion of DNA replication prior to the onset of mitosis, and chromosome segregation at mitosis (Hartwell and Weinert 1989; Murray 1992). The last checkpoint is called the "spindle assembly checkpoint". Disturbance of spindle formation arrests the cell cycle at mitosis. Drugs that inhibit microtubule function, including colchicine and nocodazole, induce mitotic arrest. While mitotic arrest has been shown to be followed by initiation of the mitochondrial pathway of apoptosis (Ibrado et al. 1998; Wang et al. 2000; Jordan 2002), the mechanism by which cell cycle perturbation is linked to the initiation of apoptosis is not yet fully understood.

Another important role of microtubules is to serve as a "railway" for particle transport within the cell. By moving on microtubular "tracks," kinesins serve a variety of functions in the cell. Blockade of axonal transport with mutant kinesins or toxins causes axonopathy (Mandelkow and Mandelkow 2002), which is the major side effect of antitumor agents. While most conventional kinesins are thought to function as vehicles that carry different cargoes (e.g. vesicles, organelles, chromosomes, protein complexes and filaments) along microtubules (Hirokawa and Takemura 2004), some kinesins are involved in mitosis (Bergnes et al. 2005). Therefore, mitotic kinesins are attractive targets for cancer chemotherapy (Bergnes et al. 2005).

3.2 Marine Toxins That Inhibit Tubulin Polymerization

3.2.1 Toxins Bind to Vinca Domain or Near Site

Although many toxins affecting tubulin polymerization have been isolated from cyanobacteria and marine sponges, one series of antineoplastic peptides was isolated from the sea hare *Dolabella auricularia* (Pettit 1997). As mentioned above, dolastatin 11 stabilizes actin filaments, whereas dolastatin 10 (Fig. 11) (Pettit et al. 1987) inhibits tubulin polymerization with an IC_{50} value of 1.2 μM (Bai et al. 1990). Another cytototoxic peptide, dolastatin 15 (Pettit et al. 1989) also inhibits tubulin polymerization with an IC_{50} value of 23 μM (Bai et al. 1992).

Halichondrin B (Fig. 11), a polyether macrolide isolated from the marine sponge *Halichondria okadai* (Hirata and Uemura 1986), was predicted to inhibit tubulin polymerization using COMPARE, an NCI computer program (Paull et al. 1992). In fact, halichondrin B inhibits tubulin polymerization with an IC_{50} value of 2 μM (Bai et al. 1991). Similarly, hemiasterlin (milnamide B) (Fig. 11), a linear peptide isolated from the marine sponge *Hemiasterella minor* (Talpir et al. 1994), blocks microtubule assembly with an IC_{50} value of 0.98 μM (Gamble et al. 1999). Several hemiasterlin analogues have been isolated from marine sponges (Crews et al. 1994; Chevallier et al. 2003).

Cryptophycin 1 (Fig. 11), a cyclic depsipeptide originally isolated as an antifungal constituent from a *Nostoc* sp. cyanobacterium (Schwartz et al. 1990), was later

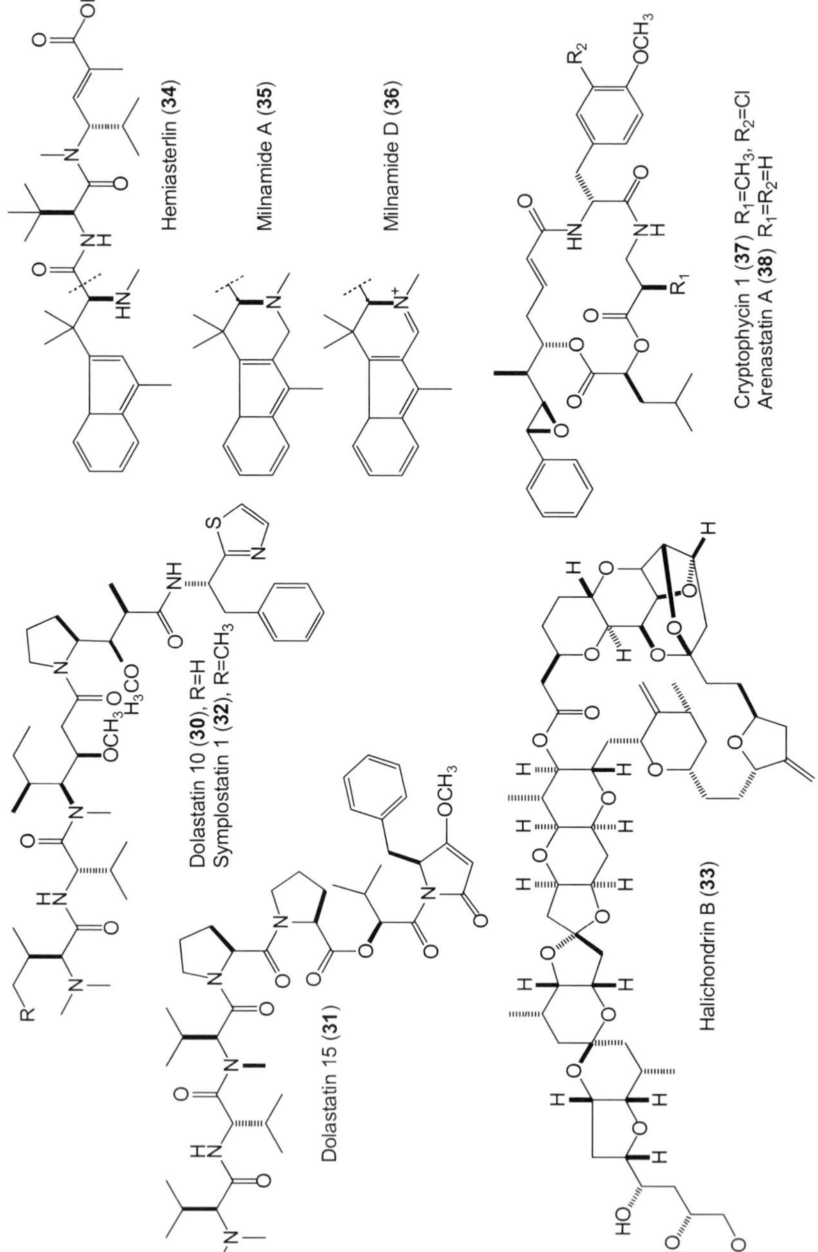

Hemiasterlin (**34**)

Milnamide A (**35**)

Milnamide D (**36**)

Cryptophycin 1 (**37**) R₁=CH₃, R₂=Cl
Arenastatin A (**38**) R₁=R₂=H

Dolastatin 10 (**30**), R=H
Symplostatin 1 (**32**), R=CH₃

Dolastatin 15 (**31**)

Halichondrin B (**33**)

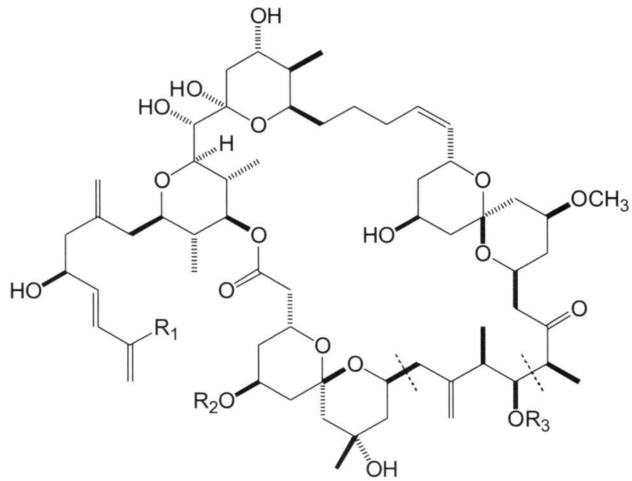

Spongistatin 1 (**39**), R_1=Cl, R_2=R_3=COCH$_3$
Spongistatin 2 (**40**), R_1=H, R_2=R_3=COCH$_3$
Spongistatin 3 (**41**), R_1=Cl, R_2=H, R_3=COCH$_3$
Spongistatin 4 (**42**), R_1=Cl, R_2=COCH$_3$, R_3=H
Spongistatin 6 (**44**), R_1=H, R_2=COCH$_3$, R_3=H

Spongistatin 5 (**43**), R_1=Cl, R_2=H
Spongistatin 7 (**45**), R_1=H, R_2=H
Spongistatin 8 (**46**), R_1=H, R_2=COCH$_3$
Spongistatin 9 (**47**), R_1=Cl, R_2=COCH$_3$

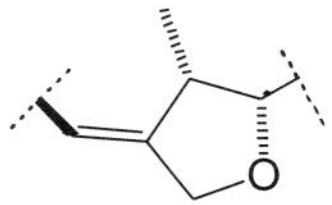

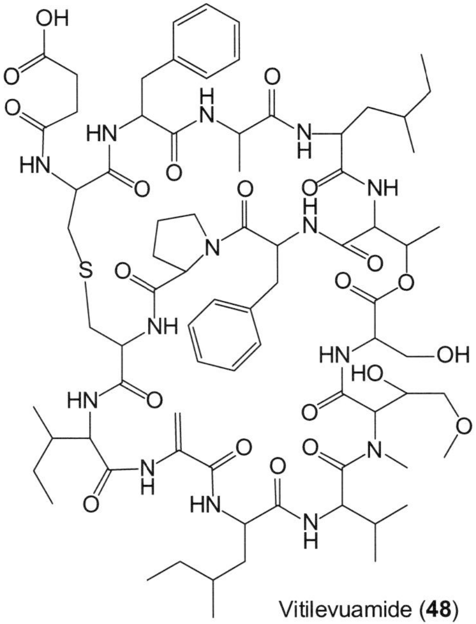

Vitilevuamide (**48**)

Fig. 11 Microtubule-depolymerizing toxins which bind to the vinca domain or near the vinca binding site

found to be highly cytotoxic ($IC_{50} < 10\,pM$ against L1210) (Smith et al. 1994). It inhibits tubulin polymerization with an IC_{50} value of $1\,\mu M$ (Kerksiek et al. 1995) by binding to the vinca domain (Bai et al. 1996; Smith and Zhang 1996), and induces mitotic arrest and apoptosis (Mooberry et al. 1997). A similar peptide named arenastatin A has been isolated from a marine sponge (Kobayashi et al. 1994a).

The binding sites for colchicine and vinblastine are entirely distinct regions of the α/β-tubulin dimer. Hemiasterlin, cryptophycin 1, and dolastatin 10 competitively bind to tubulin (Bai et al. 1999), while dolastatin 10 and halichondrin B inhibit binding of vinca alkaloids in an uncompetitive manner (Bai et al. 1990; Ludueña et al. 1993). However, the binding of dolastatin 15 was inhibited not only by cryptophycin 1 and dolastatin 10, but also by halichondrin B and vinca alkaloids. The vinca domain seems to be located in β-tubulin at the inter-dimer interface between α-tubulin and β-tubulin subunits in each protofiber (Gigant et al. 2005), suggesting that these toxins do not bind to the vinca domain itself, but the binding site(s) for these toxins consist(s) of a series of overlapping domains on the surface of tubulin (Cruz-Monserrate et al. 2003). This is consistent with the finding that hemiasterlin binds to helix H10 on α-tubulin which interacts across the interdimer interface with helix H6 of β-tubulin (Nunes et al. 2005). Furthermore, intercalation of these toxins bends protofilaments, which may explain the ring structure observed in toxin-treated tubulins (Nogales et al. 2003).

Spongistatin 1 (= altohyrtin A) (Fig. 11), a novel, strongly cytotoxic macrolide isolated from marine sponges (Pettit et al. 1993a; Kobayashi et al. 1993a), not only inhibits tubulin polymerization with an IC_{50} value of $3.6\,\mu M$ but also binding of GTP as well as of vinblastine to tubulin (Bai et al. 1993). Eight spongistatin variants are known to date (Fusetani et al. 1993; Kobayashi et al. 1993b; 1994b; Pettit et al. 1993b; 1993c; 1993d; 1994a); some of them showed similar activities (Bai et al. 1995).

Vitilevuamide (Fig. 11), a cytotoxic bicyclic pepide isolated from the tunicates *Didemnum cuculiferum* and *Polysyncranton lithostrotum*, inhibits tubulin polymerization ($IC_{50} = 2\,\mu M$) and GTP binding or vinblastine binding, but not binding of dolastatin 10 (Edler et al. 2002).

3.2.2 Others

Welwistatin (Fig. 12), an alkaloid isolated from a cyanobacterium (Stratmann et al. 1994), inhibits tubulin polymerization slowly at high concentrations and arrests SK-OV-3 cells in mitosis ($IC_{50} = 72\,nM$) without inhibiting colchicine binding and GTP hydrolysis (Zhang and Smith 1996).

Curacin A (Fig. 12), a novel metabolite of the cyanobacterium *L. majuscula* (Gerwick et al. 1994), inhibits tubulin polymerization ($IC_{50} = 4\,\mu M$) by binding to the colchicine binding site. Unlike the tubulin inhibitors mentioned above, it stimulates tubulin-dependent GTP hydrolysis similar to the toxins binding to the colchicine domain (Blokhin et al. 1995).

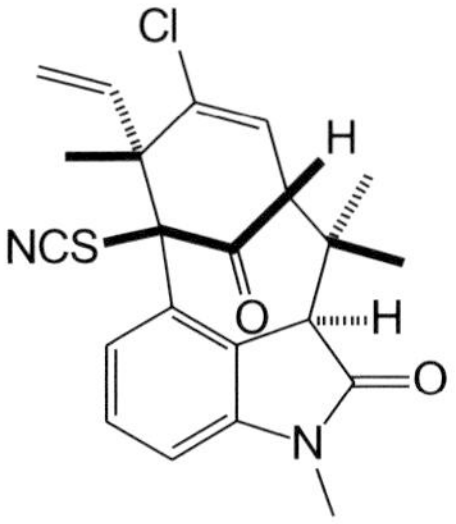

Welwistatin (**49**)

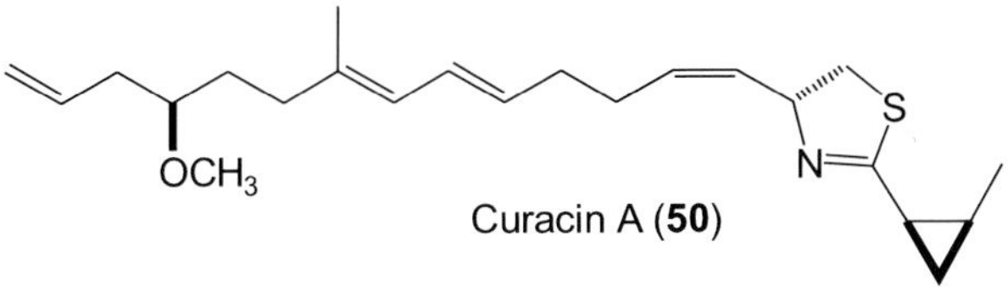

Curacin A (**50**)

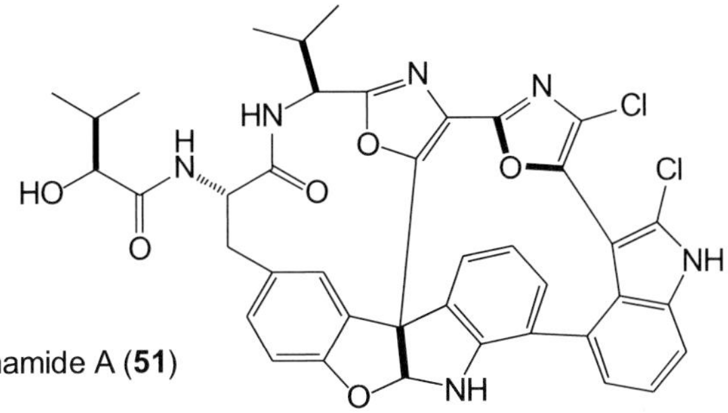

Diazonamide A (**51**)

Fig. 12 Microtubule-depolymerizing toxins which bind to sites different from the vinca domain

The story about diazonamide A (Fig. 12) is extremely dramatic. It is an unusual alkaloid isolated from the colonial ascidian *Diazona chinensis* (Lindquist et al. 1991) and its chemical structure was later revised by chemical synthesis (for review, see Ritter and Carreira 2002). COMPARE analysis predicted that it would be a tubulin toxin; subsequently it was shown to inhibit tubulin polymerization ($IC_{50} = 0.3$–$0.9\,\mu M$) and GTP binding (Cruz-Monserrate et al. 2003). However, further study revealed that its inhibitory effect on cell division is mediated by ornithine δ-amino transferase rather than tubulin (Wang et al. 2007). Strictly speaking, it is not appropriate to mention this compound here, but this story indicates that tubulin is not the only target of promising antimitotic compounds which cause G_2/M arrest and disturb microtubule networks.

3.3 Marine Toxins That Stabilize Microtubules

Several decades ago, a natural product with antitumor activity was isolated from the stem bark of the Pacific yew tree *Taxus brevifolia* and named taxol (Fig. 13) (Wani et al. 1971). Due to its weak cytotoxicity, taxol (paclitaxel) was paid little attention until its unique microtubule stabilization effect was disclosed by Horwitz and coworkers (Schiff et al. 1979). Since at the time only tubulin-depolymerizing toxins (colchicine, nocodazole and vinca alkaloids) were known, taxol provided a new tool to study tubulin functions and a new antitumor drug candidate for preclinical study (Donehower et al. 1987). However, the supply problem delayed by many years the development of taxol as a successful anticancer drug (Mann 2002).

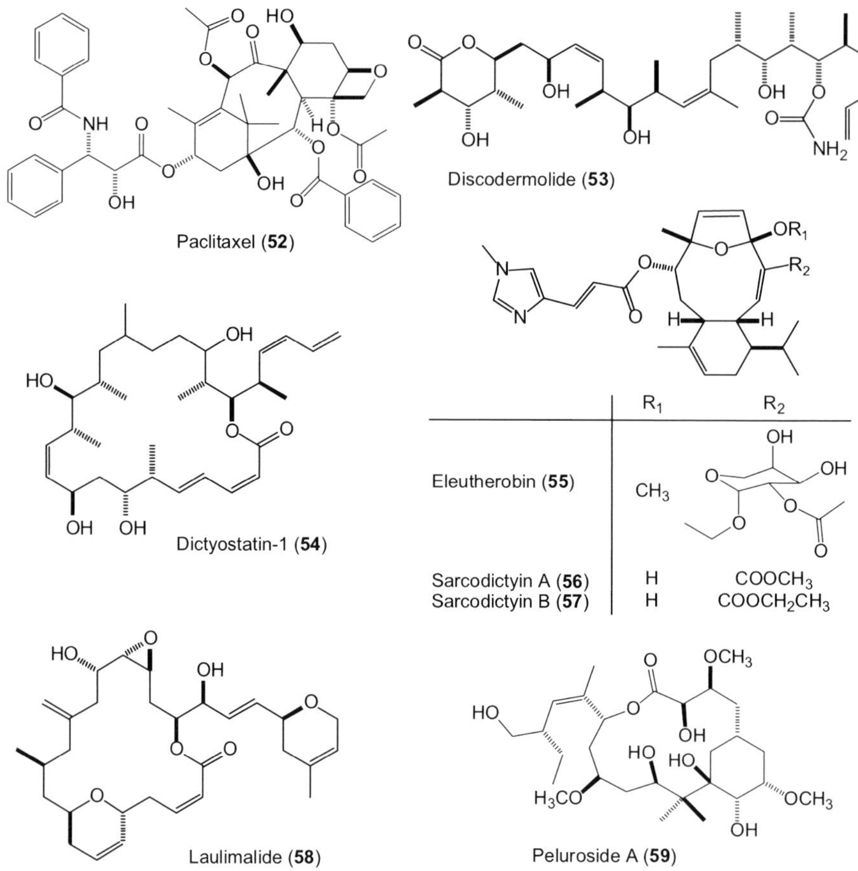

Fig. 13 Structures of marine toxins that stabilize microtubules

Following the introduction of taxol into clinical use, microtubule-stabilizing toxins of marine origin were discovered.

(+)-Discodermolide (Fig. 13), a polyketide derived from the Caribbean sponge *Discodermia dissoluta* (Gunasekera et al. 1990), was reported as an antiprolifera-tive toxin (IC_{50}=7–53 nM) arresting cell cycle at G_2/M (Longley et al. 1993; Hung et al. 1994). At first computer-automated structure evaluation programs named MultiCASE and CASE predicted interaction of discodermolide with tubulin (ter Haar et al. 1996a). Actually, it induced microtubule bundles and formation of stable tubulin polymers in vitro (IC_{50}=4.7 μM) (ter Haar et al. 1996b). Discodermolide binds competitively with paclitaxel to tubulin dimers (Hung et al. 1996). Recently, it was shown, using a photoaffinity analogue, that it binds to amino acid residues 355–359 of β-tubulin in close proximity to the taxol binding site (Xia et al. 2006).

The microtubule-stabilizing activity of dictyostatin 1 (Fig. 13), a macrolide iso-lated from a marine sponge *Spongia* sp. (Pettit et al. 1994b), was characterized nearly

10 years after its isolation (Isbrucker et al. 2003). Its mode of inhibiting microtubules was later found to be similar to that of discodermolide; it inhibits binding of discodermolide, epothilone B and paclitaxel to microtubules (Madiraju et al. 2005).

Eleutherobin (Fig. 13) isolated from a soft coral *Eleutherobia* sp. (Lindel et al. 1997) was the first marine diterpene that stabilizes microtubules (Long et al. 1998); it binds to the paclitaxel site (Hamel et al. 1999). The aglycone diterpene, sarcodictyins A and B from the Mediterranean stoloniferan coral *Sarcodictyon roseum* (D'Ambrosio et al. 1987) were less active than the glycoside (Hamel et al. 1999).

Laulimalide (= fijianolide B) (Fig. 13) of sponge origin (Corley et al. 1988; Quiñoà et al. 1988) induced tubulin assembly in vitro ($EC_{50} = 4.3 \mu M$). It arrested cell cycle of MDA-MB-435 cells at the G_2/M phase with an IC_{50} value of 5.7 nM, producing paclitaxel-like microtubule bundles (Mooberry et al. 1999). Another macrolide, peluroside A (Fig. 13) isolated from a New Zealand marine sponge *Mycale* sp. (West et al. 2000), stabilizes microtubules by binding at a different site from that of taxoids, but it shares the binding site with laulimalide (Hood et al. 2002; Gaitanos et al. 2004).

3.4 Marine Spindle Toxins As Potential Anticancer Drugs

Based on the successes and limitations of taxol as an anticancer drug, a number of tubulin-targeted marine toxins have been considered as a source of new anticancer drugs, some of which have entered clinical trials. The synthetic analogue of cryptophycin 1, LY355703 (= cryptophycin 52), has entered phase II clinical trials (Sessa et al. 2002; Stevenson et al. 2002). Tasidotin, a dolastatin 15 derivative, entered phase I as a treatment for solid tumors (Ebbinghaus et al. 2005), while LU 103793 (another synthetic analogue of dolastatin 15) proceeded to phase II for patients with metastatic breast cancer (Kerbrat et al. 2003) and advanced nonsmall-cell lung cancer (Marks et al. 2003). Dolastatin 10 has been in phase II trials in treatment of patients with advanced nonsmall-cell lung cancer (Krug et al. 2000), hormone-refractory metastatic prostate adenocarcinoma (Vaishampayan et al. 2000), metastatic melanoma (Margolin et al. 2001), advanced colorectal cancer (Saad et al. 2002), recurrent platinum-sensitive ovarian carcinoma (Hoffman et al. 2003), advanced breast cancer (Perez et al. 2005), and advanced pancreaticobiliary cancers (Kindler et al. 2005). Not only dolastatin 10 itself but also its analogue named TZT-1027 has been in phase I trials (Schoffski et al. 2004; Tamura et al. 2007).

4 Conclusion

The frontiers of biomedical science are expanded by generating and testing new hypotheses with various approaches including the use of pharmacological tools that selectively alter a particular site or process. Colchicine, a tubulin-depolymerizing compound, was isolated 120 years ago, and since then its effect, toxicity and target have been widely investigated (Dixon and Malden 1908). Searching for its target resulted in the isolation of the major component of microtubules; tubulin (Shelanski and Taylor 1967). Studies on the toxic mechanism of fungal toxins (cytochalasins) led to the theory of actin polymerization (Cooper 1987; Sampath and Pollard 1991). These tubulin- and actin-targeting toxins have been widely used as pharmacological tools to study the roles of microtubules in cellular functions. Following these successes, marine-derived cytoskeletal toxins with different modes of actions and actions at different sites were discovered. These differences are desirable since obtaining consistent results with diverse toxins reduces the possibility that results are due to a side effect. Furthermore, since the new toxins recognize different sites from the terrestrial plant toxins, we can expect that the new toxins will show less drug resistance.

Although these marine toxins are now established as research tools, still we have no clear answers to why and how invertebrates produce and store these toxins targeting actin or tubulin. Cytoskeletal proteins are conserved in a wide range of organisms; even microbes express these proteins. Nevertheless, invertebrates produce these toxins without suffering self-toxicity. We can hypothesize that the cytoskeletal proteins in marine invertebrates or the symbiotic microorganisms which really produce the toxins are lacking the binding site to toxins similar to drug-resistant cancer cells. However, this hypothesis raises another question. Many of these toxins are lipophilic and can penetrate plasma membrane of cells. In contrast with terrestrial organisms, marine invertebrates are surrounded by sea water. How do they keep toxins from discharging into sea water?

Today, other molecular biological models, such as are obtained by knocking out specific proteins, are also available to study the roles of various proteins in affecting the cellular functions of cytoskeletal proteins. However, there is a crucial dilemma to knock out tubulin or actin: it is necessary to establish the knocked-out cell clone to study the function of cells without tubulin or actin, but cells cannot proliferate without actin or tubulin! Clearly pharmacological tools still have great importance for understanding cytoskeletal dynamics.

On the other hand, the cytoskeleton is regulated differently in different positions within cells. In the case of chemotactic locomotion, the structure of the actin cytoskeleton is different in the front and in the back of moving cells. This is because different actin-binding proteins are involved at both ends. When we study cell locomotion precisely, we need new means to interfere with these cytoskeletons independently.

Now we have actin- or tubulin-targeted toxins with various mechanisms of action. For this, we have made progress in terms of efficiency (increased cell permeability of toxins), and in the matter of drug-resistance of cancer cells. We have also new inhibitors of motor proteins (blebbistatin or ispinesib), although they are not discussed in this chapter. On the other hand, inhibitors of endogenous proteins interacting with actin or tubulin are yet unexplored. During the last century, we obtained specific toxins acting on actin or tubulin. Now, in the twenty-first century, we will undoubtedly discover toxins with new mechanisms of action which can further our understanding of normal cytoskeletal function and possibly reveal how this system can be selectively altered in malignant cells to clinical advantage.

References

Allingham JS, Zampella A, D'Auria MV, Rayment I (2005) Structures of microfilament destabilizing toxins bound to actin provide insight into toxin design and activity. Proc Natl Acad Sci USA 102:14527–14532.

Allingham JS, Klenchin VA, Rayment I (2006) Actin-targeting natural products: structures, properties and mechanisms of action. Cell Mol Life Sci 63:2119–2134.

Allingham JS, Miles CO, Rayment I (2007) A structural basis for regulation of actin polymerization by pectenotoxins. J Mol Biol 371:959–970.

Bai R, Pettit GR, Hamel E (1990) Dolastatin 10, a powerful cytostatic peptide derived from a marine animal. Inhibition of tubulin polymerization mediated through the vinca alkaloid binding domain. Biochem Pharmacol 39:1941–1949.

Bai R, Paull KD, Herald CL, Malspeis L, Pettit GR, Hamel E (1991) Halichondrin B and homohalichondrin B, marine natural products binding in the vinca domain of tubulin: discovery of tubulin-based mechanism of action by analysis of differential cytotoxicity data. J Biol Chem 266:15882–15889.

Bai R, Friedman SJ, Pettit GR, Hamel E (1992) Dolastatin 15, a potent antimitotic depsipeptide derived from Dolabella auricularia: interaction with tubulin and effects on cellular microtubules. Biochem Pharmacol 43:2637–2645.

Bai R, Cichacz ZA, Herald CL, Pettit GR, Hamel E (1993) Spongistatin 1, a highly cytotoxic, sponge-derived, marine natural product that inhibits mitosis, microtubule assembly, and the binding of vinblastine to tubulin. Mol Pharmacol 44:757–766.

Bai R, Taylor GF, Cichacz ZA, Herald CL, Kepler JA, Pettit GR, Hamel E (1995) The spongistatins, potently cytotoxic inhibitors of tubulin polymerization, bind in a distinct region of the vinca domain. Biochemistry 34:9714–9721.

Bai R, Schwartz RE, Kepler JA, Pettit GR, Hamel E (1996) Characterization of the interaction of cryptophycin 1 with tubulin: Binding in the *Vinca* domain, competitive inhibition of dolastatin 10 binding, and an unusual aggregation reaction. Cancer Res 56:4398–4406.

Bai R, Durso NA, Sackett DL, Hamel E (1999) Interactions of the sponge-derived antimitotic tripeptide hemiasterlin with tubulin: comparison with dolastatin 10 and cryptophycin 1. Biochemistry 38:14302–14310.

Bai R, Verdier-Pinard P, Gangwar S, Stessman CC, McClure KJ, Sausville EA, Pettit GR, Bates RB, Hamel E (2001) Dolastatin 11, a marine depsipeptide, arrests cells at cytokinesis and induces hyperpolymerization of purified actin. Mol Pharmacol 59:462–469.

Bai R, Covell DG, Liu C, Ghosh AK, Hamel E (2002) (-)-Doliculide, a new macrocyclic depsipeptide enhancer of actin assembly. J Biol Chem 277:32165–32171.

Belmont LD, Patterson GML, Drubin DG (1999) New actin mutants allow further characterization of the nucleotide binding cleft and drub binding sites. J Cell Sci 112:1325–1336.

Bergnes G, Brejc K, Belmont L (2005) Mitotic kinesins: prospects for antimitotic drug discovery. Curr Top Med Chem 5:127–145.

Blokhin AV, Yoo H-D, Geralds RS, Nagle DG, Gerwick WH, Hamel E (1995) Characterization of the interaction of the marine cyanobacterial natural product curacin A with the colchicine site of tubulin and initial structure-activity studies with analogs. Mol Pharmacol 48:523–531.

Bubb MR, Senderowicz AMJ, Sausville EA, Duncan KLK, Korn ED (1994) Jasplakinolide, a cytotoxic natural product, induces actin polymerization and competitively inhibits the binding of phalloidin to F-actin. J Biol Chem 269:14869–14871.

Bubb MR, Spector I, Bershadsky AD, Korn ED (1995) Swinholide A is a microfilament disrupting marine toxin that stabilizes actin dimers and severs actin filaments. J Biol Chem 270:3463–3466.

Bubb MR, Spector I, Beyer BB, Fosen KM (2000) Effects of jasplakinolide on the kinetics of actin polymerization: an explanation for certain *in vivo* observations. J Biol Chem 275:5163–5170.

Carbonelli S, Zampella A, Randazzo A, Debitus C, Gomez-Paloma L (1999) Sphinxolides E-G and reidispongiolide C: four new cytotoxic macrolides from the New Caledonian Lithistida sponges *N. superstes* and *R. coerulea*. Tetrahedron 55:14665–14674.

Carmeli S, Kashman Y (1985) Structure of swinholide-a, a new macrolide from the marine sponge *Theonella swinhoei*. Tetrahedron Lett 26:511–514.

Chevallier C, Richardson AD, Edler MC, Hamel E, Harper MK, Ireland CM (2003) A new cytotoxic and tubulin-interactive milnamide derivative from a marine sponge *Cymbastela* sp. Org Lett 5:3737–3739.

Chik JK, Lindberg U, Schutt CE (1996) The structure of an open state of β-actin at 2.65 Å resolution. J Mol Biol 263:607–623.

Cooper JA (1987) Effects of cytochalasins and phalloidin on actin. J Cell Biol 105:1473–1478.

Cooper JA (1991) The role of actin polymerization in cell motility. Annu Rev Physiol 53: 585–605.

Corley DG, Herb R, Moore RE, Scheuer PJ, Paul VJ (1988) Laulimalides: new potent cytotoxic macrolides from a marine sponge and a nudibranch predator. J Org Chem 53:3644–3646.

Coué M, Brenner SL, Spector I, Korn ED (1987) Inhibition of actin polymerization by latrunculin A. FEBS Lett 213:316–318.

Crews P, Manes LV, Boehler M (1986) Jasplakinolide, a cyclodepsipeptide from the marine sponge, *Jaspis* SP. Tetrahedron Lett 27:2797–2800.

Crews P, Farias JJ, Emrich R, Keifer PA (1994) Milnamide A, an unusual cytotoxic tripeptide from the marine sponge *Auletta* cf. *constricta*. J Org Chem 59:2932–2934.

Cruz-Monserrate Z, Vervoort HC, Bai R, Newman DJ, Howell SB, Los G, Mullaney JT, Williams MD, Pettit GR, Fenical W, Hamel E (2003) Diazonamide A and a synthetic structural analog: disruptive effects on mitosis and cellular microtubules and analysis of their interactions with tubulin. Mol Pharmacol 63:1273–1280.

D'Ambrosio M, Guerriero A, Pietra F (1987) Sarcodictyin A and sarcodictyin B, novel diterpenoidic alcohols esterified by (*E*)-*N*(1)-methylurocanic acid. Isolation from the Mediterranean stolonifer *Sarcodictyon roseum*. Helv Chim Acta 70:2019–2027.

D'Auria MV, Gomez-Paloma L, Minale L, Zampella A, Verbist J-F, Roussakis C, Debitus C (1993) Three new potent cytotoxic macrolides closely related to sphinxolide from the New Caledonian sponge *Neosiphonia superstes*. Tetrahedron 49:8657–8664.

D'Auria MV, Gomez-Paloma L, Minale L, Zampella A (1994) Reidispongiolide A and B, two new potent cytotoxic macrolides from the new caledonian sponge *Reidispongia coerulea*. Tetrahedron 50:4829–4834.

Degnan BM, Hawkins CJ, Lavin MF, McCaffrey EJ, Parry DL, Watters DJ (1989) Novel cytotoxic compounds from the ascidian *Lissoclinum bistratum*. J Med Chem 32:1354–1359.

Didry D, Carlier MF, Pantaloni D (1998) Synergy between actin depolymerizing factor/cofilin and profilin in increasing actin filament turnover. J Biol Chem 273:25602–25611.

Dixon WE, Malden W (1908) Colchicine with special reference to its mode of action and effect on bone-marrow. J Physiol 37:50–76.

Doi M, Ishida T, Kobayashi M, Kitagawa I (1991) Molecular conformation of swinholide A, a potent cytotoxic dimeric macrolide from the Okinawan marine sponge *Theonella swinhoei*: X-ray crystal structure of its diketone derivative. J Org Chem 56:3629–3632.

Dominguez R (2004) Actin-binding proteins – a unifying hypothesis. Trend Biochem Sci 29: 572–578.

Donehower RC, Rowinsky EK, Grochow LB, Longnecker SM, Ettinger DS (1987) Phase I trial of taxol in patients with advanced cancer. Cancer Treat Rep 71:1171–1177.

Dos Remedios CG, Chhabra D, Kekic M, Dedova IV, Tsubakihara M, Berry DA, Nosworthy NJ (2003) Actin binding proteins: regulation of cytoskeletal microfilaments. Physiol Rev 83: 433–473.

Ebbinghaus S, Rubin E, Hersh E, Cranmer LD, Bonate PL, Fram RJ, Jekunen A, Weitman S, Hammond LA (2005) A phase I study of the dolastatin-15 analogue tasidotin (ILX651) administered intravenously daily for 5 consecutive days every 3 weeks in patients with advanced solid tumors. Clin Cancer Res 11:7807–78163.

Edler MC, Fernandez AM, Lassota P, Ireland CM, Barrows LR (2002) Inhibition of tubulin polymerization by vitilevuamide, a bicyclic marine peptide, at a site distinct from colchicine, the vinca alkaloids, and dolastatin 10. Biochem Pharmacol 63:707–715.

Estes JE, Selden LA, Gershman LC (1981) Mechanism of action of phalloidin on the polymerization of muscle actin. Biochemistry 20:708–712.

Fenteany G, Zhu S (2003) Small molecule inhibitor of actin dynamics and cell motility. Curr Top Med Chem 3:593–616.

Fukazawa Y, Saitoh Y, Ozawa F, Ohta Y, Mizuno K, Inokuchi K (2003) Hippocampal LTP is accompanied by enhanced F-actin content within the dendritic spine that is essential for late LTP maintenance in vivo. Neuron 38:447–460.

Fusetani N, Yasumuro K, Matsunaga S, Hashimoto K (1989) Mycalolides A-C, hybrid macrolides of ulapualides and halichondramide, from a sponge of the genus *Mycale*. Tetrahedron Lett 30:2809–2812.

Fusetani N, Shinoda K, Matsunaga S (1993) Bioactive marine metabolites. 48. Cinachyrolide A: a potent cytotoxic macrolide possessing two spiro ketals from marine sponge *Cinachyra* sp. J Am Chem Soc 115:3977–3981.

Gaitanos TN, Buey RM, Diaz JF, Northcote PT, Teesdale-Spittle P, Andreu JM, Miller JH (2004) Peloruside A does not bind to the taxoid site on β-tubulin and retains its activity in multidrug-resistant cell lines. Cancer Res 64:5063–5067.

Gamble WR, Durso NA, Fuller RW, Westergaard CK, Johnson TR, Sackett DL, Hamel E, Cardellina JH II, Boyd MR (1999) Cytotoxic and tubulin-interactive hemiasterlins from *Auletta* sp. and *Siphonochalina* spp. sponges. Bioorg Med Chem 7:1611–1615.

Gerwick WH, Proteau PJ, Nagle DG, Hamel E, Blokhin A, Slate DL (1994) Structure of curacin A, a novel antimitotic, antiproliferative, and brine shrimp toxic natural product from the marine cyanobacterium *Lyngbya majuscula*. J Org Chem 59:1243–1245.

Gigant B, Wang C, Ravelli RBG, Roussi F, Steinmetz MO, Curmi PA, Sobel A, Knossow M (2005) Structural basis for the regulation of tubulin by vinblastine. Nature 435:519–522.

Gouiffes D, Juge M, Grimaud N, Welin L, Sauviat MP, Barbin Y, Laurent D, Roussakis C, Henichart JP, Verbist JF (1988) Bistramide A, a new toxin from the urochordata *Lissoclinum bistratum* Sluiter: Isolation and preliminary characterization. Toxicon 26:1129–1136.

Griffiths G, Garrone B, Deacon E, Owen P, Pongracz J, Mead G, Bradwell A, Watters D, Lord J (1996) The polyether bistratene A activates protein kinase C-δ and induces growth arrest in HL60 cells. Biochem Biophys Res Commun 222:802–808.

Gunasekera SP, Gunasekera M, Longley RE, Schulte GK (1990) Discodermolide: A new bioactive polyhydroxylated lactone from the marine sponge *Discodermia dissoluta*. J Org Chem 55:4912–4915; correction (1991) J Org Chem 56:1346.

Hamel E, Sackett DL, Vourloumis D, Nicolaou KC (1999) The coral-derived natural products eleutherobin and sarcodictyins A and B: effects on the assembly of purified tubulin with and without microtubule-associated proteins and binding at the polymer taxoid site. Biochemistry 38:5490–5498.

Hartwell LH, Weinert TA (1989) Checkpoints: controls that ensure the order of cell cycle events. Science 246:629–63l.

Hirata Y, Uemura D (1986) Halichondrins – antitumor polyether macrolides from a marine sponge. Pure Appl Chem 58:701–710.

Hirata K, Muraoka S, Suenaga K, Kuroda T, Kato K, Tanaka H, Yamamoto M, Takata M, Yamada K, Kigoshi H (2006) Structure basis for antitumor effect of aplyronine A. J Mol Biol 356: 945–954.

Hirokawa N, Takemura R (2004) Kinesin superfamily proteins and their various functions and dynamics. Exp Cell Res 301:50–59.

Hoffman MA, Blessing JA, Lentz SS (2003) A phase II trial of dolastatin-10 in recurrent platinum-sensitive ovarian carcinoma: a gynecologic oncology group study. Gynecol Oncol 89:95–98.

Hood KA, West LM, Rouwe B, Northcote PT, Berridge MV, Wakefield St J, Miller JH (2002) Peloruside A, a novel antimitotic agent with paclitaxel-like microtubule-stabilizing activity. Cancer Res 62:3356–3360.

Hori M, Matsuura Y, Yoshimoto R, Ozaki H, Yasumoto T, Karaki H (1999) Actin depolymerizing action by marine toxin, pectenotoxin-2. Nippon Yakurigaku Zasshi 114 (Suppl 1): 225P–229P.

Hung DT, Nerenberg JB, Schreiber SL (1994) Distinct binding and cellular properties of synthetic (+)- and (−)-discodermolides. Chem Biol 1:67–71.

Hung DT, Chen J, Schreiber SL (1996) (+)-Discodermolide binds to microtubules in stoichiometric ratio to tubulin dimers, blocks taxol binding and results in mitotic arrest. Chem Biol 3: 287–293.

Ibrado AM, Kim CN, Bhalla K (1998) Temporal relationship of CDK1 activation and mitotic arrest to cytosolic accumulation of cytochrome C and caspase-3 activity during Taxol-induced apoptosis of human AML HL-60 cells. Leukemia 12:1930–1936.

Isbrucker RA, Cummins J, Pomponi SA, Longley RE, Wright AE (2003) Tubulin polymerizing activity of dictyostatin-1, a polyketide of marine sponge origin. Biochem Pharmacol 66: 75–82.

Ishiwata H, Nemoto T, Ojika M, Yamada K (1994a) Isolation and stereostructure of doliculide, a cytotoxic cyclodepsipeptide from the Japanese sea hare *Dolabella auricularia*. J Org Chem 59:4710–4711.

Ishiwata H, Sone H, Kigoshi H, Yamada K (1994b) Total synthesis of doliculide, a potent cytotoxic cyclodepsipeptide from the Japanese sea hare *Dolabella auricularia*. J Org Chem 59:4712–4713.

Jordan MA (2002) Mechanism of action of antitumor drugs that interact with microtubules and tubulin. Curr Med Chem Anti-Cancer Agents 2:1–17.

Jung JH, Sim CJ, Lee C-O (1995) Cytotoxic compounds from a two-sponge association. J Nat Prod 58:1722–1726.

Kashman Y, Groweiss A, Shmueli U (1980) Latrunculin, a new 2-thiazolidinone macrolide from the marine sponge *Latrunculia magifica*. Tetrahedron Lett 21:3629–3632.

Kato Y, Fusetani N, Matsunaga S, Hashimoto K, Sakai R, Higa T, Kashman Y (1987) Antitumor macrolides isolated from a marine sponge *Theonella* sp.: structure revision of misakinolide A. Tetrahedron Lett 28:6225–6228.

Kerbrat P, Dieras V, Pavlidis N, Ravaud A, Wanders J, Fumoleau P; EORTC Early Clinical Studies Group/New Drug Development Office (2003) Phase II study of LU 103793 (dolastatin analogue) in patients with metastatic breast cancer. Eur J Cancer 39:317–320.

Kerksiek K, Mejillano MR, Schwartz RE, Georg GI, Himes RH (1995) Interaction of cryptophycin 1 with tubulin and microtubules. FEBS Lett 377:59–61.

Kernan MR, Faulkner DJ (1987) Halichondramide, an antifungal macrolide from the sponge *halichondria* sp. Tetrahedron Lett 28:2809–2812.

Kigoshi H, Suenaga K, Takagi M, Akao A, Kanematsu K, Kamei N, Okugawa Y, Yamada K (2002) Cytotoxicity and actin-depolymerizing activity of aplyronine A, a potent antitumor macrolide or marine origin, and its analogs. Tetrahedron 58:1075–1102.

Kindler HL, Tothy PK, Wolff R, McCormack RA, Abbruzzese JL, Mani S, Wade-Oliver KT, Vokes EE (2005) Phase II trials of dolastatin-10 in advanced pancreaticobiliary cancers. Invest New Drugs 23:489–493.

Klenchin VA, Allingham JS, King R, Tanaka J, Marriott G, Rayment I (2003) Trisoxazole macrolide toxins mimic the binding of actin-capping proteins to actin. Nat Struct Biol 10: 1058–1063.

Klenchin VA, King R, Tanaka J, Marriott G, Rayment I (2005) Structural basis of swinholide A binding to actin. Chem Biol 12:287–291.

Kobayashi M, Tanaka J, Katori T, Matsuura M, Kitagawa I (1989) Structure of swinholide A, a potent cytotoxic macrolide from the Okinawan marine sponge *Theonella swinhoei*. Tetrahedron Lett 30:2963–2966.

Kobayashi J, Shigemori H, Ishibashi M, Yamasu T, Hirota H, Sasaki T (1991) Amphidinolides G and H: new potent cytotoxic macrolides from the cultured symbiotic dinoflagellate *Amphidinium* sp. J Org Chem 56:5221–5224.

Kobayashi J, Murata O, Shigemori H, Sasaki T (1993) Jaspisamides A-C, new cytotoxic macrolides from the Okinawan sponge *Jaspis* sp. J Nat Prod 56:787–791.

Kobayashi J, Tsuda M, Fuse H, Sasaki T, Mikami Y (1997) Halishigamides A-D, new cytotoxic oxazole-containing metabolites from Okinawan sponge *Halichondria* sp. J Nat Prod 60: 150–154.

Kobayashi M, Aoki S, Sakai H, Kawazoe K, Kihara N, Sasaki T, Kitagawa I (1993a) Altohyrtin A, a potent anti-tumor macrolide from the Okinawan marine sponge *Hyrtios altum*. Tetrahedron Lett 34:2795–2798.

Kobayashi M, Aoki S, Sakai H, Kihara N, Sasaki T, Kitagawa I (1993b) Altohyrtins B and C and 5-desacetylaltohyrtin A, potent cytotoxic macrolide congeners of altohyrtin A, from the Okinawan marine sponge *Hyrtios altum*. Chem Pharm Bull 41:989–991.

Kobayashi M, Aoki S, Ohyabu N, Kurosu M, Wang W, Kitagawa I (1994a) Arenastatin A, a potent cytotoxic depsipeptide from the Okinawan marine sponge *Dysidea arenaria*. Chem Pharm Bull 42:2196–2198.

Kobayashi M, Aoki S, Kitagawa I (1994b) Absolute stereostructures of altohyrtin A and its congeners, potent cytotoxic macrolides from the Okinawan marine sponge *Hyrtios altum*. Tetrahedron Lett 35:1243–1246.

König GM, Kehraus S, Seibert SF, Abdel-Lateff A, Müller D (2006) Natural products from marine organisms and their associated microbes. ChemBioChem 7:229–238.

Korn ED (1982) Actin polymerization and its regulation by proteins from nonmuscle cells. Physiol Rev 62:672–737.

Kouyama T, Mihashi, K (1981) Fluorimetry study of *N*-(1-pyrenyl)iodoacetamide-labelled F-actin. Local structural change of actin protomer both on polymerization and on binding of heavy meromyosin. Eur J Biochem 114:33–38.

Krug LM, Miller VA, Kalemkerian GP, Kraut MJ, Ng KK, Heelan RT, Pizzo BA, Perez W, McClean N, Kris MG (2000) Phase II study of dolastatin-10 in patients with advanced non-small-cell lung cancer. Ann Oncol 11:227–228.

Kuroda T, Suenaga K, Sasakura A, Handa T, Okamoto K, Kigoshi H (2006) Study of the interaction between actin and antitumor substance aplyronine A with a novel fluorescent photoaffinity probe. Bioconj Chem 17:524–529.

Lee JC, Timasheff SN (1977) In vitro reconstitution of calf brain microtubules: effects of solution variable. Biochemistry 16:1754–1762.

Lindel T, Jensen PR, Fenical W, Long BH, Casazza AM, Carboni J, Fairchild CR (1997) Eleutherobin, a new cytotoxin that mimics paclitaxel (Taxol) by stabilizing microtubules. J Am Chem Soc 119:8744–8745.

Lindquist N, Fenical W, Van Duyne GD, Clardy J (1991) Isolation and structure determination of diazonamides A and B, unusual cytotoxic metabolites from the marine ascidian *Diazona chinensis*. J Am Chem Soc 113:2303–2304.

Long BH, Carboni JM, Wasserman AJ, Cornell LA, Casazza AM, Jensen PR, Lindel T, Fenical W, Fairchild CR (1998) Eleutherobin, a novel cytotoxic agent that induces tubulin polymerization, is similar to paclitaxel (Taxol). Cancer Res 58:1111–1115.

Longley RE, Gunasekera SP, Faherty D, McLane J, Dumont F (1993) Immunosuppression by discodermolide. Ann NY Acad Sci 696:94–107.

Ludueña RF, Roach MC, Prasad V, Pettit GR (1993) Interaction of halichondrin B and homohalichondrin B with bovine brain tubulin. Biochem Pharmacol 45:421–427.

MacLean-Fletcher SD, Pollard TD (1980) Viscometric analysis of the gelation of acanthamoeba extracts and purification of two gelation factors. J Cell Biol 85:414–428.

Madiraju C, Edler MC, Hamel E, Raccor BS, Van Balachandran R, Zhu G, Giuliano KA, Vogt A, Shin Y, Fournier J-H, Fukui Y, Brückner AM, Curran DP, Day BW (2005) Tubulin assembly, taxoid site binding, and cellular effects of the microtubule-stabilizing agent dictyostatin. Biochemistry 44:15053–15063.

Mandelkow E, Mandelkow EM (2002) Kinesin motors and disease. Trends Cell Biol 12:585–591.

Mann J (2002) Natural products in cancer chemotherapy: past, present and future. Nat Rev Cancer 2:143–148.

Margolin K, Longmate J, Synold TW, Gandara DR, Weber J, Gonzalez R, Johansen MJ, Newman R, Baratta T, Doroshow JH (2001) Dolastatin-10 in metastatic melanoma: a phase II and pharmokinetic trial of the California Cancer Consortium. Invest New Drugs 19:335–340.

Marks RS, Graham DL, Sloan JA, Hillman S, Fishkoff S, Krook JE, Okuno SH, Mailliard JA, Fitch TR, Addo F (2003) A phase II study of the dolastatin 15 analogue LU 103793 in the treatment of advanced non-small-cell lung cancer. Am J Clin Oncol 26:336–337.

Marquez BL, Watts KS, Yokochi A, Roberts MA, Verdier-Pinard P, Jimenez JI, Hamel E, Scheuer PJ, Gerwick WH (2002) Structure and absolute stereochemistry of hectochlorin, a potent stimulator of actin assembly. J Nat Prod 65:866–871.

Marriott G, Zechel K, Jovin TM (1988) Spectroscopic and functional characterization of an environmentally sensitive fluorescent actin conjugate. Biochemistry 27:6214–6220.

Matsunaga S, Fusetani N, Hashimoto K, Koseki K, Noma M (1986) Bioactive marine metabolites. Part 13. Kabiramide C, a novel antifungal macrolide from nudibranch eggmasses. J Am Chem Soc 108:847–849.

Matsunaga S, Fusetani N, Hashimoto K, Koseki K, Noma M, Noguchi H, Sankawa U (1989) Bioactive marine metabolites. 25. Further kabiramides and halichondramides, cytotoxic macrolides embracing trisoxazole, from the *Hexabranchus* egg masses. J Org Chem 54:1360–1363.

McGough AM, Staiger CJ, Min J-K, Simonetti KD (2003) The gelsolin family of actin regulatory proteins: modular structures, versatile functions. FEBS Lett 552:75–81.

Mooberry SL, Busquets L, Tien G (1997) Induction of apoptosis by cryptophycin 1, a new anti-microtubule agent. Int J Cancer 73:440–448.

Mooberry SL, Tien G, Hernandez AH, Plubrukarn A, Davidson BS (1999) Laulimalide and iso-laulimalide, new paclitaxel-like microtubule-stabilizing agents. Cancer Res 59:653–660.

Moore RE (1981) Constituents of blue-green algae. In: Scheur PJ (ed) Marine natural products. Academic Press, New York, Vol 4, pp 1–52.

Morton WM, Ayscough KR, McLaughlin PJ (2000) Latrunculin alters the actin-monomer subunit interface to prevent polymerization. Nat Cell Biol 2:376–378.

Mundy DI, Machleidt T, Ying Y, Anderson RGW, Bloom GS (2002) Dual control of caveolar membrane traffic by microtubules and the actin cytoskeleton. J Cell Sci 115:4327–4339.

Murray AW (1992) Creative blocks: cell cycle checkpoints and feedback controls. Nature 359: 599–604.

Nagle A, Hur W, Gray NS (2006) Antimitotic agents of natural origin. Curr Drug Target 7:305–326.

Nogales E, Wang H-W, Niederstrasser H (2003) Tubulin rings: which way do they curve? Curr Opin Struct Biol 13:256–261.

Nunes M, Kaplan J, Wooters J, Hari M, Minnick AA Jr, May MK, Shi C, Musto S, Beyer C, Krishnamurthy G, Qiu Y, Loganzo F, Ayral-Kaloustian S, Zask A, Greenberger LM (2005) Two photoaffinity analogues of the tripeptide, hemiasterlin, exclusively label α-tubulin. Biochemistry 44:6844–6852.

Oda T, Crane ZD, Dicus CW, Sufi BA, Bates RB (2003) Dolastatin 11 connects two long-pitch strands in F-actin to stabilize microfilaments. J Mol Biol 328:319–324.

Oda T, Namba K, Maéda Y (2005) Position and orientation of phalloidin in F-actin determined by X-ray fiber diffraction analysis. Biophys J 88:2727–2736.

Ojika M, Kigoshi H, Yoshida Y, Ishigaki T, Nisiwaki M, Tsukada I, Arakawa M, Ekimoto H, Yamada K (2007) Aplyronine A, a potent antitumor macrolide of marine origin, and the congeners aplyronines B and C: isolation, structures, and bioactivities. Tetrahedron 63:3138–3167.

Oosawa F, Asakura S (1975) Theory of polymerization equilibrium. In: Horecker B, Kaplan NO, Marmur J, Scheraga HA (eds) Thermodynamics of the polymerization of proteins. Academic Press, London, pp 109–116.

Oosawa F, Asakura S, Hotta K, Ooi T (1959) G-F transformation of actin as a fibrous condensation. J Polymer Sci 37:323–326.

Patterson GM, Carmeli S (1992) Biological effects of tolytoxin (6-hydroxy-7-O-methyl-scytophycin b), a potent bioactive metabolite from cyanobacteria. Arch Microbiol 157: 406–410.

Patterson GM, Smith CD, Kimura LH, Britton BA, Carmeli S (1993) Action of tolytoxin on cell morphology, cytoskeletal organization, and actin polymerization. Cell Motil Cytoskel 24:39–48.

Paull KD, Lin CM, Malspeis L, Hamel E (1992) Identification of novel antimitotic agents acting at the tubulin level by computer-assisted evaluation of differential cytotoxicity data. Cancer Res 52:3892–3900.

Perez EA, Hillman DW, Fishkin PA, Krook JE, Tan WW, Kuriakose PA, Alberts SR, Dakhil SR (2005) Phase II trial of dolastatin-10 in patients with advanced breast cancer. Invest New Drugs 23:257–261.

Pettit GR (1997) The dolastatins. Prog Chem Org Nat Prod 70:1–79.

Pettit GR, Kamano Y, Herald CL, Tuinman AA, Boettner FE, Kizu H, Schmidt JM, Baczynskyj L, Tomer KB, Bontems RJ (1987) The isolation and structure of a remarkable marine animal antineoplastic constituent: Dolastatin 10. J Am Chem Soc 109:6883–6885.

Pettit GR, Kamano Y, Dufresne C, Cerny RL, Herald CL, Schmidt JM (1989) Isolation and structure of the cytostatic linear depsipeptide dolastatin 15. J Org Chem 54:6005–6006.

Pettit GR, Cichacz ZA, Gao F, Herald CL, Boyd MR (1993a) Isolation and structure of the remarkable human cancer cell growth inhibitors spongistatins 2 and 3 from an eastern Indian Ocean *Spongia* sp. J Chem Soc Chem Commun 1166–1168.

Pettit GR, Cichacz ZA, Gao F, Herald CL, Boyd MR, Schmidt JM, Hooper JNA (1993b) Isolation and structure of spongistatin 1. J Org Chem 58:1302–1304.

Pettit GR, Herald CL, Cichacz ZA, Gao F, Boyd MR, Christie ND, Schmidt JM (1993c) Antineoplastic agents. 293. The exceptional human cancer cell growth inhibitors spongistatins 6 and 7. Nat Prod Lett 3:239–244.

Pettit GR, Herald CL, Cichacz ZA, Gao F, Schmidt JM, Boyd MR, Christie ND, Boettner FE (1993d) Antineoplastic agents. 288. Isolation and structure of the powerful human cancer cell growth inhibitors spongistatins 4 and 5 from an African *Spirastrella spinispirulifera* (Porifera). J Chem Soc Chem Commun 1805–1807.

Pettit GR, Cichacz ZA, Gao F, Boyd MR, Schmidt JM (1994a) Isolation and structure of the cancer cell growth inhibitor dictyostatin 1. J Chem Soc Chem Comm 1111–1112.

Pettit GR, Cichacz ZA, Herald CL, Gao F, Boyd MR, Schmidt JM, Hamel E, Bai R (1994b) Antineoplastic agents 300. Isolation and structure of the rare human cancer inhibitory macrocyclic lactones spongistatins 8 and 9. J Chem Soc Chem Commun 1605–1606.

Quiñoà E, Kakou Y, Crews P (1988) Fijianolides, polyketide heterocyclics from a marine sponge. J Org Chem 53:3642–3644.

Ritter T, Carreira EM (2002) The diazonamides: the plot thickens. Angew Chem Int Ed 41: 2489–2495.

Rizvi SA, Tereshko V, Kossiakoff AA, Kozmin SA (2006) Structure of bistramide A-actin complex at a 1.35 Å resolution. J Am Chem Soc 128:3882–3883.

Roesener JA, Scheuer PJ (1986) Ulapualide A and B, extraordinary antitumor macrolides from nudibranch eggmasses. J Am Chem Soc 108:846–847.

Saad ED, Kraut EH, Hoff PM, Moore DF Jr, Jones D, Pazdur R, Abbruzzese JL (2002) Phase II study of dolastatin-10 as first-line treatment for advanced colorectal cancer. Am J Clin Oncol 25:451–453.

Saito S, Watabe S, Ozaki H, Fusetani N, Karaki H (1994) Mycalolide B, a novel actin depolymerizing agent. J Biol Chem 269:29710–29714.

Saito S, Watabe S, Ozaki H, Kigoshi H, Yamada K, Fusetani N, Karaki H (1996) Novel actin depolymerizing macrolide aplyronine A. J Biochem 120:552–555.

Saito S, Ozaki H, Kobayashi M, Kobayashi J, Suzuki T, Kobayashi H, Watabe S, Fusetani N Karaki H (1997) Inhibitory effects of trisoxazole macrolides isolated from marine products on actin polymerization. Jpn J Pharmacol 73 (Suppl. I):119P.

Saito S, Watabe S, Ozaki H, Kobayashi M, Suzuki T, Kobayashi H, Fusetani N, Karaki H (1998) Actin-depolymerizing effect of dimeric macrolides, bistheonellide A and swinholide A. J Biochem 123:571–578.

Saito S, Feng J, Kira A, Kobayashi J, Ohizumi Y (2004) Amphidinolide H, a novel type of actin-stabilizing agent isolated from dinoflagellate. Biochem Biophys Res Comm 320:961–965.

Sakai R, Higa T, Kashman Y (1986) Misakinolide-A, an antitumor macrolide from the marine sponge *Theonella* sp. Chem Lett 1499–1502.

Sampath P, Pollard TD (1991) Effects of cytochalasin, phalloidin, and pH on the elongation of actin filaments. Biochemistry 30:1973–1980.

Sasaki H, Ozaki H, Karaki H, Nonomura Y (2004) Actin filaments play an essential role for transport of nascent HIV-1 proteins in host cells. Biochem Biophys Res Commun 316:588–593.

Schiff PB, Fant J, Horwitz SB (1979) Promotion of microtubule assembly *in vitro* by taxol. Nature 277:665–667.

Schoffski P, Thate B, Beutel G, Bolte O, Otto D, Hofmann M, Ganser A, Jenner A, Cheverton P, Wanders J, Oguma T, Atsumi R, Satomi M (2004) Phase I and pharmacokinetic study of TZT-1027, a novel synthetic dolastatin 10 derivative, administered as a 1-hour intravenous infusion every 3 weeks in patients with advanced refractory cancer. Ann Oncol 15:671–679.

Schwartz RE, Hirsch CF, Sesin DF, Flor JE, Chartrain M, Fromtling RE, Harris GH, Salvatore MJ, Liesch JM, Yudin K (1990) Pharmaceuticals from cultured algae. J Ind Microbiol 5: 113–123.

Sessa C, Weigang-Köhler K, Pagani O, Greim G, Mor O, De Pas T, Burgess M, Weimer I, Johnson R (2002) Phase I and pharmacological studies of the cryptophycin analogue LY355703 administered on a single intermittent or weekly schedule. Eur J Cancer 38:2388–2396.

Shelanski ML, Taylor EW (1967) Isolation of a protein subunit from microtubules. J Cell Biol 34: 549–554.

Shelanski ML, Gaskin F, Cantor CR (1973) Microtubule assembly in the absence of added nucleotides. Proc Natl Acad Sci 70:765–768.

Shin J, Lee H-S, Kim J-Y, Shin HJ, Ahn J-W, Paul VJ (2004) New macrolides from the sponge *Chondrosia corticata*. J Nat Prod 67:1889–1892.

Silacci P, Mazzolai L, Gauci C, Stergiopulos N, Yin HL, Hayoz D (2004) Gelsolin superfamily proteins: key regulators of cellular functions. Cell Mol Life Sci 61:2614–2623.

Smith CD, Zhang X (1996) Mechanism of action of cryptophycin: interaction with the *Vinca* alkaloid domain of tubulin. J Biol Chem 271:6192–6198.

Smith CD, Zhang X, Mooberry SL, Patterson GM, Moore RE (1994) Cryptophycin: a new antimicrotubule agent active against drug-resistant cells. Cancer Res 54:3779–3784.

Spector I, Shochet NR, Kashman Y (1983) Latrunculins: novel marine toxins that disrupt microfilament organization in cultured cells. Science 219:493–495.

Spector I, Braet F, Shochet NR, Bubb MR (1999) New anti-actin drugs in the study of the organization and function of the actin cytoskeleton. Microsc Res Tech 47:18–37.

Statsuk AV, Bai R, Baryza JL, Verma VA, Hamel E, Wender PA, Kozmin SA (2005) Actin is the primary cellular receptor of bistramide A. Nat Chem Biol 1:383–388.

Stevenson JP, Sun W, Gallagher M, Johnson R, Vaughn D, Schuchter L, Algazy K, Hahn S, Enas N, Ellis D, Thornton D, O'Dwyer PJ (2002) Phase I trial of the cryptophycin analogue LY355703 administered as an intravenous infusion on a day 1 and 8 schedule every 21 days. Clin Cancer Res 8:2524–2529.

Stratmann K, Moore RE, Bonjouklian R, Deeter JB, Patterson GML, Shaffer S, Smith CD, Smitka TA (1994) Welwitindolinones, unusual alkaloids from the blue-green algae *Hapalosiphon welwitschii* and *Westiella intricata*. Relationship to Fischer indoles and hapalinodoles. J Am Chem Soc 116:9935–9942.

Straub FB (1942) Actin. Stud Inst Med Chem Univ Szeged II:1–15.

Suenaga K, Miya S, Kuroda T, Handa T, Kanematsu K, Sakakura A, Kigoshi H (2004) Synthesis and actin-depolymerizing activity of mycalolide analogs. Tetrahedron Lett 45:5383–5386.

Talpir R, Benayahu Y, Kashman Y, Pannell L, Schleyer M (1994) Hemiasterlin and geodiamolide TA; two new cytotoxic peptides from the marine sponge *Hemiasterella minor* (Kirkpatrick). Tetrahedron Lett 35:4453–4456.

Tamura K, Nakagawa K, Kurata T, Satoh T, Nogami T, Takeda K, Mitsuoka S, Yoshimura N, Kudoh S, Negoro S, Fukuoka M (2007) Phase I study of TZT-1027, a novel synthetic dolastatin 10 derivative and inhibitor of tubulin polymerization, which was administered to patients with advanced solid tumors on days 1 and 8 in 3-week courses. Cancer Chemother Pharmacol 60:285–293.

Tanaka J, Yan Y, Choi J, Bai J, Klenchin VA, Rayment I, Marriott G (2003) Biomolecular mimicry in the actin cytoskeleton: Mechanisms underlying the cytotoxicity of kabiramide C and related macrolides. Proc Nat Acad Sci USA 100:13851–13856.

ter Haar E, Kowalski RJ, Hamel E, Lin CM, Longley RE, Gunasekera SP, Rosenkranz HS, Day BW (1996a) Discodermolide, A cytotoxic marine agent that stabilizes microtubules more potently than taxol. Biochemistry 35:243–250.

ter Haar E, Rosenkranz HS, Hamel E, Day BW. (1996b) Computational and molecular modeling evaluation of the structural basis for tubulin polymerization inhibition by colchicine site agents. Bioorg Med Chem 4:1659–1671.

Terry DR, Spector I, Higa T, Bubb MR (1997) Misakinolide A is a marine macrolide that caps but does not sever filamentous actin. J Biol Chem 272:7841–7845.

Toonen RF, Kochubey O, de Wit H, Gulyas-Kovacs A, Konijnenburg B, Sørensen JB, Klingauf J, Verhage M (2006) Dissecting docking and tethering of secretory vesicles at the target membrane. EMBO J 25:3725–3737.

Usui T, Kazami S, Dohmae N, Mashimo Y, Kondo H, Tsuda M, Goi Terasaki A, Ohashi K, Kobayashi J, Osada H (2004) Amphidinolide H, a potent cytotoxic macrolide, covalently binds on actin subdomain 4 and stabilizes actin filament. Chem Biol 11:1269–1277.

Vaishampayan U, Glode M, Du W, Kraft A, Hudes G, Wright J, Hussain M (2000) Phase II study of dolastatin-10 in patients with hormone-refractory metastatic prostate adenocarcinoma. Clin Cancer Res 6:4205–4208.

Wang G, Shang L, Burgett AWG, Harran PG, Wang X (2007) Diazonamide toxins reveal an unexpected function for ornithine δ-amino transferase in mitotic cell division. Proc Natl Acad Sci USA 104:2068–2073.

Wang TH, Wang HS, Soong YK (2000) Paclitaxel-induced cell death: where the cell cycle and apoptosis come together. Cancer 88:2619–2628.

Wani MC, Taylor HL, Wall ME, Coggon P, McPhail AT (1971) Plant antitumor agents. VI. Isolation and structure of taxol, a novel antileukemic and antitumor agent from *Taxus brevifolia*. J Am Chem Soc 93:2325–2327.

West LM, Northcote PT, Battershill CN (2000) Peloruside A: A potent cytotoxic macrolide isolated from the New Zealand marine sponge *Mycale* sp. J Org Chem 65:445–449.

Wieland T (1977) Interaction of phallotoxins with actin. Adv Enz Reg 15:285–300.

Xia S, Kenesky CS, Rucker PV, Smith AB III, Orr GA, Horwitz SB (2006) A photoaffinity analogue of discodermolide specifically labels a peptide in β-tubulin. Biochemistry 45:11762–11775.

Yamada K, Ojika M, Ishigaki T, Yoshida Y (1993) Aplyronine A, a potent antitumor substance, and the congeners aplyronines B and C isolated from the sea hare *Aplysia kurodai*. J Am Chem Soc 115:11020–11021.

Yasumoto T, Murata M, Oshima Y, Sano M, Matsumoto GK, Clardy J (1985) Diarrhetic shellfish toxins. Tetrahedron 41:1019–1025.

Yeung K-S, Paterson I (2002) Actin-binding marine macrolides: Total synthesis and biological importance. Angew Chem Int Ed 41:4632–4653.

Zabriskie TM, Klocke JA, Ireland CM, Marcus AH, Molinski TF, Faulkner DJ, Xu C, Clardy J (1986) Jaspamide, a modified peptide from a jaspis sponge, with insecticidal and antifungal activity. J Am Chem Soc 108:3123–3124.

Zhang X, Smith CD (1996) Microtubule effects of welwistatin, a cyanobacterial indolinone that circumvents multiple drug resistance. Mol Pharmacol 49:288–294.

Zhang X, Minale L, Zampella A, Smith CD (1997) Microfilament depletion and circumvention of multiple drug resistance by sphinxolides. Cancer Res 57:3751–3758.

Carcinogenic Aspects of Protein Phosphatase 1 and 2A Inhibitors

Hirota Fujiki and Masami Suganuma

Abstract Okadaic acid is functionally a potent tumor promoter working through inhibition of protein phosphatases 1 and 2A (PP1 and PP2A), resulting in sustained phosphorylation of proteins in cells. The mechanism of tumor promotion with okadaic acid is thus completely different from that of the classic tumor promoter phorbol ester. Other potent inhibitors of PP1 and PP2A – such as dinophysistoxin-1, calyculins A–H, microcystin-LR and its derivatives, and nodularin – were isolated from marine organisms, and their structural features including the crystal structure of the PP1-inhibitor complex, tumor promoting activities, and biochemical and biological effects, are here reviewed. The compounds induced tumor promoting activity in three different organs, including mouse skin, rat glandular stomach and rat liver, initiated with three

H. Fujiki (✉)
Faculty of Pharmaceutical Sciences, Tokushima Bunri University, Yamashiro-cho, Tokushima 770-8514, Japan
e-mail: hfujiki@ph.bunri-u.ac.jp

M. Suganuma
Research Institute for Clinical Oncology, Saitama Cancer Center, Saitama 362-0806, Japan
e-mail: masami@cancer-c.pref.saitama.jp

N. Fusetani and W. Kem (eds.), *Marine Toxins as Research Tools*,
Progress in Molecular and Subcellular Biology, Marine Molecular Biotechnology 46,
DOI: 10.1007/978-3-540-87895-7, © Springer-Verlag Berlin Heidelberg 2009

different carcinogens. The results indicate that inhibition of PP1 and PP2A is a general mechanism of tumor promotion applicable to various organs. This study supports the concept of endogenous tumor promoters in human cancer development.

1   Introduction

Marine organisms are rich sources of unique compounds, and the discovery of a new compound associated with a new function can open a new era in science. Protein phosphorylation by protein kinase and protein dephosphorylation by protein phosphatase is opposite biochemical reactions, but the activities of protein kinase and protein phosphatase are cooperatively well regulated: Protein kinase and protein phosphatase modulate important biochemical and biological functions of proteins in the cells (Fig. 1). When the classic tumor promoter 12-*O*-tetradecanoylphorbol-13-acetate (TPA) is applied to initiated cells of mouse skin, TPA activates protein kinase C (PKC) and induces tumor promotion mediated through signal transduction in the cells, resulting in tumor development on mouse skin.

In contrast, inhibition by okadaic acid of protein phosphatases 1 and 2A (PP1 and PP2A), which are serine/threonine phosphatases, sustains the phosphorylation of proteins (Fig. 1), and similarly induces tumor promotion mediated through signal transduction, resulting in development of tumors on mouse skin. Although the structures of TPA and okadaic acid are different, they are potent tumor promoters that work through signal transduction and activation of gene expression. Thus,

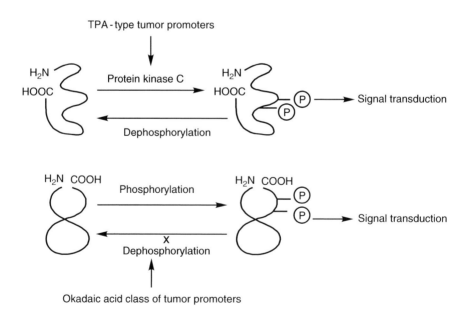

Fig. 1 Schematic illustration of mechanisms of the TPA-types and the okadaic acid class of tumor promoters

okadaic acid, which acts on the cells differently from TPA and is a potent inhibitor of PP1 and PP2A, engendered a new concept of tumor promotion. Moreover, dinophysistoxin-1 (35-methylokadaic acid), calyculin A, microcystin-LR, and nodularin – mostly isolated from marine organisms and all potent inhibitors of PP1 and PP2A – significantly extended the study of tumor promotion.

This chapter reviews PP1 and PP2A inhibitors with regard to tumor promotion in carcinogenesis. Okadaic acid, dinophysistoxin-1, and calyculin A are tumor promoters on mouse skin, and okadaic acid also induced tumor promoting activity in rat glandular stomach. Microcystin-LR and nodularin are strong tumor promoters in rat liver. Since humans are often exposed to these tumor promoters in the environment, the relationship between human cancer development and the presence of PP1 and PP2A inhibitors is a significant objective of cancer research.

In June 2006, the International Agency for Research on Cancer (IARC) in Lyon organized a meeting with scientists from eight countries to assess the carcinogenicity of microcystin-LR and nodularin in humans. In the light of the outcome of this meeting, PP1 and PP2A inhibitors became unique tools for the study of protein phosphatases, signal transduction, gene expression, tumor promotion and carcinogenesis: Investigation of their tumor promoting actions is likely to provide a better understanding of the molecular mechanisms of tumor promotion and of their effects on human health.

Since this study has a long history, supported by various research fields, we will begin by looking at historical perspectives with regard to tumor promotion by the TPA-type and non-TPA–type tumor promoters, and then demonstrate the significance of the okadaic acid class of compounds. Along with introducing a new concept of tumor promotion, we will briefly discuss how we have extended the okadaic acid pathway of tumor promotion to human cancer development under the title of Relation to Human Cancer Development.

2 Historical Perspectives

2.1 TPA-Type Tumor Promoters

How cancer cells grow from initiated cells is an important objective of carcinogenesis, the process of which is called tumor promotion and tumor progression (Fig. 2). Experimentally, a tumor promoter induces clonal growth of initiated cells, which are treated with a subthreshold dose of a carcinogen, resulting in tumor development.

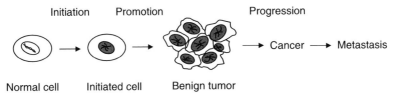

Fig. 2 Schematic illustration of multi-stage carcinogenesis, initiation, tumor promotion, and progression

Classic tumor promoters, such as 12-*O*-tetradecanoylphorbol-13-acetate (TPA) or phorbol myristate acetate (PMA), were isolated from croton oil, the seed oil of Euphorbiacea *Croton tiglium L.* in the late 1960s (Fig. 3) (Hecker 1967; Van Duuren 1969), and tumor promotion has since been studied using phorbol esters isolated from the plants.

In 1981, we first reported new tumor promoters isolated from *Streptomyces* and marine natural products other than plants, which were structurally different from TPA (Fujiki et al. 1981; Fujiki and Sugimura 1987). For example, teleocidins A (A-1 and A-2) and B (B-1, B-2, B-3 and B-4) were isolated from the actinomycete *Streptomyces mediocidicus* (Fig. 4) (Takashima and Sakai 1960). Lyngbyatoxin A was isolated from the cyanobacterium (blue-green alga) *Lyngya majuscula*, and is structurally identical to teleocidin A-1 (Fig. 4) (Cardellina et al. 1979; Fujiki et al. 1984a).

Aplysiatoxin (Fig. 5) was originally found in the digestive gland of the sea hare, *Stylocheilus longicauda* (Kato and Scheuer 1974), and later isolated from *L. majuscula* (Moore 1982).

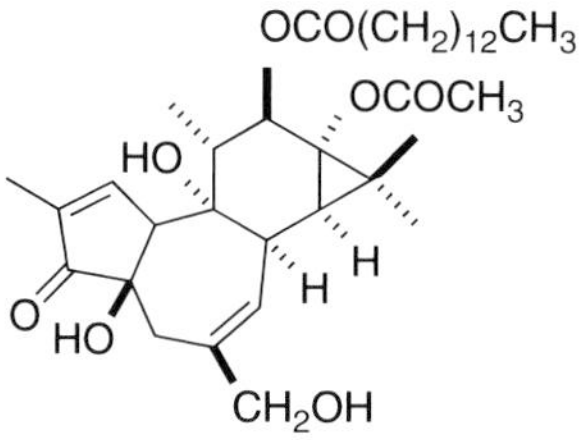

Fig. 3 Structure of TPA

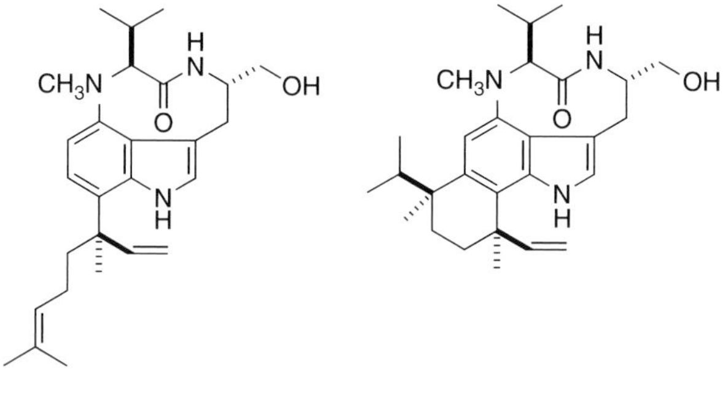

Fig. 4 Structures of teleocidin A (A-1 identical to lyngbyatoxin A and A-2) and teleocidin B (B-1, B-2, B-3 and B-4)

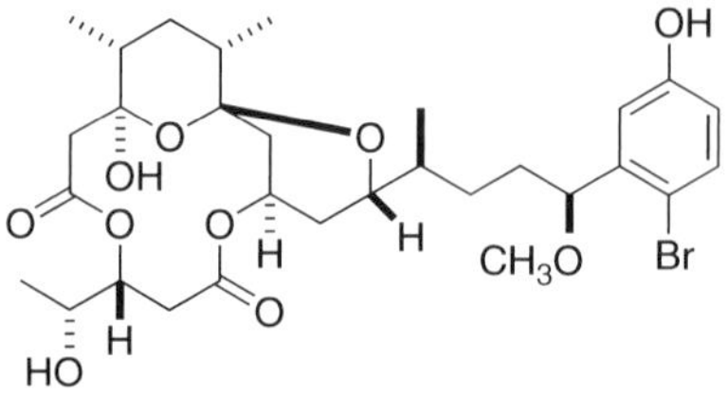

Fig. 5 Structure of aplysiatoxin

These new compounds induced tumor promotion on mouse skin initiated with 7,12-dimethylbenz[a]anthracene (DMBA) in two-stage carcinogenesis experiments (Fig. 6).

It was of great importance to note that teleocidin, lyngbyatoxin A and aplysiatoxin were biologically as active as TPA, and similarly activated PKC (Figs. 1 and 7) (Fujiki et al. 1984b), resulting in tumor promoting activity as strong as TPA on mouse skin. Because the new tumor promoters induced tumor promotion mediated through the same mechanisms of action as does TPA, we classified these two classes of compounds as the TPA-type tumor promoters (Fujiki and Sugimura 1987) (Fig. 7).

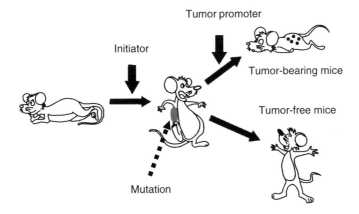

Fig. 6 Schematic illustration of two-stage carcinogenesis experiment on mouse skin

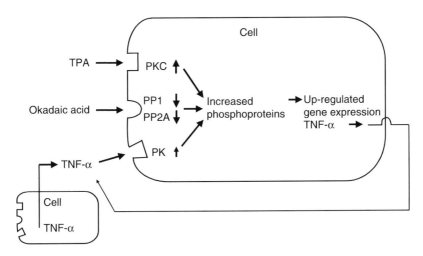

Fig. 7 Schematic illustration of tumor promotion by TPA, okadaic acid, and TNF-α in the cells

Although the study of new tumor promoters provided significant evidence on signal transduction (Nishizuka 1984), it was a challenge to overcome classical limitations in tumor promotion, such as organ and tissue specificities, which found that a single tumor promoter affected only one organ. Our original goal was to discover a general biochemical and molecular process of tumor promotion applicable to various organs, which would be the significant mechanism of cancer development in humans.

2.2 Non-TPA–Type Tumor Promoters

2.2.1 Palytoxin

To achieve our goal, we further screened for new types of tumor promoters which do not activate PKC and act differently from the TPA-types. Before going onto okadaic acid, we first showed palytoxin to be a unique tumor promoter belonging to the non-TPA–types (Fujiki et al. 1986). Palytoxin was isolated from anthozoans of the genus *Palythoa* and is a water-soluble toxin with a long, partially unsaturated aliphatic chain with interspaced cyclic ethers (Uemura et al. 1981; Moore and Bartolini 1981) (Fig. 8). Although palytoxin did not bind to the phorbol ester receptor in particulate fraction of mouse skin, or activate protein kinase C, the strong irritant activity of palytoxin suggested it as an inflammatory agent to the tissue (Fujiki et al. 1986): As expected, palytoxin induced tumor promotion on mouse

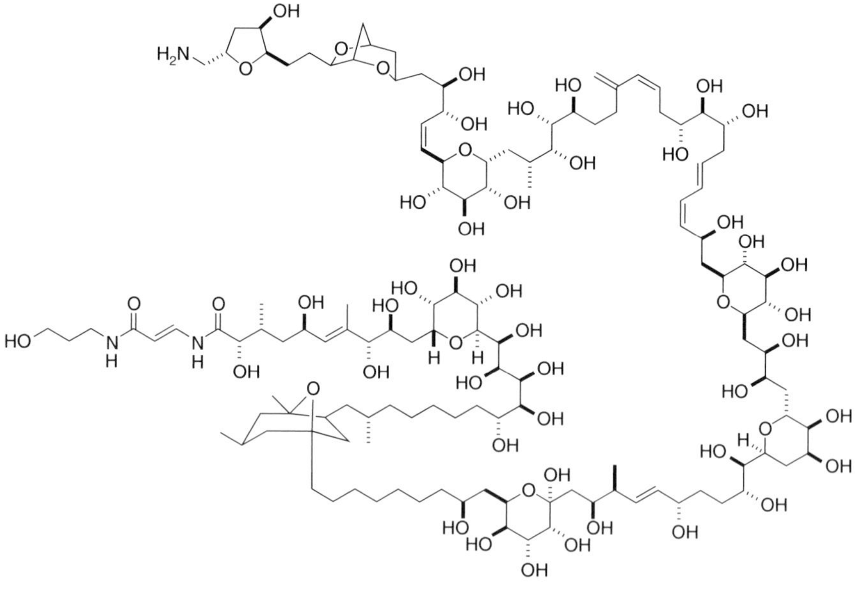

Fig. 8 Structure of palytoxin

skin initiated with DMBA. The results presented a new concept, i.e., that tumor promotion on mouse skin can be induced by the TPA-type and non-TPA–type tumor promoters (Fujiki et al. 1986).

Recently the biochemical mechanisms of palytoxin in carcinogenesis have been well investigated with regard to the Na$^+$, K$^+$-ATPase (Wattenberg 2007). Specifically, palytoxin binds to Na$^+$/K$^+$ pump, a ubiquitous P-type ATPase, to generate nonselective cation channels which are permeable to several cations, including sodium and potassium, but low for calcium (Habermann 1989; Artigas and Gadsby 2003, 2004). Subsequently, palytoxin modulates the epidermal growth factor (EGF) receptor (Wattenberg et al. 1987, 1989) and mitogen-activated protein (MAP) kinase cascade by the sodium-dependent signaling pathway (Warmka et al. 2004; Kuroki et al. 1997). The study of palytoxin encouraged us to identify mechanisms of tumor promotion that differ from that of the TPA-types (Fujiki and Sugimura 1987).

2.2.2 Okadaic Acid Class of Tumor Promoters

Yoshimasa Hirata and Kiyoyuki Yamada kindly provided us okadaic acid for our experiments, along with a note calling it "an interesting compound." Okadaic acid was isolated from the black sponge *Halichondria okadai* (Fig. 9) (Tachibana et al. 1981). The comment on "an interesting compound" referred to a pioneering paper by Shibata et al. (1982), in which they first reported that okadaic acid at higher concentrations, over 1 µM, induced smooth muscle contraction in the absence of external Ca^{2+} ion.

Based on these results, we first demonstrated that okadaic acid neither activated PKC nor bound to phorbol ester receptors in cell membrane (Fujiki et al. 1987; Suganuma et al. 1988). We also reported that incubation with okadaic acid of a mouse brain fraction containing protein kinases and protein phosphatases resulted in enhanced phosphorylation of proteins, the so-called apparent activation of protein kinases (Sassa et al. 1989). Two different research groups independently reported that okadaic acid inhibited myosin phosphatase at µM concentrations and dephosphorylation of the distinct sites in myosin light chain phosphorylated by either myosin light chain kinase or PKC (Takai et al. 1987; Erdödi et al. 1988). Okadaic acid thus came

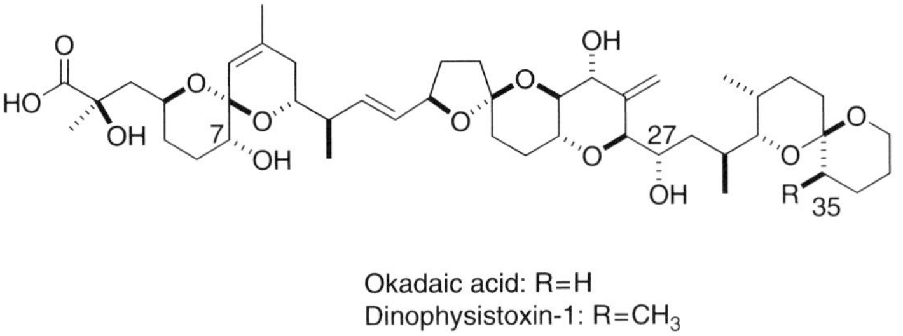

Okadaic acid: R=H
Dinophysistoxin-1: R=CH$_3$

Fig. 9 Structures of okadaic acid and dinophysistoxin-1 (35-methylokadaic acid)

to be regarded as a potent inhibitor of protein phosphatases 1 and 2A (PP1 and PP2A) (Table 1) (Bialojan and Takai 1988; Cohen 1989; Suganuma et al. 1992a).

Okadaic acid and dinophysistoxin-1, the latter of which was isolated from the hepatopancreas of the mussel *Mytilus edulis* (Fig. 9) (Murata et al. 1982), were shown to be tumor promoters on mouse skin that were as potent as the TPA-types (Suganuma et al. 1988; Fujiki et al. 1988). The next exciting news was that okadaic acid in drinking water induced tumor promotion in rat glandular stomach initiated with *N*-methyl-*N*'-nitro-*N*-nitrosoguanidine (Suganuma et al. 1992b): The results with okadaic acid thus caused revision of a classic theory of tumor promotion.

Since the tumor promoting activity of okadaic acid and dinophysistoxin-1 appeared to be linked to inhibition of PP1 and PP2A, we screened further inhibitors of PP1 and PP2A, utilizing an assay based on inhibition of specific [27-^{3}H]okadaic acid binding to the particulate fraction of mouse skin. We subsequently identified other compounds – calyculins (Fig. 10), microcystins (Fig. 11), and nodularin (Fig. 12) – which bind to the okadaic acid receptors. Calyculin A, isolated from the marine sponge *Discodermia calyx* (Fig. 10) (Kato et al. 1986), became an additional member of the okadaic acid class of tumor promoters on mouse skin.

Then microcystin-LR isolated from colonial and filamentous cyanobacterium (blue green alga) *Microcystis aeruginosa* (Fig. 11) (Carmichael and Mahmood 1984) and nodularin isolated from the brackish water cyanobacterium *Nodularia spumigena* (Fig. 12) (Rinehart et al. 1988; Carmichael 1988) were confirmed as

Table 1 Inhibition of PP1 and PP2A by the okadaic acid class of compounds

Compounds	IC$_{50}$ (nM)	
	Inhibition of PP1	Inhibition of PP2A
Okadaic acid	3.4	0.07
Calyculin A	0.3	0.13
Microcystin-LR	0.1	0.10

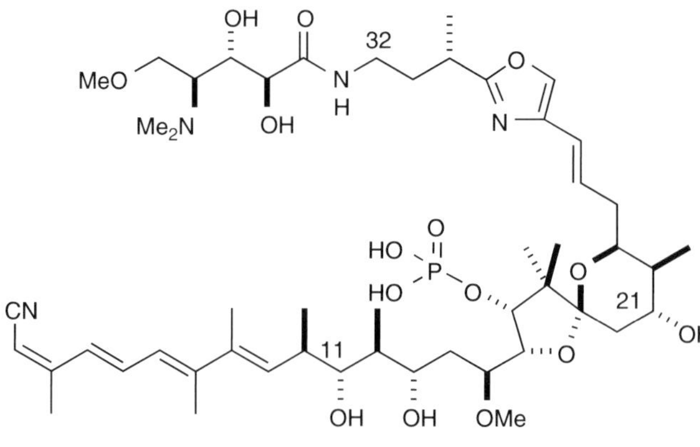

Fig. 10 Structure of calyculin A

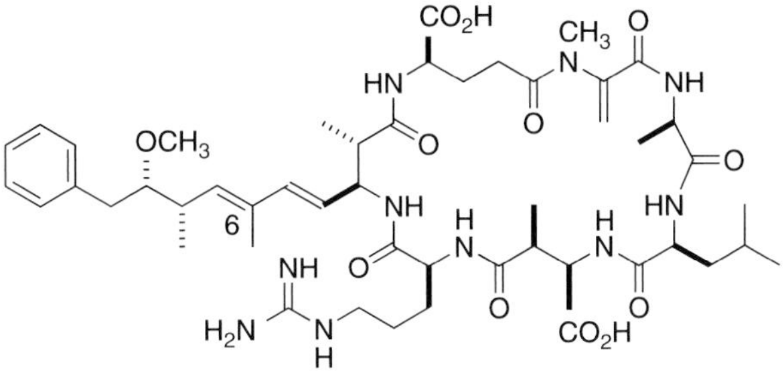

Fig. 11 Structure of microcystin-LR

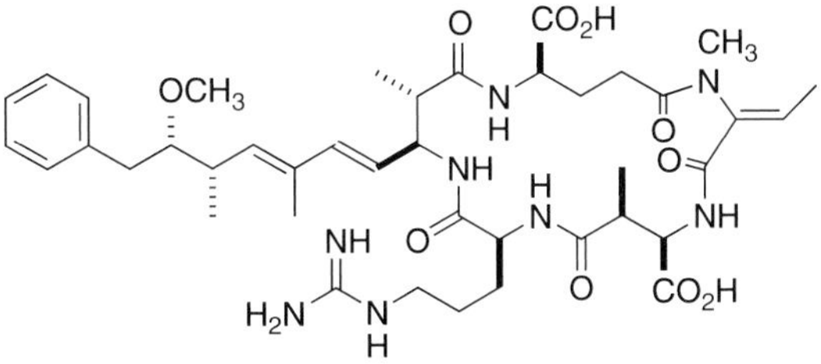

Fig. 12 Structure of nodularin

tumor promoters in rat liver initiated with diethylnitrosamine (Nishiwaki-Matsushima et al. 1992; Ohta et al. 1994).

Although their structures are not related, PP1 and PP2A inhibitors induced tumor promotion in three different organs initiated with three different carcinogens (Fig. 7). Thus, we classified the compounds as the okadaic acid class of tumor promoters (Fujiki and Suganuma 1993).

Considering these findings, we conclude that the okadaic acid-inhibitable phosphatase pathway is a widely distributed and potentially important mechanism for tumor promotion in various organs. This chapter reviews the okadaic acid class of compounds derived from marine natural products, including okadaic acid, dinophysistoxin-1, calyculins A–H (Matsunaga and Fusetani 1991; Matsunaga et al. 1991, 1997), microcystin-LR and nodularin, looking at unique biochemical features and tumor promotion.

3 Inhibitors of PP1 and PP2A

3.1 Structural Features and Inhibitory Activity

3.1.1 The Okadaic Acids

To begin with, let's look at the origin and structural uniqueness of the okadaic acid class of compounds. Okadaic acid is a polyether compound of a C_{38} fatty acid, isolated from the black sponge *Halichondria okadai* (Fig. 9). "Okadai" comes from the name of a Japanese educator, Dr. Yaichiro Okada: When the structure was elucidated by Paul Scheuer's group at the University of Hawaii in 1981 (Tachibana et al. 1981), Paul Scheuer named the compound okadaic acid in honor of Dr. Okada.

Although dinophysistoxin-1 (35-methylokadaic acid) was first isolated from the hepatopancreas of the mussel *Mytilus edulis* as a causative agent of diarrhetic shellfish poisoning (Fig. 9) (Murata et al. 1982), okadaic acid and dinophysistoxin-1 are now known to be synthesized by marine dinoflagellates of the genus *Dinophysis* and to be accumulated in the digestive glands of mussels or scallops (Murakami et al. 1982; Yasumoto et al. 1985). However, the origin of okadaic acid in the black sponge *H. okadai* is still unknown.

To study the structure-function relationships of okadaic acid, the specific binding of [27-³H]okadaic acid (14 Ci/mmol) to PP1 and PP2A contained in both particulate and cytosolic fractions of mouse skin was determined: The binding activity was then compared with inhibition of PP1 and PP2A, and with induction of ornithine decarboxylase in mouse skin (Suganuma et al. 1989). Results with numerous semi-synthesized derivatives of okadaic acid showed that the carboxyl group as well as the four hydroxyl groups at C-2, C-7, C-24 and C-27 of okadaic acid are important for the activity (Nishiwaki et al. 1990a). Among semi-synthesized compounds, okadaic acid spiroketal II, okadaic acid glycol, and okadaic acid spiroketal I appeared to be similarly active, while 7-*O*-docosahexaenoylokadaic acid, glycookadaic acid, okadanol, and 7-*O*-palmitoylokadaic acid were considerably active. Okadaic acid tetramethyl ether and okadaic acid methyl ester tetramethyl ether were inactive in various assays. Acanthifolicin, which was isolated from the marine sponge *Pandaros acanthifolium* and confirmed as a new episulfide-containing polyether carboxylic acid, was predicted to be an additional member of the okadaic acid class of compounds (Schmitz et al. 1981).

Various research groups independently reported the inhibitory activity of the okadaic acid class of compounds using purified catalytic subunits of PP1 and PP2A (Hescheler et al. 1988; Ishihara et al. 1989; Honkanen et al. 1990; MacKintosh et al. 1990). Our previous studies with the radioactive photoaffinity probe [³H] methyl 7-*O*-(4-azidobenzoyl) okadaate suggested that okadaic acid, calyculin A and microcystins bind to the catalytic subunits of PP1 and PP2A contained in the partially purified enzyme fractions (Nishiwaki et al. 1990b). Since the particulate and cytosolic enzyme fractions contain different proportions of PP1 and PP2A – and the catalytic activity of the protein phosphatases varies due to association with regulatory components – we determined the relative inhibitory doses of the okadaic acid class of compounds toward the purified catalytic subunits of PP1 and PP2A at the same time (Table 1) (Suganuma et al. 1992a).

For our experiments, the catalytic subunit of PP1 was isolated from rabbit skeletal muscle, and that of PP2A was isolated from human erythrocytes (Brautigan and Shriner 1988). The dose-response for inhibitory activity of the catalytic subunit of PP1 showed the order of potency as microcystin-LR > calyculin A > okadaic acid, whereas that for inhibition of PP2A was in the order okadaic acid > microcystin-LR > calyculin A. Table 1 shows that the IC_{50} values for inhibition by calyculin A and microcystin-LR are in the same range (0.1–0.3 nM) for both PP1 and PP2A. By contrast, okadaic acid was a significantly stronger inhibitor of PP2A than PP1 (IC_{50} values of 0.07 nM and 3.4 nM, respectively) (Suganuma et al. 1992a). The results indicated that the three okadaic acid classes of compounds are all potent inhibitors of PP1 and PP2A, but not of recombinant rat brain protein tyrosine phosphatase 1.

3.1.2 Calyculin A

Calyculin A was isolated from lipophilic extract of the marine sponge *Discodermia calyx*, which was the major constituent (Fig. 10). The compound was a strong inhibitor of development of starfish embryos as well as strong cytotoxic compound against L1210 leukemia cells (Kato et al. 1986), and showed both contraction of intact and skinned fibers and inhibition of the endogenous phosphatase of smooth muscle myosin B (Ishihara et al. 1989). Calyculin A is a novel spiroketal of an unprecedented skeleton containing phosphate, oxazole, nitrile, and amides (Fig. 10). Calyculin C is the homologue of calyculins A and B, and calyculins D–H are geometrical isomers of either calyculin A or C. These calyculins were isolated from the same sponge (Kato et al. 1988; Matsunaga et al. 1991). Calyculins A, B, E, and F are geometrical isomers of olefins of the hydrophobic tail, i.e., the tetraene portion (Matsunaga et al. 1991). However, when calyculins were prepared avoiding light, and their enzyme activity was determined in the dark, calyculins B, E, and F showed significantly less activity than did calyculin A (Wakimoto et al. 2002) (Table 2).

In addition, inhibitory activity of seventeen natural and chemically modified calyculin derivatives were further studied against PP1 (Matsunaga et al. 1997; Wakimoto et al. 2002). Dephosphonocalyculin A, for example, was found to be an inactive form of calyculin A (Table 2) isolated from the same sponge, suggesting that the phosphate group is absolutely essential for the inhibition of the enzymes and that some structural parts other than the phosphate group participate in binding to the enzymes (Wakimoto et al. 2002). Studies on the structure-activity relationships of calyculins revealed that C-17 phosphate, C-13 hydroxyl group, and tetraene moieties are essential for inhibition of PP1 and PP2A (Wakimoto et al. 2002).

Table 2 Inhibitory activity of calyculins toward PP1 and PP2A

Calyculins	IC_{50} (nM)	
	PP1	PP2A
A	8.2	1.0
B	112	9.0
C	29	2.6
E	185	14
F	55	7.5
Dephosphonocalyculin A	>10,000	>10,000

Table 3 Enzyme inhibitory and cytotoxic activities of calyculin A and its derivatives

	IC_{50}		
	PP1	PP2A	P388 cells
Compounds		nM	ng/ml
Calyculin A	8.2	1.0	0.170
Hemicalyculin A	14	1.0	450
C1/C34-calyculin A	9.0	1.1	40
Dephosphonocalyculin A	>10,000	>10,000	>5,000
11,13-O-isopropylidene-calyculin A	>10,000	>10,000	2,000

The enzyme inhibitory activities of PP1 and PP2A, and the cytotoxic activity of P388 leukemia cells by calyculin A and its derivatives, were also compared (Table 3). Calyculin A, hemicalyculin A and C1/C34-calyculin A had comparable enzyme inhibitory activities, while dephosphonocalyculin A and 11,13-O-isopropylidene-calyculin A were inactive derivatives. Hemicalyculin A lacked two γ-amino acid units, and C1/C34-calyculin A lacked only the dimethylamino group. The cytotoxicity of calyculins was in the reverse order of their acidity (Wakimoto et al. 2002).

3.1.3 Microcystins

Microcystins isolated from colonial and filamentous cyanobacteria – especially *Microcystis aeruginosa*, *M. viridis*, *Anabena flos-aquae*, and *Oscillatoria agardhii* – induce severe intrahepatic hemorrhages and liver necrosis at low concentrations in rats and mice (Botes et al. 1984; Carmichael 1988). The waterblooms of toxic cyanobacteria (blue-green algae), found in eutrophic freshwater municipal and residential water supplies, are an increasing environmental hazard in several areas of the world (Francis 1878; Carmichael 1988; Falconer et al. 1998). Microcystins are cyclic heptapeptides containing two variable L-amino acids and an unusual amino acid, 3-amino-9-methoxy-10-phenyl-2,6,8-trimethyl-deca-4,6-dienoic acid (Adda) (Fig. 11) (Rinehart et al. 1988). Microcystins differ primarily in the two variable amino acids and are named using a nomenclature based on these two L-amino acids. Microcystin-LR contains leucine and arginine in the variable positions (Fig. 11).

In 1987, three microcystins – -LR, -YR, and -RR – were provided to us by Kenichi Harada and Mariyo Watanabe, and in collaboration with them we studied their mechanisms of action (Harada et al. 1988; Watanabe et al. 1988). Three microcystins, along with nodularin, inhibited both *in vitro* protein phosphatase activity present in a cytosolic fraction and specific [³H]okadaic acid binding to cytosolic and particulate fractions of mouse liver, and this resulted in an increase of phosphoproteins with similar ED_{50} values (Yoshizawa et al. 1990). Microcystin-LR inhibited PP1 and PP2A with Ki values below 0.1 nM, and PP2B 1,000-times less potently (MacKintosh et al. 1990); it was also of great interest to find that microcystin-LR inhibited both PP1 and PP2A with the same potency, whereas okadaic acid inhibited

PP2A much more strongly than PP1 (Suganuma et al. 1992a) (Table 1). Seven inhibitors of PP1, including microcystins-LL, -LV, -LM, -LF and -LZ (Z represents an unknown hydrophobic amino acid), were purified from blooms of *M. aeruginosa*, and they inhibited PP1 with IC_{50} values of 0.06–0.4 nM (Craig et al. 1993).

Geometrical isomers at C-6 in the Adda molecule of microcystins-LR and -RR, named 6(*Z*)-Adda microcystins-LR and -RR, were isolated as minor components with maternal microcystins from *Microcystis* spp (Fig. 11) (Harada et al. 1990; Namikoshi et al. 1992). The structure-activiy relationships between 6(*E*)-Adda microcystin-LR and 6(*Z*)-Adda microcystin-LR were significantly different: 6(*Z*)-Adda microcystin-LR inhibited the PP1 activity and released the alanine aminotransferase (ALT) from rat liver into serum 10–100 times more weakly than did maternal microcystin-LR (Nishiwaki-Matsushima et al. 1991). The antibodies specific for PP1 and PP2A revealed that, in hepatocytes, microcystin forms secondary covalent bonds with the C-terminal of PP1 and PP2A catalytic subunits (Runnegar et al. 1995).

3.1.4 Nodularin

Nodularin was isolated from the toxic brackish-water cyanobacterium *Nodularia spumigena* as a hepatotoxic compound (Rinehart et al. 1988; Carmichael 1988). Nodularin is a monocyclic pentapeptide which contains Adda but lacks one of the L and D amino acids found in the microcystins (Fig. 12). Nodularin is a new inhibitor of protein phosphatases, as is microcystin-LR. Using computer-assisted molecular modeling of nodularin and microcystin-LR, these two compounds were shown to have a similar molecular orientation with respect to the Adda portion, the peptide ring, and the two acidic groups (Taylor et al. 1992; Quinn et al. 1993). Nodularin did not bind covalently to PP1 or PP2A, whereas microcystin-LR interacted covalently with both (Bagu et al. 1997).

A similar compound to nodularin, motuporin, was isolated from the marine sponge *Theonella swinhei* collected in Papua New Guinea (de Silva et al. 1992). Although motuporin contains valine instead of the arginine in nodularin, these compounds are equipotent inhibitors of PP1 and PP2A (Craig et al. 1996). The results are consistent with the report that the arginine residue in microcystin-LR does not significantly contribute to biological activity, based on a comparison of enzyme inhibition and receptor binding activity (Nishiwaki-Matsushima et al. 1992). The two compounds are often referred to as nodularin-R and nodularin-V (motuporin) in the literature.

3.2 Crystal Structure of PP1-Inhibitor Complex

Since okadaic acid, calyculin A and microcystin-LR have similar, high potencies for inhibition of PP1 and PP2A, they were assumed to have some common functional molecular structures that are responsible for this activity, before the crystal structures of PP1-inhibitor complexes were reported. These are: formation of an

Fig. 13 Flexible cavity of okadaic acid. Conformation was determined based on interpretation of NMR data (Uemura and Hirata 1989)

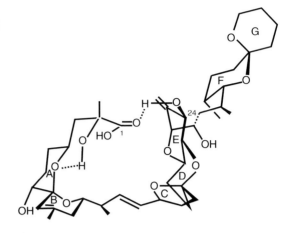

intramolecular hydrogen bond between C-1 carbonyl and C-24 hydroxyl groups of okadaic acid is assumed (Uemura and Hirata 1989) (Fig. 13); hydrogen bonds between phosphate ester and various parts of calyculin A are present (Fig. 10) (Kato et al. 1988); and microcystins and nodularin are cyclic peptides (Figs. 11 and 12) (Taylor et al. 1996). These points have been strongly supported by computer-assisted molecular modeling (Quinn et al. 1993), so we concluded that okadaic acid, calyculin A and microcystin-LR belong functionally to the same okadaic acid class (Fujiki and Suganuma 1993).

In 1995, Goldberg et al. first reported the three-dimensional structure of the catalytic subunit of PP1 in a complex with microcystin-LR. The catalytic subunit of PP1 consists of three main grooves: acidic, hydrophobic, and C-terminal grooves. The active site lies at one of the interfaces between an α-helical domain and a β-sheet domain, and the two metal ions and two conserved arginine residues (Arg 96 and Arg 221) together create a region of positive electrostatic potential at the active site of the enzyme. Egloff et al. (1995) investigated the nature and function of the two metal ions (Mn^{2+} and Fe^{2+} or Fe^{3+}) and reported that metal ions promote hydrolysis reactions by activating water molecules and stabilizing nucleophilic hydroxide and oxide ions. Figure 14 shows the molecular surface of the catalytic subunit of PP1-microcystin-LR complex (Goldberg et al. 1995), and similar crystal structures of the molecular complex of catalytic subunit of PP1-okadaic acid (Maynes et al. 2001), PP1-calyculin A (Kita et al. 2002), and PP1-motuporin (Maynes et al. 2006) have also been reported.

3.2.1 Okadaic Acid

The structures of okadaic acid and okadaic acid complexed with PP1 have very similar overall conformations. As Uemura and Hirata (1989) suggested (Fig. 13), a hydrogen bond formed by a cyclic structure between the C-1 acid and the C-24

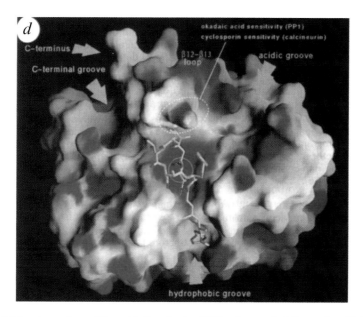

Fig. 14 Molecular surface of the catalytic subunit of PP1–microcystin-LR complex (Goldberg et al. 1995)

hydroxyl group was confirmed in both free okadaic acid and the PP1-okadaic acid complex (Maynes et al. 2001). The acid motif in okadaic acid accepts a hydrogen bond from the hydroxyl group of Tyr 272 in PP1, and the esterification or removal of the acidic moiety in okadaic acid results in elimination of its inhibitory activity. The spiroketal moiety at the other end binds to the hydrophobic groove on the surface of the enzyme. Tyr 206 and Ile 130 in the hydrophobic groove appear to be the two most important residues in this interaction because of their proximity to the hydrophobic segment of okadaic acid. Other hydrogen bonding interactions occur between Arg 96 and the C-2 hydroxyl group of okadaic acid, and between Arg 221 and the C-24 hydroxyl group of okadaic acid. The other hydrophobic interactions that occur are in the C-4 to C-16 region of okadaic acid and Phe 276 and Val 250 of the catalytic subunit. The hydrophobic ring of Phe 276 may inhibit entry of okadaic acid into the active site. Moreover, a significant difference between PP2A and PP1 inhibition by okadaic acid was also revealed by crystal structural analysis, i.e., PP2A has a hydrophobic cage that better accommodates the hydrophobic end of okadaic acid, whereas PP1 contains an open-ended groove (Xing et al. 2006).

3.2.2 Calyculin A

The crystal structure of the PP1-calyculin A complex was determined at 2.0 Å resolution (Kita et al. 2002). The overall and secondary structures of PP1-calyculin A complex are similar to those of PP1-okadaic acid and PP1-microcystin-LR complexes. Calyculin A is located in two of the three grooves, the hydrophobic and the acidic grooves on the molecular surface. Although calyculin A is docked on PP1 in the binding mode of an extended form, both the crystal structure (Kato et al. 1986) and solution structure of free calyculin A exhibiting intramolecular interactions (Volter et al. 1999) are different from the structure in the complex. The pathway of the C-terminal groove on the surface of PP1-calyculin A complex was closed by Cys 273 belonging to the β12–β13 loop. However, Tyr 272 in β12–β13 loop is essential for the association of PP1 with calyculin A and for the inhibition of the enzyme, based on the results with the site-directed mutagenesis (Zhang et al. 1996). At the active site, the metals, ligand atoms, water molecules, and the phosphate group of calyculin A formed a close network (Kita et al. 2002). Moreover, the loss of the phosphate group and the modification of C-13 hydroxyl group abolished the inhibitory activity. The crystal structure of the complex clearly showed the significance of the salt bridges between the phosphate in calyculin A and the two arginine residues 96 and 221 in the substrate recognition site. The entire study on the PP1-calyculin A complex indicates that the hydrophobic groove plays the dominant role in binding calyculin A, the acidic groove plays a supporting role, and the C-terminal groove functions in molecular recognition (Kita et al. 2002).

3.2.3 Microcystin-LR

The crystal structure of PP1-microcystin-LR complex at 2.1 Å was initially determined by Goldberg et al. (1995), as reported above. They found that microcystin-LR binds to the catalytic subunit of PP1 through the three regions of the surface, and the long hydrophobic Adda side chain was packed into the hydrophobic groove. Although covalent linkage between a methyldehydroalanine residue in microcystin-LR and the Sγ of Cys 273 is present, it is not the primary cause of inhibition of PP1 by microcystin-LR (Maynes et al. 2001). The carboxyl group of D-*erythro*-β-methylaspartic acid was also found to interact with Tyr 134 (Maynes et al. 2001, 2006). Interaction between microcystin-LR and the C-terminal groove occurs at the loop connecting the β12–β13, where the leucine side chain in microcystin-LR is packed close to the side chain of Tyr 272 (Goldberg et al. 1995).

It is worthwhile to note the difference in the inhibitory activity of okadaic acid and of microcystin-LR against PP1 (Table 1), based on the crystal structural analysis of the catalytic subunit of PP1. The reason that the activity of microcystin-LR is higher than that of okadaic acid is assumed to be the presence of a hydrogen bond between Arg 96 in PP1 and D-*erythro*-β-methylaspartic acid in microcystin-LR, an interaction not found in PP1-okadaic acid complex, and this interaction may have accounted for the 100-fold greater inhibition of PP1 by microcystin-LR compared with okadaic acid. The covalent binding of microcystin-LR to catalytic subunits of PP1 and PP2A was confirmed by Goldberg et al. (1995) and Bagu et al. (1997), but the covalent linkage of microcystin-LR was thought not to be essential for inhibition of PP1, since nodularin, including motuporin, did not bind covalently to PP1 and PP2A (Maynes et al. 2006). Recently the crystal structure of the PP2A holoenzyme – microcystin-LR complex was reported (Xu et al. 2006), but the results will not be dealt with here.

4 Tumor Promotion and Carcinogenic Activities

4.1 Okadaic Acid

4.1.1 Tumor Promotion on Mouse Skin

Recent studies indicate that the okadaic acid class of compounds bind to their own receptors – catalytic subunits of PP1 and PP2A – and inhibit their activities, resulting in the sustained phosphorylation of proteins. The mimicking of the activation of protein kinases by the TPA-type tumor promoters suggests the presence of tumor promoting activity by the okadaic acid class of compounds, since treatments with compounds of the okadaic acid class and of the TPA-types similarly induced irritation of mouse ear and ornithine decarboxylase (ODC) on mouse skin, which is a marker enzyme of cell proliferation. Two-stage carcinogenesis experiments on mouse skin were conducted using the same procedure as with the TPA-types (Fig. 6), and we will now discuss the results with okadaic acid and dinophysistoxin-1.

Experimentally, initiation was carried out by single application of 100 μg of 7,12-dimethylbenz[a]anthracene (DMBA) on the back of 8-week-old female CD-1 mice (Fujiki et al. 1982). After 1 week, okadaic acid (0.1, 1.0, 5.0, and 10 μg, corresponding to 0.12, 1.2, 6.2, and 12 nmol) in 0.1 ml of acetone was applied to the skin twice a week until week 30. As a control experiment, two groups treated with DMBA alone and with okadaic acid alone were also included (Fig. 15). Each group

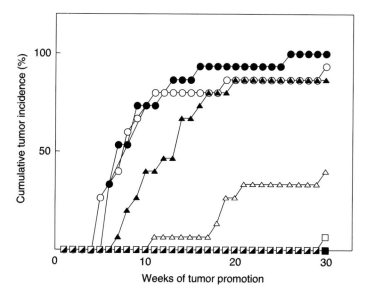

Fig. 15 Tumor promoting activity of okadaic acid. The groups were treated with DMBA plus okadaic acid (0.1 μg per application, △; 1 μg, ▲; 5 μg, ○; and 10 μg, ●), DMBA plus okadaic acid tetramethyl ether (1 μg, ■), okadaic acid alone (10 μg, □), and DMBA alone (X half-shaded blocks under the ■ and □)

consisted of 15 mice, and the numbers of tumors of 1 mm or more in diameter were counted weekly. Dinophysistoxin-1 (1.0, and 5.0 µg, corresponding to 1.2 and 6.2 nmol) was similarly applied to mouse skin in separate experiments.

Tumor promoting activity was evaluated by the percentage of tumor-bearing mice and the average number of tumors per mouse, every week. As Fig. 15 shows, the group treated with DMBA and okadaic acid showed gradual increase in the percentage of tumor-bearing mice (and the average number of tumors per mouse), whereas the groups treated with DMBA alone and with okadaic acid alone did not significantly develop tumors; and an inactive okadaic acid tetramethyl ether did not produce any tumors on mouse skin. The results showed that repeated applications of okadaic acid induced clonal growth of the initiated cells, demonstrating the tumor promoting activity of okadaic acid. Since dinophysistoxin-1 showed tumor promoting activity equipotent with okadaic acid, the results are summarized in Table 4.

It was previously reported that a single application of DMBA results in activation of the c-Harvey-*ras* gene with a mutation at the second nucleotide of codon 61 (CAA → CTA), and that TPA induces tumors mediated through clonal growth of initiated cells containing the mutation on mouse skin (Fig. 2) (Balmain and Pragnell 1983; Quintanilla et al. 1986). Moreover, DNA isolated from tumors in the two groups treated with DMBA and okadaic acid, and with DMBA and dinophysistoxin-1, contained the same mutation as determined by the polymerase chain reaction and DNA sequencing (Table 4) (Fujiki et al. 1989). Thus, the study with the okadaic acid class of compounds confirmed that DMBA induces a mutation in an oncogene of mouse skin cells, and that a tumor promoter clonally proliferates the growth of an initiated cell, developing to tumor.

The tumor is usually at first benign (papilloma), but later becomes a malignant (carcinoma) with numerous associated genetic changes. In fact, histological examination of skin tumors produced by DMBA and okadaic acid revealed that the percentage incidences of papillomas and squamous cell carcinomas were 92.3% and 5.1% (Suganuma et al. 1988). Since the experiment with tumor promotion on mouse skin is time consuming and painstaking work, the inhibitory activity of the okadaic acid class of compounds toward PP1 and PP2A can serve as a tentative evaluator of their tumor promoting activity. In this sense, acanthifolicin is assumed to be an additional tumor promoter, in addition to okadaic acid and dinophysistoxin-1.

Table 4 Tumor promoting activities of the okadaic acid class of compounds and activation of c-H-*ras* gene

Okadaic acid class tumor promoters	Amounts per application (nmol)	Maximal percentage of tumor bearing mice in week 30	Average no. of tumors per mouse	Mutation of the second nucleotide in codon 61
Okadaic acid	1.2	86.7	7.2	A → T
Dinophysistoxin-1	1.2	100.0	8.5	A → T
Calyculin A	1.0	93.3	4.3	A → T

4.1.2   Carcinogenic Effects on Rat Glandular Stomach

Evidence that okadaic acid and dinophysistoxin-1 are causative agents of diarrhetic shellfish poisoning in humans encouraged us to look at the gastrointestinal tract as another target organ in addition to mouse skin. Since TPA induces tumor promotion only in squamous cells of the skin, we were interested in studying the tumor promoting activity of okadaic acid in rat glandular stomach. The reason we selected the rat glandular stomach as important target for our study is that these tumors are histologically similar to tumors of human stomach cancer. This experiment would provide a test of the classical theory of tissue and organ specificity of tumor promoters.

4.1.3   Biochemical and Biological Effects on Glandular Stomach

[^{3}H]Okadaic acid distribution using two different routes – po and ip – was studied, and [^{3}H]okadaic acid ($14\,\mu$Ci/$0.2\,$ml sesame oil) was orally administered into the stomach of mice. Most of the radioactivity (>77%) was found in the contents of the gastrointestinal tract 3 h after intubation – 4% was 19 h later – suggesting that some of the okadaic acid had interacted with cell membrane; radioactivity of 30% was found in the feces 19 h later (unpublished results). When [^{3}H]okadaic acid ($28\,\mu$Ci/$0.2\,$ml saline solution) was ip injected into mice, most of the radioactivity (33.3%) was found in the contents of the gastrointestinal tract 3 h after intubation – 5% was found 19 h later. However, 27.4% of the radioactivity was found in the liver 3 h after ip injection, suggesting that ip-administered [^{3}H]okadaic acid was excreted through hepatobiliary circulation (unpublished results).

Before proceeding to the rodent carcinogenesis experiments, the increase of labeling indices and induction of ornithine decarboxylase by okadaic acid were studied in mucosa of rat glandular stomach (Yuasa et al. 1994; Fujiki et al. 1995). A single administration of okadaic acid at three different doses – 1, 10, and $50\,\mu$g – into rat stomach induced significant enhancement of BrdU labeling indices in various parts of the digestive tract dose-dependently. Table 5 shows that $10\,\mu$g okadaic acid

Table 5 BrdU labeling indices in mucosa of rat gastrointestinal tract

Organs	Control (average of 3 rats)	Okadaic acid ($10\,\mu$g) (average of 5 rats)
Forestomach	8.7 ± 0.8	$17.7 \pm 3.7^*$
Glandular stomach		
Fundus	2.5 ± 0.8	$4.7 \pm 0.8^*$
Pylorus	3.1 ± 0.4	$5.4 \pm 0.9^*$
Small intestine		
Jejunum	16.0 ± 1.6	$23.7 \pm 2.3^*$
Ileum	16.0 ± 2.1	$25.5 \pm 3.3^*$
Colon		
Proximal	8.1 ± 1.5	$16.9 \pm 1.7^*$
Distal	8.4 ± 0.5	$16.1 \pm 0.7^*$

$^*p < .01$

increased the labeling indices in mucosa of the gastrointestinal tract, including forestomach, glandular stomach, small intestine and colon, by almost twofold, compared with those of control (Yuasa et al. 1994).

The maximum ODC induction in rat glandular stomach by a single administration of 10–30 μg okadaic acid was obtained 4 h after intubation: 10 μg okadaic acid induced 140 pmol CO_2/30 min of incubation/mg of protein in the supernatant of the mucosa isolated from the glandular stomach. ODC is a rate-limiting enzyme for polyamine biosynthesis and is also an indication that an okadaic acid signal has reached the gene level. Thus, induction of ODC with okadaic acid is a significant marker of tumor promotion. When the cytosolic fraction of rat glandular stomach was subjected to DEAE cellulose column chromatography, PP1 and PP2A were found in the separate fractions (Suganuma et al. 1992b), which indicated that okadaic acid also induced ODC gene expression in the glandular stomach mediated through the okadaic acid pathway.

4.1.4 Tumor Promotion in Rat Glandular Stomach

Experimentally, initiation of a two-stage carcinogenesis experiment in rat glandular stomach of male SD rats was achieved by adding N-methyl-N'-nitro-N-nitrosoguanidine (MNNG) to the drinking water at a concentration of 100 mg/L for the first 8 weeks. One week after initiation, a solution of okadaic acid was given orally to the rats, i.e. about 10 μg okadaic acid/rat/day for 47 weeks from weeks 9 to 55 of the experiment, and subsequently about 20 μg okadaic acid/rat/day for 17 weeks from weeks 56 to 72. Two control groups treated with MNNG alone and with okadaic acid alone were included in this experiment. Tumor promoting activity in the glandular stomach was expressed as neoplastic changes, which included adenomatous hyperplasia and adenocarcinoma.

Table 6 summarizes the percentage of neoplastic change-bearing rats and the average number of neoplastic changes per rat of each group at week 72 of the experiment. The group treated with MNNG and okadaic acid showed 75.0% neoplastic change-bearing rats, whereas the group treated with MNNG alone showed 46.4%, and the group treated with okadaic acid alone showed 0%. The average number of neoplastic changes per rat in the groups treated with MNNG and okadaic

Table 6 Tumor promoting activity of okadaic acid in two-stage carcinogenesis experiment in rat glandular stomach

Treatment	No. of surviving rats at week 72	No. of rats with neoplastic changes (%)	Average no. of neoplastic changes/rat
MNNG + Okadaic acid	16	12 (75.0)[*]	1.1 ± 0.9
MNNG	28	13 (46.4)	0.6 ± 0.8
Okadaic acid	9	0 (0)	0

[*]$p < .05$

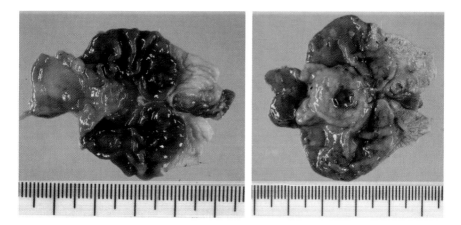

Fig. 16 A neoplastic change in the glandular stomach of the group treated with MNNG as initiator alone (left), and carcinoma in the glandular stomach of the group treated with MNNG followed by okadaic acid (right)

acid, with MNNG alone, and with okadaic acid alone were 1.1 ± 0.9, 0.6 ± 0.8 and 0, respectively (Suganuma et al. 1992b).

Figure 16 shows the glandular stomach bearing a small neoplastic change of the group treated with MNNG alone (left) and that bearing large adenocarcinoma of the group treated with MNNG and okadaic acid (right). These pictures indicate that okadaic acid as a tumor promoter enhanced the carcinogenic activity of MNNG as an initiator.

In this case, an adenomatous hyperplasia was defined as a non-invasive proliferative glandular lesion associated with slight cellular atypia and extension into the stomach lumen, which is comparable to papilloma in mouse skin carcinogenesis. The results showed that the treatment with okadaic acid enhanced the neoplastic changes in the rat glandular stomach initiated with MNNG ($p < 0.05$) (Suganuma et al. 1992b). No experiment with dinophysistoxin-1 was conducted, because the biochemical activity of dinophysistoxin-1 is the same as that of okadaic acid.

4.2   Calyculin A

4.2.1   Tumor Promotion on Mouse Skin

Calyculin A ($1.0\,\mu g$, $1.0\,nmol$) was similarly applied to the mouse skin initiated with DMBA, as previously described for okadaic acid. The percentage of tumor-bearing mice and the average number of tumors per mouse in week 30 are summarized in Table 4 (Suganuma et al. 1990). DNA isolated from tumors treated with DMBA and calyculin A contained the mutation of the second nucleotide of codon 61 (Table 4) (Fujiki et al. 1989).

4.3   Microcystin-LR

4.3.1   Tumor Promotion in Rat Liver

To demonstrate the tumor promoting activity of microcystin-LR in the liver, micro-cystin-LR was given by repeated ip injections into male Fischer 344 rats initiated with diethylnitrosamine (DEN), a liver carcinogen and liver initiator, using the standard procedure for two-stage carcinogenesis experiments: an initiator followed by a tumor promoter. The tumor promoting activity of microcystin-LR in separate experiments was similarly estimated by induction of glutathione *S*-transferase placental form (*GST-P*)-positive foci in rat liver (Nishiwaki-Matsushima et al. 1992) (Table 7), and in particular by three parameters: the number of *GST-P* positive foci/liver (No./cm^2), area of foci/liver (mm^2/cm^2) and volume of foci/liver (v/v%) (Table 8) (Ohta et al. 1994). The results showed, for the first time, that microcystin-LR is a new liver tumor promoter – not a liver carcinogen.

Tumor promoting activity of microcystin-LR with ip injections of 1 or 10 μg/kg was confirmed in male Fischer 344 rats by two-stage carcinogenesis experiments initiated with DEN and aflatoxin B1 followed by partial hepatectomy (Table 7) (Sekijima et al. 1999). Repeated ip injections of microcystin-LR – 100 times over 28 weeks – without an initiator induced neoplastic nodules – probably benign tumors in ICR mouse liver – suggesting that microcystin-LR has very weak carcinogenic activity (Table 7) (Ito et al. 1997).

4.3.2   Biochemical and Biological Effects

Microcystin-LR induced numerous biochemical and molecular changes in the cells: 1.0 μM microcystins-LR and -YR induced morphological and biochemical changes in rat primary cultured hepatocytes (Yoshizawa et al. 1990), but microcystin-YR at concentrations up to 9.6 μM did not induce any effects in human fibroblasts. Effects were induced only by microinjection of high concentrations of 670 μM microcystin-YR (Matsushima et al. 1990), which suggested that microcystin does not easily penetrate into the fibroblasts and the other cells, whereas okadaic acid induces effects at much lower concentrations. This reflects the tissue specificity of the agents, but when microcystins are given ip or po to test animals, massive intrahepatic hemorrhages and necrosis of hepatocytes are histologically observed.

Moreover, a significant study on the incorporation of [^{3}H]dihydromicrocystin-LR into the liver of mice found great differences between ip and po administrations: ip-administration resulted in the highest uptake into the liver – 71.5 ± 6.9% of the total administered radioactivity after 1 h – whereas po-administration resulted in 0.68% uptake after 6 h, and less than 0.5% uptake after 6 days (Nishiwaki et al. 1994). Microcystin-LR (35 μg/kg) ip injected into mice was detected in serum (8.7 ng/ml) and liver (45 ng/ml) 15 min later by direct competitive enzyme-linked immunosorbent assay (ELISA): The peak in serum (62.7 ng/ml) was 2 h after injection and that in liver cytosol (250 ng/ml) was 12 h after (Lin and Chu 1994).

Table 7 Tumor promoting activity and/or carcinogenicity of microcystin-LR

Groups	Initiators	Microcystin-LR (μg/kg) ip, times	Subject species	Partial hepatectomy	Biomarker	Estimation	References
1	DEN	10, ip 12 times	Fischer 344 male rats	+	GST-P foci	Tumor promoter	Nishiwaki-Matsushima et al. (1992)
2	DEN	25, ip 20 times	Fischer 344 male rats	−	GST-P foci	Tumor promoter	Ohta et al. (1994)
3	DEN	10, ip 12 times	Fischer 344 male rats	+	GST-P foci	Tumor promoter	Sekijima et al. (1999)
4	DEN + aflatoxin B$_1$	10, ip 12 times	Fischer 344 male rats	+	GST-P foci	Tumor promoter	Sekijima et al. (1999)
5	−	20, ip 100 times	ICR mice	−	Neoplastic nodules	Weak carcinogen	Ito et al. (1997)

Table 8 Induction of *GST-P* positive foci by microcystin-LR

Groups	DEN	Microcystin-LR (μg/kg) ip 20 times	Effective No. of rats	No. of foci/liver (No./cm^2)	Area of foci/liver (mm^2/cm^2)	Volume of foci/liver (v/v%)
1	+	–	20	10.0 ± 2.9[a]	0.18 ± 0.07	0.37 ± 0.18
2	–	–	5	0	0	0
3	+	25	18	95.7 ± 27.8[*]	4.74 ± 2.23[*]	8.55 ± 4.04[*]
4	–	25	17	1.6 ± 1.4	0.02 ± 0.02	0.04 ± 0.03

[a]Mean ± SD
[*]$p < .005$

Microcystin-LR up-regulated expression of tumor necrosis factor-α (TNF-α) after its ip administration (Sueoka et al. 1997). Microcystin-LR also up-regulated expression of early response genes, such as *c-jun, jun B, jun D, c-fos,* and *fos B,* in rat liver after its ip-administration, and in primary cultured rat hepatocytes 6 h after treatment (Sueoka et al. 1997). Microcystin-LR at concentrations of 0.1, 0.3 and 1.0 μg/ml stimulated the synthesis of TNF-α and interleukin-1β in rat peritoneal macrophages *in vitro*, and the supernatants of macrophages stimulated with microcystin-LR induced electrogenic secretion in rabbit ileal mucosa (Rocha et al. 2000). Subsequent evidence using TNF-α-deficient mice confirmed that TNF-α is an endogenous tumor promoter and a cancer mediator in the target organs of tumor promotion (Suganuma et al. 1999).

4.4 Nodularin

4.4.1 Tumor Promotion in Rat Liver

The tumor promoting activity of nodularin was investigated using a two-stage carcinogenesis experiment in Fischer 344 male rats initiated with DEN followed by repeated ip injections of nodularin. Nodularin showed strong tumor promoting activity in rat liver with regard to the parameters, including number of *GST-P* positive foci/liver (No./cm^2), area of foci/liver (mm^2/cm^2) and volume of foci/liver (v/v%) (Table 9). It was of great interest to note that treatment with DEN followed by nodularin administered 25 μg/kg ip 20 times induced 71.75 ± 18.78% volume of foci per liver, and nodularin alone induced weakly *GST-P* positive foci as strongly as DEN alone did (Ohta et al. 1994).

Figure 17 shows the liver of the group treated with DEN alone - where initiating activity in the liver (left) was found - and that of the group treated with DEN followed by nodularin, which shows large numbers of nodular foci (right). Thus, compared with the tumor promoting activity of microcystin-LR (Table 8), nodularin has both stronger tumor promoting activity than microcystin-LR and initiating activity as strong as that of DEN, suggesting that nodularin is a new liver carcinogen. The

Table 9 Induction of *GST-P* positive foci by nodularin

Groups	DEN	Nodularin (μg/kg) ip 20 times	Effective No. of rats	No. of foci/ liver (No./cm²)	Area of foci/ liver (mm²/cm²)	Volume of foci/liver (v/v%)
1	+	–	20	10.0 ± 2.9[a]	0.18 ± 0.07	0.37 ± 0.18
2	–	–	5	0	0	0
3	+	25	20	106.0 ± 22.6*	39.87 ± 10.51*	71.75 ± 18.78*
4	–	25	16	6.3 ± 7.3	0.49 ± 0.89	0.92 ± 1.58

[a]Mean ± SD
*$p < .005$

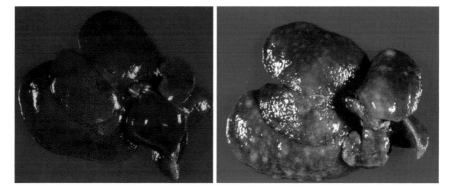

Fig. 17 Liver of the group treated with DEN as initiator alone (left), and that of the group treated with DEN followed by nodularin (25 μg/kg) ip 20 times

results indicated for the first time that nodularin is a new liver tumor promoter with strong initiating activity (Table 9).

The tumor promoting activity of nodularin was confirmed by a two-stage carcinogenesis experiment in Fischer 344 male rats initiated with DEN (Song et al. 1999). The activity of PP1 and PP2A was significantly inhibited in the hyperplastic nodules of rat liver (Lim et al. 2001). The transforming growth factor-β (TGF-β) family consists of three highly homologous genes in humans: *TGF-β1, TGF-β2, and TGF-β3*. Increased TGF-β1 gene expression was shown by Northern blot analysis of total cellular RNAs isolated from rat liver during the two-stage carcinogenesis experiments (Lim et al. 1999). And treatment with DEN plus nodularin induced a high level of TGF-β1 gene expression with concomitant loss of its receptor expression in the enzyme-altered hepatocytes, suggesting an escape from TGF-β1-induced apoptosis during the early stage of carcinogenesis (Lim 2003). In addition, nodularin significantly induced peroxisomal palmitoyl-CoA oxidase and cytochrome P-450 4A1 expression in nodules of rat liver (Lim et al. 2001).

4.4.2 Biochemical and Biological Effects

Nodularin induced up-regulated expression of *tumor necrosis factor-α* (*TNF-α*) gene and early response genes in rat liver after its ip administration: The up-regulated expression of early response genes, such as *c-jun*, *jun B*, *jun D*, *c-fos*, *fos B* and *fra*-1, and that of *TNF-α* gene were observed in primary cultured rat hepatocytes treated with 0.1 μM nodularin 6 h after treatment. In contrast, TPA, which is not a liver tumor promoter, induced no apparent expression of the *TNF-α* gene in primary cultured rat hepatocytes (Sueoka et al. 1997).

Nodularin induced free-radical formation, reduction of glutathione levels and DNA damage in rat hepatocytes. The DNA damage induced by nodularin increased at both 24 and 48 h, in contrast to the effects of microcystin-LR, suggesting that nodularin has stronger carcinogenic effects than does microcystin-LR (Maatouk et al. 2004). Nodularin significantly enhanced 8-oxo-7,8-dihydro-2'-deoxyguanosine (8-oxo-dG) as a marker of DNA oxidative damage in primary rat hepatocytes and *in vivo* in rat liver cells (Bouaïcha et al. 2005). Nodularin at non-cytotoxic low concentrations (2 and 10 ng/ml) induced significant increases in intracellular reduced glutathione (GSH) levels and production of reactive oxygen species (ROS) in freshly isolated rat hepatocytes (Bouaïcha and Maatouk 2004).

4.5 Evaluation of Human Risk of Microcystin-LR and Nodularin

Cyanobacteria contain numerous toxins, which are classified into five different groups - hepatotoxin, neurotoxin, cytotoxin, dermatotoxin, and irritant toxin (Rao et al. 2002; Wiegand and Pflugmacher 2005). Extracts of cyanobacteria – blue green algae blooms collected from various sources – induced liver damage in mice that was associated with increased mortality (Briand et al. 2003), damage for enterocytes isolated from chicken small intestine (Falconer et al. 1992), and decreased body weight in BALB/c mice (Shen et al. 2003). The toxic effects of microcystin-LR on carp, rainbow trout and loach have also been studied, since the occurrence of fish kills is associated with toxic cyanobacterial blooms (Fischer and Dietrich 2000).

The tumor promoting activity of microcystin-LR may be involved in human liver cancer from polluted drinking water. The incidence of primary liver cancer in Qidong County, Peoples' Republic of China, where people drink pond and ditch water, was about eight times higher than that in populations who drink well water (Yu 1989). It was also reported that the microcystins are not affected by normal chlorination, flocculation, and filtration procedures used by water treatment facilities (Krishnamurthy et al. 1989). Thus, in Haimen, another high mortality area of primary liver cancer in China, the content of intake microcystins is estimated at 255.6 μg/day/person, and the average content of microcystins was 164.9 ± 93.2 ng/L in pond-ditch water and 188.7 ± 20.8 ng/L in irrigation water (Yu 1995; Yu et al. 1998).

The Guideline for Canadian Drinking Water Quality (Federal-Provincial-Territorial Committee on Drinking Water 2002) sets the maximum acceptable concentration for microcystin-LR at 1.5 µg/L, and the provisional WHO-guideline for drinking water, as well as the Oregon Health Division for blue green algal supplements, currently use microcystin-LR 40 µg/kg/body weight/day as the tolerable daily intake values (Dietrich and Hoeger 2005).

Based on the results of numerous studies with microcystins and nodularin, it is appropriate to evaluate their risk of carcinogenicity for humans. The WHO International Agency for Research on Cancer (IARC) Lyon organized a meeting in June 2006 to assess the carcinogenicity of microcystin-LR and nodularin for humans, and assessment was considered from three perspectives: (1) Carcinogenicity in rodents: The tumor promoting activity of microcystin-LR through ip repeated injections was significantly evaluated, but the activity was only shown with ip injections. The results with microcystin-LR of long-term animal carcinogenesis experiments were not reported, probably due to the high cost of the compound; (2) Epidemiological evidence: The incidence of primary liver cancer in Qidong and Haimen Counties, Peoples' Republic of China was discussed in relation to drinking pond and ditch water containing microcystins. However, evidence of carcinogenicity was inadequate, due to fewer detailed studies; (3) Unique mechanisms of action of the compounds: The potent tumor promoting activity with microcystin-LR and nodularin was for the first time shown to be related to potent inhibition of PP1 and PP2A, associated with up-regulated expression of various important genes. Study of the compounds opened up new vistas in the science of carcinogenesis, and working group concluded that microcystin-LR is "possibly carcinogenic to humans" (group 2B) and that nodularin is "not classifiable as to carcinogenicity" (group 3) (Grosse et al. 2006).

5 Relation to Human Cancer Development

In 1993, we had the fortunate opportunity to publish a review article in *Advances in Cancer Research* entitled "Tumor promotion by inhibitors of protein phosphatases 1 and 2A: The okadaic acid class of compounds" (Fujiki and Suganuma 1993). In this paper, we considered three possibilities related to human cancer: (1) Exposure to the okadaic acid class compounds; (2) Involvement of endogenous protein inhibitors of PP1, inhibitor-1 and -2, in the cells; (3) How the effects of okadaic acid pathway mimic those of cytokines, such as tumor necrosis factor-α (TNF-α) and interleukin-1 (IL-1) (Fig. 7). Today the third aspect seems most worthy of further study.

Our recent study revealed a fascinating link between chemical tumor promoters and cytokines as endogenous tumor promoters: TNF-α-deficient mice were refractory to tumor promotion of okadaic acid on mouse skin, suggesting that TNF-α is the essential cytokine in tumor promotion (Suganuma et al. 1999); IL-1 was also proposed as an additional endogenous tumor promoter (Suganuma et al. 2002).

These cytokines are strongly linked to inflammation in cancer microenvironment, so carcinogenesis in humans is now understood as a disease affected by TNF-α up-regulation and NF-κB activation (Fujiki and Suganuma 2005). These are results that may help us provide a partial answer to the essential question: How can we prevent cancer in humans?

6 Conclusion

This chapter presents a new concept of tumor promotion, the okadaic acid pathway, indicating that inhibition of protein phosphatases 1 and 2A is a general biochemical mechanism of tumor promotion applicable to the skin, the glandular stomach and the liver. Thus, the okadaic acid pathway overcomes the classical limitation of tumor promotion, based on tissue and organ specificities. Study with the okadaic acid class of tumor promoters could open the way to pinpointing some mechanisms of cancer development in humans.

Acknowledgements This work was supported by Scientific Research on Priority Areas for Cancer Research from the Ministry of Education, Culture, Sports, Science, and Technology, Japan: Encouragement of Young Scientists from Japan Society for the Promotion of Science of Japan: Comprehensive Research on Aging and Health, and Cancer Research from the Ministry of Health, Labor, and Welfare, Japan: the Smoking Research Fund. We thank Dr. Takashi Sugimura, President emeritus at the National Cancer Center, for his encouragement during the course of this work, Dr. Nobuhiro Fusetani for his stimulating collaboration and careful reading of our manuscript, and Dr. Shigeki Matsunaga for providing the important information. To Drs. D. Uemura, K. Yamada, late Y. Hirata, T. Yasumoto, M. Murata, M. Tatematsu, K-I. Harada, late M.F. Watanabe, and the many scientists from abroad, including Drs. late P.J. Scheuer, late R.E. Moore, F.J. Schmidt, W.W. Carmichael, R.J. Quinn, C. Taylor, I.K. Lim, T.J. Park, S.-Z. Yu, M.R. Rosner, E.V. Wattenberg, Y. Grosse and V. Cogliano, we express gratitude for their fruitful discussion. We also thank Japanese and American scientists cited in the references for generous collaborations.

References

Artigas P, Gadsby DC (2003) Na$^+$/K$^+$-pump ligands modulate gating of palytoxin-induced ion channels. Proc Natl Acad Sci USA 100:501–505.

Artigas P, Gadsby DC (2004) Large diameter of palytoxin-induced Na$^+$/K$^+$-pump channels and modulation of palytoxin interaction by Na$^+$/K$^+$-pump ligands. J Gen Physiol 123:357–376.

Bagu JR, Sykes BD, Craig MM, Holmes CFB (1997) A molecular basis for different interactions of marine toxins with protein phosphatase-1. Molecular models for bound motuporin, microcystins, okadaic acid, and calyculin A. J Biol Chem 272:5087–5097.

Balmain A, Pragnell IB (1983) Mouse skin carcinomas induced *in vivo* by chemical carcinogens have a transforming Harvey-*ras* oncogene. Nature 303:2–74.

Bialojan C, Takai A (1988) Inhibitory effect of a marine-sponge toxin, okadaic acid, on protein phosphatases. Specificity and kinetics. Biochem J 256:283–290.

Botes DP, Tuinman AA, Wesseles PL, Viljoen CC, Kruger H, Williams DH, Santikarn S, Smith RJ, Hammond SJ (1984) The structure of cyanoginosin-LA, a cyclic heptapeptide toxin from the cyanobacterium *Microcystis aeruginosa*. J Chem Soc Perkin Trans 1:2311–2319.

Bouaïcha N, Maatouk I (2004) Microcystin-LR and nodularin induce intracellular glutathione alteration, reactive oxygen species production and lipid peroxidation in primary cultured rat hepatocytes. Toxicol Lett 148:53–63.

Bouaïcha N, Maatouk I, Plessis M-J, Périn F (2005) Genotoxic potential of microcystin-LR and nodularin *in vitro* in primary cultured rat hepatocytes and *in vivo* in rat liver. Environ Toxicol 20:341–347.

Brautigan DL, Shriner CL (1988) Methods to distinguish various types of protein phosphatase activity. Meth Enzym 159:339–346.

Briand JF, Jacquet S, Bernard C, Humbert JF (2003) Health hazards for terrestrial vertebrates from toxic cyanobacteria in surface water ecosystems. Vet Res 34:361–377.

Cardellina II JH, Marner F-J, Moore RE (1979) Seaweed dermatitis: structure of lyngbyatoxin A. Science 204:193–195.

Carmichael WW (1988) Toxins of fresh water cyanobacteria. In: Tu A (ed) Handbook of natural toxins. Dekker, Inc, New York and Basel, Vol 3, pp. 121–147.

Carmichael WW, Mahmood NA (1984) The structure of cyanoginosin-LA, a cyclic heptapeptide toxin from the cyanobacterium *Microcystis aeruginosa*. In: Ragelis EP (ed) Seafood toxins. American Chemical Society, Washington, DC, pp. 377–389.

Cohen P (1989) The structure and regulation of protein phosphatases. Annu Rev Biochem 58:453–508.

Craig M, McCready TL, Luu HA, Smillie MA, Dubord P, Holmes CFB (1993) Identification and characterization of hydrophobic microcystins in Canadian freshwater cyanobacteria. Toxicon 31:1541–1549.

Craig M, Luu HA, McCready TL, Williams D, Andersen RJ, Holmes CF (1996) Molecular mechanisms underlying the interaction of motuporin and microcystins with type-1 and type-2A protein phosphatases. Biochem Cell Biol 74:569–578.

de Silva ED, Williams DE, Andersen RJ, Klix H, Holmes CFB, Allen TM (1992) Motuporin, a potent protein phosphatase inhibitor isolated from the Papua New Guinea sponge *Theonella swinhoei* Gray. Tetrahedron Lett 33:1561–1564.

Dietrich D, Hoeger S (2005) Guidance values for microcystins in water and cyanobacterial supplement products (blue-green algal supplements): a reasonable or misguided approach? Toxicol Appl Pharmacol 203:273–289.

Egloff M-P, Cohen PTW, Reinemer P, Barford D (1995) Crystal structure of the catalytic subunit of human protein phosphatase 1 and its complex with tungstate. J Mol Biol 254:942–959.

Erdödi F, Rokolya A, Di Salvo J, Bárány M, Bárány K (1988) Effect of okadaic acid on phosphorylation-dephosphorylation of myosin light chain in aortic smooth muscle homogenate. Biochem Biophys Res Commun 153:156–161.

Falconer IR, Dornbusch M, Moran G, Yeung SK (1992) Effect of the cyanobacterial (blue-green algal) toxins from *Microcystis aeruginosa* on isolated enterocytes from the chicken small intestine. Toxicon 30:790–793.

Falconer IR, Smith JV, Jackson ARB, Jones A, Runnegar MTC (1998) Oral toxicity of a bloom of the cyanobacterium *Microcystis aeruginosa* administered to mice over periods up to 1 year. J Toxicol Environ Health 24:291–305.

Federal-Provincial-Territorial Committe on Drinking Water (2002) Cyanobacterial toxins – microcystin-LR. In: Guidelines for Canadian drinking water quality: supporting documantation. pp. 1–22.

Fischer WJ, Dietrich DR (2000) Pathological and biochemical characterization of microcystin-induced hepatopancreas and kidney damage in carp (*Cyprinus carpio*). Toxicol Appl Pharmacol 164:73–81.

Francis G (1878) Poisonous Australian lake. Nature 18:11–12.

Fujiki H, Sugimura T (1987) New classes of tumor promoters: teleocidin, aplysiatoxin, and palytoxin. Adv Cancer Res 49:223–264.

Fujiki H, Suganuma M (1993) Tumor promotion by inhibitors of protein phosphatases 1 and 2A: the okadaic acid class of compounds. Adv Cancer Res 61:143–194.

Fujiki H, Suganuma M (2005) Translational research on TNF-α as an endogenous tumor promoter and green tea as cancer preventive in humans. J Environ Sci Health 23:3–30.

Fujiki H, Mori M, Nakayasu M, Terada M, Sugimura T, Moore RE (1981) Indole alkaloids: dihydroteleocidin B, teleocidin, and lyngbyatoxin A as members of a new class of tumor promoters. Proc Natl Acad Sci USA 78:3872–3876.

Fujiki H, Suganuma M, Matsukura N, Sugimura T, Takayama S (1982) Teleocidin from *Streptomyces* is a potent promoter of mouse skin carcinogenesis. Carcinogen 3:895–898.

Fujiki H, Suganuma M, Hakii H, Bartolini G, Moore RE, Takayama S, Sugimura T (1984a) A two-stage mouse skin carcinogenesis study of lyngbyatoxin A. J Cancer Res Clin Oncol 108:174–176.

Fujiki H, Tanaka Y, Miyake R, Kikkawa U, Nishizuka Y, Sugimura T (1984b) Activation of calcium-activated, phospholipid-dependent protein kinase (protein kinase C) by new classes of tumor promoters: teleocidin and debromoaplysiatoxin. Biochem Biophys Res Commun 120:339–343.

Fujiki H, Suganuma M, Nakayasu M, Hakii H, Horiuchi T, Takayama S, Sugimura T (1986) Palytoxin is a non-12-*O*-tetradecanoylphorbol-13-acetate type tumor promoter in two-stage mouse skin carcinogenesis. Carcinogen 7:707–710.

Fujiki H, Suganuma M, Suguri H, Yoshizawa S, Ojika M, Wakamatsu K, Yamada K, Sugimura T (1987) Induction of ornithine decarboxylase activity in mouse skin by a possible tumor promoter, okadaic acid. Proc Japan Acad Ser B 63:51–53.

Fujiki H, Suganuma M, Suguri H, Yoshizawa S, Takagi K, Uda N, Wakamatsu K, Yamada K, Murata M, Yasumoto T, Sugimura T (1988) Diarrhetic shellfish toxin, dinophysistoxin-1, is a potent tumor promoter on mouse skin. Jpn J Cancer Res 79:1089–1093.

Fujiki H, Suganuma M, Yoshizawa S, Kanazawa H, Sugimura T, Manam S, Kahn SM, Jiang W, Hoshina S, Weinstein IB (1989) Codon 61 mutations in the c-Harvey-*ras* gene in mouse skin tumors induced by 7,12-dimethylbenz[*a*]anthracene plus okadaic acid class tumor promoters. Mol Carcinogen 2:184–187.

Fujiki H, Suganuma M, Komori A, Sueoka E, Okabe S, Sueoka N, Aida M, Kozu T, Hamano A, Sakai Y, Tatematsu M (1995) Tumor promotion and prevention of stomach cancer. In: Nishi M, Sugano H, Takahashi T (eds) 1st International gastric cancer congress. Monduzzi Editore, Bologna, pp 19–25.

Goldberg J, Huang H-b, Kwon Y-g, Greengard P, Nairn AC, Kuriyan J (1995) Three-dimensional structure of the catalytic subunit of protein serine/threonine phosphatase-1. Nature 376:745–753.

Grosse Y, Baan R, Straif K, Secretan B, Ghissassi FEl, Cogliano V (2006) Carcinogenicity of nitrate, nitrite, and cyanobacterial peptide toxins. Lancet Oncol 7:628–629.

Habermann E (1989) Palytoxin acts through Na$^+$/K$^+$-ATPase. Toxicon 27:1171–1187.

Harada K-I, Matsuura K, Suzuki M, Oka H, Watanabe MF, Oishi S, Dahlem AM, Beasley VR, Carmichael WW (1988) Chemical analysis of toxic peptides from cyanobacteria by reversed phase high performance liquid chromatography. J Chromatogr 448:275–283.

Harada K-I, Matsuura K, Suzuki M, Watanabe MF, Oishi S, Dahlem AM, Beasley VR, Carmichael WW (1990) Isolation and characterization of the minor components associated with microcystins LR and RR in the cyanobacterium (blue-green algae). Toxicon 28:55–64.

Hecker E (1967) Phorbol esters from croton oil – Chemical nature and biological activities. Naturwissenschaften 54:282–284.

Hescheler J, Mieskes G, Rüegg JC, Takai A, Trautwein W (1988) Effects of a protein phosphatase inhibitor, okadaic acid, on membrane currents of isolated guinea-pig cardiac myocytes. Pflugers Arch 412:248–252.

Honkanen RE, Zwiller J, Moore RE, Daily SL, Khatra BS, Dukelow M, Boynton AL (1990) Characterization of microcystin-LR, a potent inhibitor of type 1 and type 2A protein phosphatases. J Biol Chem 265:19401–19404.

Ishihara H, Martin BL, Brautigan DL, Karaki H, Ozaki H, Kato Y, Fusetani N, Watabe S, Hashimoto K, Uemura D, Hartshorne DJ (1989) Calyculin A and okadaic acid: inhibitors of protein phosphatase activity. Biochem Biophys Res Commun 159:871–877.

Ito E, Kondo F, Terao K, Harada K-i (1997) Neoplastic nodular formation in mouse liver induced by repeated intraperitoneal injections of microcystin-LR. Toxicon 35:1453–1457.

Kato Y, Scheuer PJ (1974) Aplysiatoxin and debromoaplysiatoxin, constituents of the marine mollusk *Stylocheilus longicauda*. J Am Chem Soc 96:2245–2246.

Kato Y, Fusetani N, Matsunaga S, Hashimoto K, Fujita S, Furuya T (1986) Calyculin A, a novel antitumor metabolite from the marine sponge *Discodermia calyx*. J Am Chem Soc 108:2780–2781.

Kato Y, Fusetani N, Matsunaga S, Hashimoto K, Koseki K (1988) Isolation and structure elucidation of calyculins B, C, and D, novel antitumor metabolites, from the marine sponge *Discodermia calyx*. J Org Chem 53:3930–3932.

Kita A, Matsunaga S, Takai A, Kataiwa H, Wakimoto T, Fusetani N, Isobe M, Miki K (2002) Crystal structure of the complex between calyculin A and the catalytic subunit of protein phosphatase 1. Structure 10:715–724.

Krishnamurthy T, Szafraniec L, Hunt DF, Shabanowitz J, Yates III JR, Hauer CR, Carmichael WW, Skulberg O, Codd GA, Missler S (1989) Structural characterization of toxic cyclic peptides from blue-green algae by tandem mass spectrometry. Proc Natl Acad Sci USA 86:770–774.

Kuroki DW, Minden A, Sanchez I, Wattenberg EV (1997) Regulation of a c-Jun amino-terminal kinase/stress-activated protein kinase cascade by a sodium-dependent signal transduction pathway. J Biol Chem 272:23905–23911.

Lim IK (2003) Erratum to "Spectrum of molecular changes during hepatocarcinogenesis induced by DEN and other chemicals in Fischer 344 male rats." Mech Age Dev 124:697–708.

Lim IK, Park SC, Song KY, Park TJ, Lee MS, Kim S-J, Hyun BH (1999) Regulation of selection of liver nodules initiated with *N*-nitrosodiethylamine and promoted with nodularin injections in Fischer 344 male rats by reciprocal expression of transforming growth factor-β1 and its receptors. Mol Carcinogen 26:83–92.

Lim IK, Park TJ, Park SC, Yoon G, Kwak CS, Lee MS, Song KY, Choi YK, Hyun BH (2001) Selective left-lobe atrophy by nodularin treatment accompanied by reduced protein phosphatase 1/2A and increased peroxisome proliferation in rat liver. Int J Cancer 91:32–40.

Lin J-R, Chu FS (1994) Kinetics of distribution of microcystin LR in serum and liver cytosol of mice: an immunochemical analysis. J Agric Food Chem 42:1035–1040.

Maatouk I, Bouaïcha N, Plessis MJ, Périn F (2004) Detection by ^{32}P-postlabelling of 8-oxo-7,8-dihydro-2'-deoxyguanosine in DNA as biomarker of microcystin-LR- and nodularin-induced DNA damage *in vitro* in primary cultured rat hepatocytes and *in vivo* in rat liver. Mutat Res 564:9–20.

MacKintosh C, Beattie KA, Klumpp S, Cohen P, Codd GA (1990) Cyanobacterial microcystin-LR is a potent and specific inhibitor of protein phosphatases 1 and 2A from both mammals and higher plants. FEBS Lett 264:187–192.

Matsunaga S, Fusetani N (1991) Absolute stereochemistry of the calyculins, potent inhibitors of protein phosphatases 1 and 2A. Tetrahedron Lett 32:5605–5606.

Matsunaga S, Fujiki H, Sakata D, Fusetani N (1991) Calyculins E, F, G, and H, additional inhibitors of protein phosphatases 1 and 2A, from the marine sponge *Discodermia calyx*. Tetrahedron 47:2999–3006.

Matsunaga S, Wakimoto T, Fusetani N (1997) Isolation of four new calyculins from the marine sponge *Discodermia calyx*. J Org Chem 62:2640–2642.

Matsushima R, Yoshizawa S, Watanabe MF, Harada K-I, Furusawa M, Carmichael WW, Fujiki H (1990) *In vitro* and *in vivo* effects of protein phosphatase inhibitors, microcystins and nodularin, on mouse skin and fibroblasts. Biochem Biophys Res Commun 171:867–874.

Maynes JT, Bateman KS, Cherney MM, Das AK, Luu HA, Holmes CFB, James MNG (2001) Crystal structure of the tumor-promoter okadaic acid bound to protein phosphatase-1. J Biol Chem 276:44078–44882.

Maynes JT, Luu HA, Cherney MM, Andersen RJ, Williams D, Holmes CFB, James MNG (2006) Crystal structures of protein phosphatase-1 bound to motuporin and dihydromicrocystin-LA: elucidation of the mechanism of enzyme inhibition by cyanobacterial toxins. J Mol Biol 356:111–120.

Moore RE (1982) Toxins, anticancer agents, and tumor promoters from marine prokaryotes. Pure Appl Chem 54:1919–1934.

Moore RE, Bartolini G (1981) Structure of palytoxin. J Am Chem Soc 103:2491–2494.

Murakami Y, Oshima Y, Yasumoto T (1982) Identification of okadaic acid as a toxic component of a marine dinoflagellate *Prorocentrum lima*. Bull Jpn Soc Sci Fish 48:69–72.

Murata M, Shimatani M, Sugitani H, Oshima Y, Yasumoto T (1982) Isolation and structural elucidation of the causative toxin of the diarrhetic shellfish poisoning. Bull Jpn Soc Sci Fish 48:549–552.

Namikoshi M, Rinehart KL, Sakai R, Stotts RR, Dahlem AM, Beasley VR, Carmichael WW, Evans WR (1992) Identification of 12 hepatotoxins from a Homer Lake bloom of the cyanobacteria *Microcystis aeruginosa*, *Microcystis viridis*, and *Microcystis wesenbergii*: nine new microcystins. J Org Chem 57:866–872.

Nishiwaki S, Fujiki H, Suganuma M, Furuya-Suguri H, Matsushima R, Iida Y, Ojika M, Yamada K, Uemura D, Yasumoto T, Schmitz FJ, Sugimura T (1990a) Structure-activity relationship within a series of okadaic acid derivatives. Carcinogen 11:1837–1841.

Nishiwaki S, Fujiki H, Suganuma M, Ojika M, Yamada K, Sugimura T (1990b) Photoaffinity labeling of protein phosphatase 2A, the receptor for a tumor promoter okadaic acid, by [27–^{3}H] methyl 7-*O*-(4-azidobenzoyl) okadaate. Biochem Biophys Res Commun 170:1359–1364.

Nishiwaki-Matsushima R, Nishiwaki S, Ohta T, Yoshizawa S, Suganuma M, Harada K-I, Watanabe MF, Fujiki H (1991) Structure-function relationships of microcystins, liver tumor promoters, in interaction with protein phosphatase. Jpn J Cancer Res 82:993–996.

Nishiwaki-Matsushima R, Ohta T, Nishiwaki S, Suganuma M, Kohyama K, Ishikawa T, Carmichael WW, Fujiki H (1992) Liver tumor promotion by the cyanobacterial cyclic peptide toxin microcystin-LR. J Cancer Res Clin Oncol 118:420–424.

Nishiwaki R, Ohta T, Sueoka E, Suganuma M, Harada K-I, Watanabe MF, Fujiki H (1994) Two significant aspects of microcystin-LR: specific binding and liver specificity. Cancer Lett 83:283–289.

Nishizuka Y (1984) The role of protein kinase C in cell surface signal transduction and tumour promotion. Nature 308:693–698.

Ohta T, Sueoka E, Iida N, Komori A, Suganuma M, Nishiwaki R, Tatematsu M, Kim SJ, Carmichael WW, Fujiki H (1994) Nodularin, a potent inhibitor of protein phosphatases 1 and 2A, is a new environmental carcinogen in male F344 rat liver. Cancer Res 54:6402–6406.

Quinn RJ, Taylor C, Suganuma M, Fujiki H (1993) The conserved acid binding domain model of inhibitors of protein phosphatases 1 and 2A: molecular modelling aspects. Bioorg Med Chem Lett 3:1029–1034.

Quintanilla M, Brown K, Ramsden M, Balmain A (1986) Carcinogen-specific mutation and amplification of Ha-*ras* during mouse skin carcinogenesis. Nature 322:78–80.

Rao PVL, Gupta N, Bhaskar ASB, Jayaraj R (2002) Toxins and bioactive compounds from cyanobacteria and their implications on human health. J Environ Biol 23:215–224.

Rinehart KL, Harada K-i, Namikoshi M, Chen C, Harvis CA, Munro MHG, Blunt JW, Mulligan PE, Beasley VR, Dahlem AM, Carmichael WW (1988) Nodularin, microcystin, and the configuration of Adda. J Am Chem Soc 110:8557–8558.

Rocha MFG, Sidrim JJC, Soares AM, Jimenez GC, Guerrant RL, Ribeiro RA, Lima AAM (2000) Supernatants from macrophages stimulated with microcystin-LR induce electrogenic intestinal response in rabbit ileum. Pharmacol Toxicol 87:46–51.

Runnegar M, Berndt N, Kong S-M, Lee EYC, Zhang L (1995) *In vivo* and *in vitro* binding of microcystin to protein phosphatases 1 and 2A. Biochem Biophys Res Commun 216:162–169.

Sassa T, Richter WW, Uda N, Suganuma M, Suguri H, Yoshizawa S, Hirota M, Fujiki H (1989) Apparent "activation" of protein kinases by okadaic acid class tumor promoters. Biochem Biophys Res Commun 159:939–944.

Schmitz FJ, Prasad RS, Gopichand Y, Hossain MB, van der Helm D, Schmidt P (1981) Acanthifolicin, a new episulfide-containing polyether carboxylic acid from extracts of the marine sponge *Pandaros acanthifolium*. J Am Chem Soc 103:2467–2469.

Sekijima M, Tsutsumi T, Yoshida T, Harada T, Tashiro F, Chen G, Yu S-Z, Ueno Y (1999) Enhancement of glutathione S-transferase placental-form positive liver cell foci development by microcystin-LR in aflatoxin B$_1$-initiated rats. Carcinogen 20:161–165.

Shen PP, Zhao SW, Zheng WJ, Hua ZC, Shi Q, Liu ZT (2003) Effects of cyanobacteria bloom extract on some parameters of immune function in mice. Toxicol Lett 143:27–36.

Shibata S, Ishida Y, Kitano H, Ohizumi Y, Habon J, Tsukitani Y, Kikuchi H (1982) Contractile effects of okadaic acid, a novel ionophore-like substance from black sponge, on isolated smooth muscles under the condition of Ca deficiency. J Pharmacol Exp Ther 223:135–143.

Song KY, Lim IK, Park SC, Lee SO, Park HS, Choi YK, Hyun BH (1999) Effect of nodularin on the expression of glutathione S-transferase placental form and proliferating cell nuclear antigen in N-nitrosodiethylamine initiated hepatocarcinogenesis in the male Fischer 344 rat. Carcinogen 20:1541–1548.

Sueoka E, Sueoka N, Okabe S, Kozu T, Komori A, Ohta T, Suganuma M, Kim SJ, Lim IK, Fujiki H (1997) Expression of the tumor necrosis factor α gene and early response genes by nodularin, a liver tumor promoter, in primary cultured rat hepatocytes. J Cancer Res Clin Oncol 123:413–419.

Suganuma M, Fujiki H, Suguri H, Yoshizawa S, Hirota M, Nakayasu M, Ojika M, Wakamatsu K, Yamada K, Sugimura T (1988) Okadaic acid: an additional non-phorbol-12-tetradecanoate-13-acetate-type tumor promoter. Proc Natl Acad Sci USA 85:1768–1771.

Suganuma M, Suttajit M, Suguri H, Ojika M, Yamada K, Fujiki H (1989) Specific binding of okadaic acid, a new tumor promoter in mouse skin. FEBS Lett 250:615–618.

Suganuma M, Fujiki H, Furuya-Suguri H, Yoshizawa S, Yasumoto S, Kato Y, Fusetani N, Sugimura T (1990) Calyculin A, an inhibitor of protein phosphatases, a potent tumor promoter on CD-1 mouse skin. Cancer Res 50:3521–3525.

Suganuma M, Fujiki H, Okabe S, Nishiwaki S, Brautigan D, Ingebritsen TS, Rosner MR (1992a) Structurally different members of the okadaic acid class selectively inhibit protein serine/threonine but not tyrosine phosphatase activity. Toxicon 30:873–878.

Suganuma M, Tatematsu M, Yatsunami J, Yoshizawa S, Okabe S, Uemura D, Fujiki H (1992b) An alternative theory of tissue specificity by tumor promotion of okadaic acid in glandular stomach of SD rats. Carcinogen 13:1841–1845.

Suganuma M, Okabe S, Marino MW, Sakai A, Sueoka E, Fujiki H (1999) Essential role of tumor necrosis factor α (TNF-α) in tumor promotion as revealed by TNF-α-deficient mice. Cancer Res 59:4516–4518.

Suganuma M, Okabe S, Kurusu M, Iida N, Ohshima S, Saeki Y, Kishimoto T, Fujiki H (2002) Discrete roles of cytokines, TNF-α, IL-1, IL-6 in tumor promotion and cell transformation. Int J Oncol 20:131–136.

Tachibana K, Scheuer PJ, Tsukitani Y, Kikuchi H, Van Engen D, Clardy J, Gopichand Y, Schmitz FJ (1981) Okadaic acid, a cytotoxic polyether from two marine sponges of the genus Halichondria. J Am Chem Soc 103:2469–2471.

Takai A, Bialojan C, Troschka M, Ruegg JC (1987) Smooth muscle myosin phosphatase inhibition and force enhancement by black sponge toxin. FEBS Lett 217:81–84.

Takashima M, Sakai H (1960) A new toxic substance, teleocidin, produced by Streptomyces. Part I. Production, isolation and chemical studies. Bull Agric Chem Soc Jpn 24:647–651.

Taylor C, Quinn RJ, McCulloch R, Nishiwaki-Matsushima R, Fujiki H (1992) An alternative computer model of the 3-dimensional structures of microcystin-LR and nodularin rationalising their interactions with protein phosphatases 1 and 2A. Bioorg Med Chem Lett 2:299–302.

Taylor C, Quinn RJ, Suganuma M, Fujiki H (1996) Inhibition of protein phosphatase 2A by cyclic peptides modelled on the microcystin ring. Bioorg Med Chem Lett 6:2113–2116.

Uemura D, Hirata Y (1989) Antitumor polyethers from marine animals. In: Rahman AU (ed) Studies in natural products chemistry. Elsevier, Amsterdam, Vol 5, pp. 377–401.

Uemura D, Ueda K, Hirata Y (1981) Further studies on palytoxin. II. Structure of palytoxin. Tetrahedron Lett 22:2781–2784.

Van Duuren BL (1969) Tumor-promoting agents in two-stage carcinogenesis. Prog Exp Tumor Res 11:31–68.

Volter KE, Pierens GK, Quinn RJ, Wakimoto T, Matsunaga S, Fusetani N (1999) The solution structures of calyculin A and dephosphonocalyculin A by NMR. Bioorg Med Chem Lett 9:717–722.

Wakimoto T, Matsunaga S, Takai A, Fusetani N (2002) Insight into binding of calyculin A to protein phosphatase 1: isolation of hemicalyculin A and chemical transformation of calyculin A. Chem Biol 9:309–319.

Warmka JK, Mauro LJ, Wattenberg EV (2004) Mitogen-activated protein kinase phosphatase-3 is a tumor promoter target in initiated cells that express oncogenic Ras. J Biol Chem 279:33085–33092.

Watanabe MF, Oishi S, Harda K-i, Matsuura K, Kawai H, Suzuki M (1988) Toxins contained in *Microcystis* species of cyanobacteria (blue-green algae). Toxicon 26:1017–1025.

Wattenberg EV (2007) Palytoxin: exploiting a novel skin tumor promoter to explore signal transduction and carcinogenesis. Am J Physiol Cell Physiol 292:C24–C32.

Wattenberg EV, Fujiki H, Rosner MR (1987) Heterologous regulation of the epidermal growth factor receptor by palytoxin, a non-12-*O*-tetradecanoyl phorbol-13-acetate-type tumor promoter. Cancer Res 47:4618–4622.

Wattenberg EV, Byron KI, Villereal ML, Fujiki H, Rosner MR (1989) Sodium as a mediator of non-phorbol tumor promoter action, Down-signal-regulated kinase transmits palytoxin-stimulated signals leading to altered gene expression in mouse keratinocytes. Toxicol Appl Pharmacol 185:8–17.

Wiegand C, Pflugmacher S (2005) Ecotoxicological effects of selected cyanobacterial secondary metabolites: a short review. Toxicol Appl Pharmacol 203:201–218.

Xing Y, Xu Y, Chen Y, Jeffrey PD, Chao Y, Lin Z, Li Z, Strack S, Stock JB, Shi Y (2006) Structure of protein phosphatase 2A core enzyme bound to tumor-inducing toxins. Cell 127:341–353.

Xu Y, Xing Y, Chen Y, Chao Y, Lin Z, Fan E, Yu JW, Strack S, Jeffrey PD, Shi Y (2006) Structure of the protein phosphatase 2A holoenzyme. Cell 127:1239–1251.

Yasumoto T, Murata M, Oshima Y, Sano M, Matsumoto GK, Clardy J (1985) Diarrhetic shellfish toxins. Tetrahedron 41:1019–1025.

Yoshizawa S, Matsushima R, Watanabe MF, Harada K-i, Ichihara A, Carmichael WW, Fujiki H (1990) Inhibition of protein phosphatases by microcystin and nodularin associated with hepatotoxicity. J Cancer Res Clin Oncol 116:609–614.

Yu S-Z (1989) Drinking water and primary liver cancer. In: Tang ZY, Wu MC, Xia SS (eds) Primary liver cancer. Springer-Verlag, Berlin, pp. 20–30.

Yu S-Z (1995) Primary prevention of hepatocellular carcinoma. J Gastroenterol Hepatol 10:674–682.

Yu S-Z, Chen G, Zhi X-L, Li J (1998) Primary liver cancer: natural toxins and prevention in China. J Toxicol Sci 23:143–147.

Yuasa H, Yoshida K, Iwata H, Nakanishi H, Suganuma M, Tatematsu M (1994) Increase of labeling indices in gastrointestinal mucosae of mice and rats by compounds of the okadaic acid type. J Cancer Res Clin Oncol 120:208–212.

Zhang L, Zhang Z, Long F, Lee EY (1996) Tyrosine-272 is involved in the inhibition of protein phosphatase-1 by multiple toxins. Biochemistry 35:1606–1611.

Index